Kant's Theory of Biology

Kant's Theory of Biology

Edited by Ina Goy and Eric Watkins

DE GRUYTER

ISBN 978-3-11-048171-6
e-ISBN 978-3-11-022579-2

Library of Congress Cataloging-in-Publication Data
A CIP catalog record for this book has been applied for at the Library of Congress.

Bibliographic information published by the Deutsche Nationalbibliothek
The Deutsche Nationalbibliothek lists this publication in the Deutsche Nationalbibliographie;
detailed bibliographic data are available on the internet at http://dnb.dnb.de

© 2014 Walter de Gruyter GmbH, Berlin/Boston
Cover image: The cover image is from the "Mangrove Humming Bird" plate in John James
Audubon's (1785–1851) "Birds of America" (1827–1838) series. Reprinted with permission of
the Archives Service Center, University of Pittsburgh, 7500 Thomas Blvd., Pittsburgh, PA 15260,
United States.
Printing: Hubert & Co. GmbH & Co. KG, Göttingen
♾ Printed on acid-free paper
Printed in Germany

www.degruyter.com

Contents

Citations and Abbreviations

1 Citations

Citations to Kant's writings give an abbreviated English title of the work and the volume and page numbers of the *Akademie-Ausgabe*, except in the case of the *Critique of Pure Reason*, for which the page numbers of the first (A) or second (B) edition are given. Translations quoted are from *The Cambridge Edition of the Works of Immanuel Kant*, except where noted. The list below gives the *Akademie-Ausgabe* (AA) volume and page numbers for each abbreviated work. Where possible, the title of the *Cambridge Edition* (CE) volume in which a translation appears is also listed, along with the page numbers of the translation. Bibliographical information for the *Akademie-Ausgabe* and the volumes of the *Cambridge Edition* can be found in the bibliography of this volume.

2 Abbreviations

Anthropology   *Anthropology from a Pragmatic Point of View*
AA VII 119 – 333
CE *Anthropology, History, and Education*, 231 – 429

Argument   *The Only Possible Argument in Support of a Demonstration of the Existence of God*
AA II 65 – 163
CE *Theoretical Philosophy 1755 – 1770*, 111 – 201

Correspondence   *Correspondence*
AA X–XIII
CE *Correspondence* (selection)

CprR   *Critique of Practical Reason*
AA V 1–164
CE *Practical Philosophy*, 139 – 271

CPR A   *Critique of Pure Reason* (1st Edition)
AA IV 5 – 252
CE *Critique of Pure Reason*

CPR B   *Critique of Pure Reason* (2nd Edition)
AA III 2 – 552
CE *Critique of Pure Reason*

CPJ	*Critique of the Power of Judgment* AA V 165–486 CE *Critique of the Power of Judgment*, 53–346
Dreams	*Dreams of a Spirit-Seer Elucidated by Dreams of Metaphysics* AA II 317–73 CE *Theoretical Philosophy 1755–1770*, 305–59
Faculties	*The Conflict of the Faculties* AA VII 5–116 CE *Religion and Rational Theology*, 239–327
First Introduction	The first, unpublished introduction to the *Critique of the Power of Judgment* AA XX 195–251 CE *Critique of the Power of Judgment*, 3–51
Human Race	*Determination of the Concept of a Human Race* AA VIII 91–106 CE *Anthropology, History, and Education*, 145–59
Lect. Log. Busolt	Notes to lectures on logic given by Kant in 1789 or 1790, attributed to Gotthilf Christoph Wilhelm Busolt AA XXIV/1.2 603–86
Lect. Log. Dohna	Notes to lectures on logic given by Kant in 1792, attributed to Graf Heinrich Ludwig Adolph zu Dohna-Wundlacken AA XXIV/1.2 687–784 CE *Lectures on Logic*, 425–516
Lect. Met. Dohna	Notes to lectures on metaphysics given by Kant in 1792–3, attributed to Graf Heinrich Ludwig Adolph zu Dohna-Wundlacken AA XXVIII/2.1 611–702 CE *Lectures on Metaphysics*, 355–91 (selection)
Lect. Met. Herder	Notes to lectures on metaphysics given by Kant in the early 1760s, attributed to Johann Gottfried Herder AA XXVIII/1 1–166 CE Lectures on Metaphysics, 1–16 (selection)
Lect. Met. K$_2$	Notes to lectures on metaphysics given by Kant and dated in the early 1790s, author unknown AA XXVIII/2.1 705–816 CE *Lectures on Metaphysics*, 393–413 (selection)
Lect. Met. L$_1$	Notes to lectures on metaphysics given by Kant probably in the mid 1770s, author unknown

AA XXVIII/1 167–350
CE *Lectures on Metaphysics*, 19–106 (selection)

Lect. Met. Vigilantius Notes to lectures on metaphysics given by Kant and dated 1794–5, attributed to Johann Friedrich Vigilantius
AA XXIX/1.2 941–1040
CE *Lectures on Metaphysics*, 415–506

Lect. Moral Phil. Collins Notes to lectures on moral philosophy given by Kant probably in the mid 1770s (though dated 1784–5), attributed to Georg Ludwig Collins
AA XXVII/1 237–471
CE *Lectures on Ethics*, 37–222

Lect. Moral Phil. Mrong. Notes to lectures on moral philosophy given by Kant probably in the mid 1770s (but copied later), attributed to Christoph Coelestin Mrongovius
AA XXVII/2.2 1395–581

Lect. Nat. Law Feyerabend Notes to lectures on natural law given by Kant probably in 1784, attributed to Gottlieb Feyerabend
AA XXVII/2.2 1317–94

Lect. Phys. Danzig Notes to lectures on physics given by Kant probably in 1785, attributed to Christoph Coelestin Mrongovius
AA XXIX/1.1 92–169

Lect. Rat. Theol. Pölitz Notes to lectures on theology given by Kant probably 1783–4, attributed to Karl Heinrich Ludwig Pölitz
AA XXVIII/2.2 998–1126
CE *Religion and Rational Theology*, 341–446

Lect. Rat. Theol. Volckmann Notes to lectures on theology given by Kant probably 1783–4, attributed to Johann Wilhelm Volckmann
AA XXVIII/2.2 1127–225

Logic *The Jäsche Logic*
AA IX 1–150
CE *Lectures on Logic*, 521–640

MM *Metaphysics of Morals*
AA VI 205–493
CE *Practical Philosophy*, 365–492, 509–603

Notes and Fragments *Notes and Fragments*
AA XIV–XXIII
CE *Notes and Fragments*

On the Common Saying	*On the Common Saying: That May Be Correct in Theory, but it is of no Use in Practice* AA VIII 275 – 313 CE *Practical Philosophy*, 279 – 309
OP	*Opus Postumum* AA XXI and XXII CE *Opus Postumum*, 3 – 256
Progress	*What Real Progress has Metaphysics Made in Germany Since the Time of Leibniz and Wolff?* AA XX 257 – 332 CE *Theoretical Philosophy after 1781*, 351 – 424
Prolegomena	*Prolegomena to Any Future Metaphysic that Will be Able to Come Forward as Science* AA IV 255 – 383 CE *Theoretical Philosophy after 1781*, 51 – 169
Races	*Of the Different Races of Human Beings* AA II 429 – 43 CE *Anthropology, History, and Education*, 84 – 97
Religion	*Religion within the Boundaries of Mere Reason* AA VI 3 – 202 CE *Religion and Rational Theology*, 57 – 215
Review of Herder	*Review of J. G. Herder's Ideas for the Philosophy of the History of Humanity* AA VIII 45 – 66 CE *Anthropology, History, and Education*, 124 – 42
Teleological Principles	*On the Use of Teleological Principles in Philosophy* AA VIII 159 – 84 CE *Anthropology, History, and Education*, 195 – 218
Theory of Heavens	*Universal Natural History and Theory of the Heavens or Essay on the Constitution and the Mechanical Origin of the Whole Universe According to Newtonian Principles* AA I 217 – 368 CE *Natural Science*, 191 – 308
Universal History	*Idea for a Universal History with a Cosmopolitan Aim* AA VIII 17 – 31 CE *Anthropology, History, and Education*, 108 – 20

Ina Goy and Eric Watkins
Introduction

1 Kant's Theory of Biology

1.1 Historical Background

Whereas early modern advocates of experimental philosophy, Cartesian mechanism, and Newtonian mathematical physics avoided positing final causes and teleological explanations, many philosophers and natural researchers in the seventeenth and eighteenth centuries believed that efficient causes and non-teleological explanations were insufficient to explain the processes that regularly occurred in nerves and muscles, and in plant and animal generation, and thus tried to reinstate final causes and teleological explanations. For example, the physico-theological accounts of John Ray (1627–1705) and William Derham (1657–1735) shaped natural philosophy in England, while Christian Wolff's (1679–1754) deistic teleology was influential in Germany.

In the life sciences (at this time a field of research intersecting with medicine, anatomy, physiology, and physics) preformationist theories of the generation of living beings dominated the debate of the seventeenth and first half of the eighteenth century, whereas in the second half of the eighteenth century epigenetic accounts increasingly gained support. Defenders of preformation claimed—consistent with the creation narrative of the Old Testament—that the ultimate principle of an organism was a divine, preformed germ that contained en miniature all the predispositions of a prospective living being. Among them, ovists thought that the preformed germ was the female egg while animalculists identified it as the male sperm.

In the first half of the eighteenth century, the detection of the self-reproduction and regeneration of polyps[1] and the generation of deformed offspring provided counterexamples to the creationist narrative and fostered doubts about the explanatory reach of the doctrine of preformation. The appearance of the characteristics of both mother and father in inter-species hybrids (such as mules) also seemed to contradict the idea of a one-sided heredity of either the mother's (ovism) or the father's characteristics (animalculism) in newborns. These con-

---

[1] The experiments with polyps are famously reported in Abraham Trembley's (1710–1784) treatise "Mémoires pour servir à l'histoire d'un genre de polypes d'eau douce", published in 1744.

cerns paved the way for epigenetic accounts of generation in the second half of the eighteenth century. Defenders of this new approach argued that organic life begins with unstructured matter and self-organizing powers. Whereas earlier epigenetic accounts described these forces mechanically, in Newtonian terms of attraction and repulsion, later accounts held that self-organizing powers must be vitalistic.

Immanuel Kant (1724–1804), who lived through the development of both preformationist and epigenetic accounts, was familiar with the classificatory schemes of theories of organized beings and the specific vocabulary that natural researchers used in his time. But he did not straightforwardly adopt these schemes and terms. When discussing possible explanations of plant and animal generation, for instance in §81 of the *CPJ*, he focused on the natural and divine contributions to generation and treated both "*individual preformation*" and "*generic preformation*" (which he equated with "*epigenesis*" [*CPJ* V 423.2–4]) as different forms of prestabilist as opposed to occasionalist views of creation (*CPJ* V 422.22). Kant praised epigenesis for relying less on God for the organization of the forms of nature. He adopted terms that were widely used in the preformation-epigenesis debate, such as 'germs', 'educt', 'product', 'evolution', 'involution', but he did not characterize his own position as either preformationist or epigenetic, nor, for that matter, as some blend of the two. Instead, Kant classified theories of generation as subspecies of what he called occasionalism and prestabilism (ibid.). This classification expresses a specific interest in organized beings that seems for him to have been inseparably connected to metaphysical and theological claims. One of the major historical scholarly controversies concerning Kant's views on generation focuses on the extent to which his account tends to be preformationist or epigenetic, or an alternative to both (see Zammito 2007, Fisher 2007, Fisher and Goy in this volume).

Kant rejected the ancient view of a *generatio aequivoca*, a theory that explains the generation of organized beings occurs from "the mechanism of crude, unorganized matter" such as mud or slime. He also criticized the lack of empirical evidence for the *generatio univoca heteronyma*: theories that explain the generation of organized beings of one kind out of organized beings of another, such as the generation of land animals out of amphibians, and of amphibians out of aquatic animals. He thought that the *generatio univoca homonyma*, i.e., the generation of organized beings out of organized beings of the same kind, was most widely confirmed by experience (*CPJ* V 419.26–38 and 420.34–6).

Although apparently familiar with many natural researchers of his time, Kant seldom refers to them by name or discusses their accounts in detail. Kant's works contain almost no extended discussion of specific individual's accounts of natural research in the life sciences of his time. One exception is a brief review of Johann Gottfried Herder's (1744–1803) "Ideas Toward a Philosophy of the History of Man" in 1785 (see Ameriks 2011, Zammito's articles, and Zuckert in this volume). A second is his response to Johann Georg Adam Forster's (1754–1794) objections to his own account of the origin of human races (see Forster 1786) in the third of his three short essays on human races. A third exception is his reaction to Johann Friedrich Blumenbach's (1752–1840) theory of epigenesis: Kant praises it in his *Teleological Principles* (VIII 180.31–5), in §81 of the *CPJ* (V 424.7–34), and in several other passages from the 1790s[2]. He even exchanged two letters with Blumenbach in 1790 (*Correspondence* XI 184.30–185.25, 211.1–23). But generally Kant refrains from explicitly referring to contemporaries in these fields. This is all the more remarkable given that it was standard practice in the literature on organized beings at the time to present a new account or a modification of an existing one only after providing extensive quotations, excerpts, paraphrases and comments on earlier theories. Kant's ambivalent relationship to Blumenbach is another long-standing focus of historical scholarly debates (see Lenoir 1980, 1981, and 1982, Richards 2000 and 2002, Look 2006, Zammito 2012).

Kant also makes brief allusions to Pierre-Louis Moureau de Maupertuis' (1698–1759) "laws of desire and aversion" and to George-Louis Leclerc de Buffon's (1707–1788) "internal forms" in his early essay *Argument* (II 115.4–8). He quotes with meticulous accuracy from the first volume of Buffon's "Natural History", which, in B. J. Zink's German translation, contained Buffon's account of organized beings from the second volume of the "Natural History". He praises Hermann Samuel Reimarus' (1694–1768) physico-theology (*CPJ* V 476.36) and mentions Charles Bonnet's (1720–1793) idea of a natural chain of living beings in several places (*Teleological Principles* VIII 180.2, 31–5). Kant also speaks highly of the influential taxonomical ideas of Carl Linnaeus (1707–1778), lauding his "principle of the persistence of the character of the pollinating parts of plants", without which "the systematic *description of nature* of the vegetable kingdom would not have been ordered and enlarged in so praiseworthy a manner" (*Teleological Principles* VIII 161.18–21, *CPJ* V 427.4). Although a variety of learned remarks attest to Kant's interest in these studies, none of them directly impacted his writings on organized beings during the critical period to the extent that, for instance, disagreements with Gottfried Wilhelm Leibniz's (1646–1716) or

---

[2] Kant mentions Blumenbach in his *Lect. Met. K₂* (XXVIII/2.1 762.21), in notes on Soemmering's work (*Correspondence* XIII 400.7), in his essay *Faculties* (VII 89.5) and in his writing *Anthropology* (VII 299.15), in *Notes and Fragments* (XIV 619.4) on physical geography, and in his latest notes (*OP* XXI 180.27).

Isaac Newton's (1643–1727) theories had shaped his works in natural philosophy in his early years.

1.2 The Development of Kant's Thoughts about Organized Beings

The "Introductions" and the second half of the *CPJ* contain Kant's richest and most sophisticated account of organized beings, but his interest in the life sciences reaches far back to the beginnings of his career (Fisher's doctoral dissertation summarizes this long path). Remarks from 1755 confirm that Kant was aware of an intractable difference between organisms and objects that can be fully explained mechanically. However, he did not formulate a specific law for biological entities until much later. Though he was familiar with most theories of organisms of his own time, he adhered to none of them; and he passed over the problem of the inexplicability of organisms in his philosophical system for more than three decades, even as he presented that system as complete. It was not until 1788 that Kant discovered that the teleological lawfulness of organized beings could allow him to explain organisms theoretically while at the same time enabling him to find a place for them within his larger project of transcendental philosophy. Kant presented the most mature version of his doctrine of the teleology of organic nature in the second half of the *CPJ* in 1790, and he continued to revise his views on organized beings until the end of his productive career.

In his early *Theory of Heavens* (1755) Kant describes the cosmos as arising according to mechanical laws and uses this mechanical cosmology in a physico-theological proof for the existence of God. For the first time, he admits that he is unable to explain organized beings in terms of mechanical laws:

> Are we in a position to say: *Give me matter and I will show you how a caterpillar can be created?* Do we not get stuck at the first step due to ignorance about the true inner nature of the object and the complexity of the diversity contained in it? It should therefore not be thought strange if I dare to say that we will understand the formation of all the heavenly bodies, the cause of their motion, in short, the origin of the whole present constitution of the universe sooner than the creation of a single plant or caterpillar becomes clearly and completely known on mechanical grounds (*Theory of Heavens* I 230.16–26).

In the early *Argument* essay (1763) Kant presents his views on philosophical arguments for the existence of God. He elaborates on the possible relations between God and the laws and powers of nature at length and responded to some of the positions of the preformation-epigenesis debate of his time. In his essays on human races in 1775, 1785, and 1788, Kant discusses generation and the heredity of human beings and explains the origin and development of human races. He defends the view that there are four races that arose from "*a single first [human] phylum*" (*Human Race* VIII 98.36). The phylum is said to have contained the predispositions for all races in preformed germs, and to have developed into four distinct races under the influence of different climates and environments. In these essays, Kant also incorporates terms similar to those used in the epigenetic and preformationist debates, in which mechanism was no longer dominant.

In his review of Johann Gottfried Herder's (1744–1803) "Ideas", written in 1785, Kant criticizes Herder's account of the generation of organisms. Herder defends the existence of a single "animating force that organizes everything [...] in such a way that the schema of the perfection of this organization is supposed to be the human being, to which all earthly creatures, from the lowest stage on, approach" (*Review Herder* VIII 52.24–7). Although Kant seems to agree with the notion of an organic power as "a principle of life" that "modifies *itself* internally in accordance with differences of external circumstances" (*Review Herder* VIII 62.26–7) he disagrees with the notion of the "unity of the organic force [...] which, as self-forming in regard to the manifoldness of all organic creatures, [...] is supposed to constitute the entire distinctiveness of its many genera and species". In his eyes this notion lies "outside the field of the observational doctrine of nature" and is too speculative (*Review Herder* VIII 54.28–34). Instead he mentions "germs" and "original dispositions" of a self-forming faculty that would account for the diversity of organized beings (*Review Herder* VIII 62.34). Kant also objects to Herder's idea that the organization of a human being presupposes and contains the organizations of all other beings.

Kant's *Metaphysical Foundations* (1786) proposes a special metaphysics, in particular of "*rational physics*" or "*[r]ational [p]hysiology*" (*CPR* B 874–5), which is primarily concerned with moving powers (attraction and repulsion) and mechanical laws. In this work—while reflecting on the external mechanical causation of motion and rest as states of matter—Kant admits that he is unable to explain the phenomenon of life, understood as the capacity of a substance to determine itself to act or change from an inner principle (*Metaphysical Foundations* IV 544.1–30).

Finally, the appendices to the "Transcendental Dialectic" in the *CPR* (1781/7) express Kant's search for the appropriate order of organized beings beyond mechanical laws: in these sections of the *CPR* Kant discusses reason's attempt to form a regulative teleological order and to find unity among the a priori universal laws of nature. Similarly, in the third *Critique* Kant describes reason's attempt to discover a teleological order and unity among the empirical laws of nature.

Kant's search for the specific lawfulness of organized beings reached its turning point in the insight that organized beings can be characterized by means of physical teleological laws that at the same time can serve as transcendental principles in his critical philosophy. In the *Teleological Principles* in 1788 Kant writes:

> Since the concept of an organized being already includes that it is some matter in which everything is mutually related to each other as end and means, which can only be thought as a *system of final causes*, and since therefore their possibility only leaves the teleological but not the physical-mechanical mode of explanation, at least as far as *human* reason is concerned, there can be no investigation in physics about the origin of all organization itself. The answer to this question [...] would lie *outside* of natural science *in metaphysics* (*Teleological Principles* VIII 179.8 – 18).

Kant argues that an organized being is "a material being which is possible only through the relation of everything contained in it to each other as end and means", and that a "basic power that is effectuated through an organization has to be thought as a cause effective according to *ends*" (*Teleological Principles* VIII 181.1–7). Two years later, Kant articulates a comprehensive metaphysical account of organized beings in both "Introductions" to the *CPJ* along with its second half. He would return to the life sciences in his last remarks in the *OP* during the years 1796–1804.

Kant's theory of biology in the *CPJ* not only belongs to a series of writings concerned with natural philosophy, and in particular, the explanation of organized beings, but as part of the third of his three *Critiques* it also fulfills a specific function within his critical project. In the third *Critique* Kant tries to find the "bridge" (*CPJ* V 195.16) or "transition" (*CPJ* V 176.14) between two realms of philosophy. The theoretical philosophy of the first *Critique* deals with the sensible realm of nature, represented through the laws of understanding, while the practical philosophy of the second *Critique* concerns the supersensible realm of freedom, represented through the law of pure practical reason and the moral law in its more teleological expressions, such as the Formula of Humanity and the kingdom of ends. The third *Critique* is supposed to connect both parts of philosophy in the unity of a single philosophical system:

> although there is an incalculable gulf fixed between the domain of the concept of nature, as the sensible, and the domain of the concept of freedom, as the supersensible, so that from the former to the latter [...] no transition is possible [...]: yet the latter *should* have an influence on the former, namely the concept of freedom should make the end that is imposed by its laws real in the sensible world; and nature must consequently also be able to be conceived in such a way that the lawfulness of its form is at least in agreement with the possibility of the ends that are to be realized in it in accordance with the laws of freedom (*CPJ* V 175.36 – 176.9).

The problem of the transition between the two realms of nature and freedom and their order is not primarily concerned with the agreement between the universally valid, a priori necessary laws of understanding (the universal laws of nature) and the universally valid, a priori necessary law of pure practical reason (the moral law): in the *Groundwork* (IV 421.18–20) Kant had already asserted an agreement between both kinds of laws in the second variant of the formula of the practical law[3]. Instead the transition concerns the possibility of a rational order of the contingent empirical manifold of nature, as represented in the empirical laws of nature (*CPJ* V 183.22–184.21). The contingency of the empirical order of nature could conflict with the necessary a priori order of pure practical reason.

An agreement between theoretical and practical laws is needed for the possibility of the realization of moral maxims (that follow from the law of pure practical reason) in the seemingly contingent manifoldness of the empirical forms of nature. It is needed both for the possibility of individual moral actions that are supposed to take place in nature, and for the possibility of the highest moral good, i.e., for an ideal world in which the totality of all moral maxims would have to take place in nature. Both thoughts presuppose that not only the universal structures of nature, but also the seemingly contingent empirical manifold of nature can be represented as a rational order. Organized beings and physical teleological laws account for the possibility of a rational order of the empirical manifold of nature since physical teleological laws contain the notion of a natural purpose, which is an idea of reason. The a priori unity of a natural purpose in physical teleological laws stands in a systematic connection with moral teleological laws since they both belong to the noumenal world. It would be useful if future research could explain the influence of Kant's changing views on organized nature upon the possibility of the realization of the practical law in nature and the possibility of the highest moral good.

1.3 Kant's Account of Organized Beings in the *CPJ* and Related Scholarly Debates

Kant's account of organized beings unfolds in four major parts of the text of the *CPJ*: in the first and abbreviated second "Introduction(s)", the "Analytic" and the

---

3 The second formula of the practical law says that one should "*act as if the maxim of [one's] action were to become by [one's] will a universal law of nature*".

"Dialectic", and in parts of the "Methodology of the Teleological Power of Judgment".

The prevailing concern in the "Introductions" is the clarification of the "transcendental principle" of the "formal purposiveness of nature" (*CPJ* V 181.13–5), a principle of the reflecting power of judgment that characterizes organized beings. Kant discusses it in particular in section V of the "Introduction". He never provides this principle with a precise formulation, which invites diverse interpretations of its sentence structure and meaning. But Kant at least gives several explicit descriptions of its function and systematic place: it is a principle that brings a necessary unity into the otherwise contingent empirical manifoldness of nature (*CPJ* V 183.14–184.10). It helps us to understand why we apply physical teleological maxims when we consider nature, such as 'in an organized product of nature nothing is in vain or purposeless' (*CPJ* V 376.13–4), 'nature takes the shortest way', 'nature makes no leaps, either in the sequence of its changes or in the juxtaposition of specifically different forms', 'the great multiplicity of its empirical laws is nevertheless unity under a few principles', and the like (*CPJ* V 182.19–25). Kant claims that we must presuppose the transcendental principle of the formal purposiveness of nature since otherwise "no thoroughgoing interconnection of empirical cognitions into a whole of experience would take place" (*CPJ* V 183.30–1).

Further remarks in the "Introductions" could be read as suggesting that the principle of the formal purposiveness of nature is the underlying principle for all physical teleological laws. These laws are maxims of the reflecting power of judgment that are "grounded on a principle a priori", although "we discover the end of nature solely through experience" (*First Introduction* XX 239.27–30). Kant seems to indicate that physical teleological judgments have two components: they contain an a priori principle of unity, on the one hand ('search for the purposive unity in the empirical manifold of nature!'), and they contain an empirical concept that we achieve through experience, on the other hand ('search for this unity under the empirical concept 'for flying'!'). In addition, physical teleological laws are thought to be final causal laws since a natural purpose ('for flying') is understood as a concept that brings about the united form of the organized being. It is the "concept of an object" which "contains the ground of *the reality* of this object" (*CPJ* V 180.31–2, italics added).

In the "Analytic of the Teleological Power of Judgment" Kant develops further claims about organized beings. One central statement is that organized beings are machines and mechanical aggregates, but not *mere* machines and mechanical aggregates. Organized beings can be explained to some extent based on mechanical laws and "*motive* power" (*CPJ* V 374.22), but cannot be explained based on mechanical laws and *motive* powers alone. Understanding them requires final causal or physical teleological laws and a *"formative* power" (*CPJ* V 374.23). Different from mechanical machines like watches, which are brought about as aggregates by an external engineer, organized beings like trees are generated by an internal principle as self-organized purposive wholes.

Kant's notions of mechanism and moving powers are a matter of controversy among scholars. He seems to presuppose these terms based either on discussions in his own earlier writings or on the historical context of his time; but he never explicitly defines them in the *CPJ*. Several scholars present catalogues of possible meanings of these terms (for example, McLaughlin [1989, 138–41/ 1990, 152–6, and McLaughlin in this volume]; Ginsborg [2001, 238–43]; Quarfood [2004, 196–205]; Zuckert [2007, 101–7]). One question is whether Kant's notion of mechanical laws reflects the causal law in the second analogy of experience in Kant's *CPR*, or the (dynamical and) mechanical laws in Kant's *Metaphysical Foundations*, or empirical laws more broadly. Another scholarly debate concerns the meaning of physical teleological laws and the nature of the formative power, i.e., the laws and powers that characterize organized beings *as such*. This analysis is made particularly difficult due to the lack of explicit examples of physical teleological laws and due to the rare occurrences of the term 'formative power' (van den Berg 2009, Frigo 2009, Goy 2012). The most significant occurrence of latter term seems to be a sentence in §65 (*CPJ* V 374.21–6) stating that the formative power is a self-organizing power that, as part of Kant's account of final causation, is responsible for the generation of the intentional, end-directed form of organized beings. For possible formulations of physical teleological laws one could begin with the antithesis of the regulative version of the antinomy (*CPJ* V 387.6–9). Further debates concern the relation of parts to parts and of parts to whole in an organized being, understood as a self-organized whole as opposed to a mechanical aggregate. Cheung (2009) provides new insights into the famous tree example on the basis of its historical context.

In the "Analytic" Kant claims that both mechanical and physical teleological laws characterize organized beings whereby only the latter are distinctive of organized beings as such. In the "Dialectic of the Teleological Power of Judgment" (§§69–78) Kant points out that an antinomy can arise within the power of judgment between higher-order mechanical and higher-order physical teleological laws, which unify the empirical laws involving organized beings. Kant attempts to show how the competition between these conflicting principles can be resolved.

The "Antinomy of Teleological Judgment" (§§69–78) is one of the most controversial passages of the *CPJ*. One controversy concerns what the conflict is about (see Frank/Zanetti [2001, 1288–9] and Quarfood [2004, 160–6]). It could be seen as a conflict between regulative maxims of the reflecting power of judgment, i.e., a conflict between alternative forms of explanation, as most current interpreters think (Allison [2003], Ginsborg [2006], Goy [forthcoming], Watkins [2009], Breitenbach [2009, 109–31], McLaughlin in this volume). Or it could be seen as a conflict between constitutive principles of the determining power of judgment, i.e., a conflict between the structures of organized beings, as an earlier generation of interpreters like Erich Adickes, Alfred Cyril Ewing, Ernst Cassirer, and Rudolf Eisler thought (for brief surveys of both views see McLaughlin [1989, 125–37], Watkins [2008, 254]). Quarfood (2004, 160–208 and in this volume) suggests yet a third reading of the intended conflict. He thinks that the antinomy consist in the confusion of a pair of regulative maxims of the reflecting power of judgment with a pair of constitutive principles of the determining power of judgment. He argues that although mechanism and teleology are regulative maxims, we are inclined to take them constitutively, that is, we are inclined to ascribe ontological commitments to those claims. A fourth possible reading of the antinomy is that the antinomy consists in a conflict between a constitutive principle of the determining power of judgment and a regulative maxim of the reflecting power of judgment.

Another major controversy concerns the resolution of the antinomy. Several suggestions have been made in the literature. In Quarfood's reading the resolution consists in the detection of confusions between both kinds of laws as regulative and constitutive principles and in the avoidance of the ontological claims entailed by the constitutive principles. Other interpreters suggest that the resolution of the antinomy is reached in a supersensible ground that contains the original unity of both kinds of laws. For Förster (2002a und b; 2008; 2011, 149–60) it is an intuitive understanding; others have called it an indeterminate ground of nature, or a conjunction of both. A different line of interpretation claims that the resolution of the antinomy consists in a unifying principle for our human judgment, such as the hierarchy of the two kinds of laws (see Ginsborg [2006, 461–2], Breitenbach [2009, 124–31]). Further questions concern the function of §§72–3— Kant's discussion of dogmatic teleologies of nature—within the overall argument of the antinomy.

In the final section of the *CPJ*, the "Methodology of the Teleological Power of Judgment", Kant describes the systematic order between organized beings (including human beings) and their natural purposes, and human beings as noumenal beings and their moral purposes. With regard to the relation between organized beings as natural ends and the human beings as natural and moral end(s) Kant claims that organized beings as natural purposes are ends in themselves that can serve as means and as ends for other organized beings. But they can serve as means only for the noumenal aspect of the human being and its moral purpose. The moral purpose of human beings as noumenal beings is the final end of nature since it is the only unconditioned end in the world (see Guyer in this volume, Höffe [2008]). Kant then clarifies the relation between organized beings and their natural and moral orders and the divine being (critical notions of God), arguing that the divine being guarantees the original unity of the natural and moral orders in the world (see Goy in this volume). §§79–91 of the *CPJ* in particular have not received sufficient critical attention so far.

2 Contributions to This Volume: An Overview

2.1 Part One: Kant's Theory of Biology and Research on Nature in the Seventeenth and Eighteenth Centuries

Part One consists of three essays that consider Kant's views on a range of topics in biology prior to 1790 and thus describe the immediate historical background to Kant's most explicit views as they are developed in the *CPJ*. In "Metaphysics and Physiology in Kant's Attitude Towards Theories of Preformation", Mark Fisher considers Kant's complex attitude throughout his pre-critical and critical periods toward the theories of organic generation that were prevalent among his predecessors (such as Leibniz, Malebranche, and Maupertuis), namely preformation and epigenesis. Fisher asserts that Kant, relying primarily on philosophical argument rather than empirical investigation, reconceives of these theories such that they do not necessarily exclude each other, and emphasizes the distinction between occasionalism and prestabilism. Specifically, Fisher maintains that Kant rejects the *individual* preformation of preexistence theories (i.e., preformation theories maintaining that individual members of each plant and animal kind are co-created), on the grounds that they are as inconsistent with the scientific method as direct divine intervention into nature would be. However, Fisher claims that rejecting this kind of view does not prevent Kant from accepting a theory of *generic* preformation (i.e., preformation with respect to kinds). Fisher then shows that Kant's version of generic preformation is perfectly consistent with an epigenetic theory that is committed to a fundamental generative power, which involves an immaterial and thus unknowable principle. In this way, Fisher illustrates how Kant can account for the vital functioning of organic bodies (by way of an epigenetic account of generation) but without undermining a natural-causal account of the production of bodies from other bodies of the same kind (which presupposes generic preformation).

In her "Epigenetic Theories: Caspar Friedrich Wolff and Immanuel Kant", Ina Goy investigates Caspar Friedrich Wolff's epigenetic account of organisms, describing in detail the essential force he asserts, the inorganic (mechanistic) and organic processes for which it is responsible, and the part-whole relations that it produces. She then considers Kant's various uses of the term 'formative power', arguing that he employs the term with two different meanings, one epistemological, the other biological. With this background in hand, Goy provides a detailed interpretation of a crucial passage in §65 of the *CPJ*, in which Kant attributes a formative power to organisms. Specifically, she lays out its basic elements (which distinguish it from machines and matter as such) and then offers an interpretation of Kant's claim that the formative power propagates itself, according to which it generates a certain organization in matter and preserves *itself* as an organic power. Goy concludes by taking issue with prior interpretations of Wolff's influence on Kant by arguing that although Kant's invocation of formative power has less to do with Blumenbach than is often supposed, Wolff's account does not resolve the main problem of explaining generation and thus is not of use to Kant to such an end. At the same time, according to Goy, Wolff's account remains systematically important for Kant, because Wolff does invoke the part-whole relation to account for organisms, which was not standard at the time, and Kant is close to him on that point. However, Kant disagrees with Wolff about the precise way in which the part-whole relation obtains in organisms, since he does not divide the process of production into an inorganic and an organic phase and posit the former as occurring prior to the latter. As a result, though Kant is systematically close to Wolff, his ultimate view still differs from Wolff's basic account in fundamental ways.

Rachel Zuckert's essay, "Organisms and Metaphysics: Kant's First Herder Review", discusses Kant's reasons, expressed in his first review of Herder's *Ideas towards a Philosophy of the History of Man* in 1785, for rejecting his former student's proposal of a single organic force. Zuckert argues that Kant's objections are based not on his own metaphysical assumptions about the difference between human beings and other organic life forms or on the distinctiveness of organic beings as contrasted with inorganic matter (which might be just as dogmatic as the views of Herder that he is criticizing), but rather on appropriately critical epistemological grounds. Specifically, she asserts that Kant rejects Herder's position because it cannot be supported on the basis of inductive generalizations or the direct observation of nature, nor can it be articulated in the form of universal scientific laws governing natural kinds. For, on Kant's view, in proposing a single life-force Herder is in effect postulating an unconditioned entity that necessarily goes beyond any evidence or experience we could have, which makes it an inadmissible dogmatic metaphysical assertion. Zuckert then suggests that Kant's view during this period is somewhat different from that of the *CPR* and that of the *CPJ*, such that it represents a transitional period for Kant. In the first Herder review Kant thinks that there can be legitimate determinative teleological judgment concerning organisms and that regulative principles can help us to discover teleological *laws*, whereas Kant is silent on these points in the first *Critique* and then comes to reject them later in the *CPJ*, though he does not, as a result of this final shift, become any more sympathetic to Herder's organic force. What these shifting constellations reveal, according to Zuckert, is that the significant question that Kant ultimately faces is whether Herder's concept of a single organic force or his own concept of purposiveness is most satisfactory when serving as a regulative principle that guides our scientific investigation.

2.2 Part Two: Kant's Theory of Biology—Commentaries on the "Critique of the Teleological Power of Judgment" and Other Writings

Part Two is by far the most substantive part of this volume, consisting of ten essays that present Kant's views as expressed in specific sections of the "Critique of the Teleological Power of Judgement" in the *CPJ* and in the *OP*. Luca Illetterati discusses how to understand organisms as natural ends, which is the central topic of sections §§61–8, in "Teleological Judgment: Between Technique and Nature". Illetterati argues that Kant attempts to find a coherent justification for a teleological explanation of organic life, one that does not require either backwards causation, which is metaphysically suspect, or the assumption of extrinsic purposiveness, which is plausible in the case of artifacts whose cause lies in the intentions of a designer, but not in the case of natural organisms. For natural organisms display a structure that is analogous to the technical-practical behavior of agents, but do not have intentions (except in humans) and would therefore have to have a different internal principle of self-organization. Illetterati argues that these reasons lead Kant to try to resolve the problem confronting our understanding of organisms by attributing to them a regulative rather than a constitutive status. Illetterati suggests, however, that Kant's solution is characterized by a lingering tension. On the one hand, Kant explicitly contrasts artifacts and organisms, since the former require, whereas the latter exclude external purposiveness. On the other hand, acknowledging that the principle of internal purposiveness can be attributed only a regulative status shows that we cannot grasp their true ontological status, since the only way we have of understanding it is by way of analogy with our technical-practical knowledge.

In "Kant's Account of Biological Causation", Predrag Šustar focuses on Kant's analysis in §§64–6 of the unique causal structure instantiated in natural ends (specifically of organisms) as what is both cause and effect of itself, arguing for two claims. The first claim is that Kant is committed to a dispositionalist view of biological causation. According to this view, biological processes are to be understood, fundamentally, in terms of powers, capacities, or dispositions and their necessary manifestations under appropriate circumstances and not in terms of Humean discrete events that are only loosely or contingently connected. Šustar also evaluates the strengths and shortcomings of Kant's account by comparing and contrasting it with an example from contemporary molecular biology, arguing that the case of photosynthesis reveals that Kant needs to appeal to a formative power and not simply a motive force. Šustar's second claim is that Kant's analysis is consistent with either eliminativism or deflationism about traditional metaphysical systems (such as hylozoism or animism), which has an influence on his dispositionalism about biological causation in general.

In "Nature in General as a System of Ends", Eric Watkins focuses on several passages in §§66–7 that assert that reflection on organisms necessarily leads to two claims about nature in general. The first claim is that not only organisms, but in fact *every* thing in nature must also be judged teleologically, while the second is that nature as a whole is a) a system of purposes that b) has a purpose itself. Watkins argues that Kant's distinctive conception of reason as a faculty that searches for the unconditioned condition of all conditioned objects is crucial to understanding Kant's arguments for each of these claims. For characterized in this way, reason has a legitimate interest not only in the inner form of organisms (with its reciprocal causal ties both among parts and between the parts and the whole they constitute), but also in the *external* conditions on these organisms and in the *purpose* of the *existence* of objects in nature. However, reason, in seeking further conditions in the systematic connections that occur within nature, also seeks a final unconditional purpose for nature as a whole. Such a purpose must of necessity lie outside of nature, and later on in the *CPJ* Kant identifies it with human beings, though understood not as natural organisms, but rather as free and rational noumenal agents.

In "Biological Purposiveness and Analogical Reflection", Angela Breitenbach considers the peculiar status of teleological judgments about organisms, which depend in some way on the particular experiences of such objects that we have, but without making any determinate claims about those objects. Breitenbach argues that such teleological judgments can be elucidated by clarifying their analogical status. In particular, she argues that the analogy Kant draws between organisms (as natural ends) and the capacity of practical reason (rather than artifacts in general) can explain two especially unusual phenomena. First, when we find that we cannot explain a certain class of objects purely mechanically, thinking about those objects as if they were directed at their own ends can guide further scientific inquiry into the object. For example, even if a bird's feathers cannot be explained purely mechanistically, thinking of them as directed to the end of flying can help us to direct our research into their weight, composition, and shape (rather than, say, their color). Second, and more important, the analogy explains why our encounter with living beings gives rise to the peculiar kind of experience that we have of organisms not merely as not mechanically explicable but also as eliciting the notion of objective natural ends, which one might otherwise never have reason to bring into conjunction with nature given that this concept is not, for all we know, a constitutive principle for natural objects. Specifically, Breitenbach argues that Kant takes the analogy to ground a *symbolic* representation of the purposiveness of living beings such that it makes the representation of something as a living being possible not by drawing out *existing* similarities between the apparent purposiveness of living beings and our own intentional activity, but by *projecting* thoughts that we associate with reason's intentional activity onto our consideration of organic nature, though without thereby assuming or asserting that nature must be that way. (This case is thus analogous to the aesthetic case in which our non-conceptual reflective response to an object gives rise to an experience of aesthetic pleasure but without representing the object as having the property "beauty".)

In "Mechanical Explanation in the 'Critique of the Teleological Power of Judgment'", Peter McLaughlin considers what Kant means when he refers, in the "Antinomy of Teleological Judgment", to the mechanism of nature that is necessary but still only a regulative maxim that reflects a peculiarity of our understanding. After canvassing a range of possible meanings of the term "mechanism" in this context, he argues that it has a reductionist meaning, according to which wholes (and their properties) are to be explained on the basis of their parts (and the interactions among them). McLaughlin then responds to several important objections that have been raised against this proposal in the literature. While some of these objections focus on specific features of the part-whole relationship that define mechanistic causality, the most serious objection, raised by Ginsborg, calls into question the very idea that mechanistic causality should be understood in terms of part-whole relations, on the grounds that machines are just as mechanistically inexplicable as organisms are. McLaughlin responds to this last objection by arguing that machines and organisms are mechanistically inexplicable for different reasons. Machines are mechanically inexplicable because they involve a form of concept-mediated causality, involving the actions of an embodied mind, whereas organisms seem to involve a causality *sui generis* that we cannot recognize as real. For this reason we must treat them as if they were, in part, ideal, in the sense that we know that the idea of the whole is not a cause, or real ground, of the organism, but rather simply a ground of cognition of the organism, or a sort of marker. McLaughlin then explores two different ways in which one might try to understand the peculiar status of mechanism as necessary but still only regulative rather than constitutive. One possibility would be to appeal to the specifically spatial character of the part-whole relation in bodies, which would thus involve a relation between intuition and the understanding, while the other possibility focuses on the essentially compositional nature of the understanding itself. McLaughlin does not find either of these possibilities to be particularly well-supported (either textually or philosophically) and thus ends on a skeptical note, namely that we still lack a satisfactory account of how the maxim of mechanism is genuinely necessary but merely regulative.

Marcel Quarfood, in "The Antinomy of Teleological Judgment: What It Is and How It Is Solved", argues for a novel interpretation of the "Antinomy of the Teleological Power of Judgment" and how Kant intends to solve it. Specifically, rather than choosing between standard interpretations of the antinomy available in the literature that focus exclusively on either the constitutive or the regulative principles that Kant formulates, Quarfood proposes that the antinomy arises when one takes the regulative maxims (which, he thinks, are not contradictory) as implying the constitutive principles (which are, in fact, inconsistent though not susceptible of direct proof). After explaining how his interpretation make sense of why Kant discusses the "dogmatic" systems described in §72, Quarfood argues that the regulative principles do not involve a contradiction because they are what Kant calls "disparate principles" (*CPJ* 391.11–5), which differ without being contradictorily opposed. He also argues that we are tempted to switch from the perfectly legitimate regulative, but disparate principles to the contradictory and thus illegitimate constitutive principles due to the heautonomy of reflective judgment. In the course of developing his interpretation, Quarfood offers an interpretation of the proofs of the regulative principles and an analysis of how transcendental idealism underlies the temptation to take the regulative principles as constitutive, providing in this way a comprehensive interpretation of the most fundamental features of Kant's antinomy of teleological judgment.

In "Purposiveness, Necessity, and Contingency", Philippe Huneman addresses the concepts of contingency and necessity in the context of Kant's discussion in §§76–7 of the *CPJ*. Specifically, after providing a brief description of Kant's account of organisms as natural purposes, Huneman explains what it means for purposiveness to embody the "lawfulness of the contingent as such" by noting how biological phenomena are contingent with respect to the necessity of the laws of mechanics, but still possess a kind of lawfulness (that allows one to distinguish, for example, a well-formed chick from a monstrous outgrowth, despite the fact that both follow necessarily from eggs according to the laws of nature). Huneman then examines Kant's transcendental genealogy of the very concept of purposiveness as necessarily embedded in the finiteness of our power of knowledge (§77). He concludes by showing how this elucidation is consistent with Kant's doctrine of modalities and by providing a justification for the solution of the antinomy of teleological judgment. The central idea behind Huneman's argument here is that what appears contingent to a being like ourselves, endowed with a discursive understanding that represents possibility and necessity, and a sensible faculty of intuition that represents existence, is none the less thought by us to have a kind of necessity, yet one that we cannot grasp directly, but rather only indirectly, by postulating of a different, intuitive kind of understanding, which, in effect, cancels the difference between mechanism and teleology.

Ina Goy discusses Kant's attitude towards the so-called argument from design in "Kant's Theory of Biology and the Argument from Design", which Kant treats in several passages in §§65, 75, 85 of the *CPJ*. After first reconstructing three different versions of the argument from design by Aquinas, Hume, and Paley and raising several objections to the argument, Goy considers various ways in which one might see Kant as making an argument from design in the "Analytic of the Teleological Power of Judgement": one based on the similarities between the production of art and the generation of natural things, a second based on the notion of the supersensible that Kant introduces in §§66–7, and a third that draws on the potential source of organic matter and specific kinds of formative powers. She then turns to various arguments one might locate in the "Dialectic of the Teleological Power of Judgement": one based on his use of the term 'technique of nature' and a second in the explicit presupposition of an intentional designer, though only as a subjective principle for the reflecting power of judgment. Finally, Goy considers Kant's explicit statements about different theistic proofs in the "Methodology of the Teleological Power of Judgment", including the 'physico-theological' proof and the moral argument that he had first developed in the *CprR*. Goy concludes by responding to possible difficulties of her interpretation.

In "Freedom, Happiness, and Nature: Kant's Moral Teleology", Paul Guyer argues that two paradoxes concerning freedom and teleology that naturally arise on the basis of passages in §§83–4 and §§86–7, can be resolved. The first paradox is based on two conflicting claims about the status of freedom. According to the first and second *Critiques* Kant views freedom as a *non-natural* capacity, but according to the *CPJ* freedom is supposed to serve as the final end of *nature*. The second paradox concerns Kant's account of happiness. According to the second *Critique*, happiness is included in the final end of nature in the form of the highest good. In the *CPJ*, by contrast, freedom, not happiness, is viewed as the final end of nature. Guyer resolves the first paradox by noting that freedom is not the final end *within* nature since nature cannot produce freedom. The final end within nature is culture, which serves as the preparation for freedom. Freedom, by contrast, is the normative final end of nature, which is distinct from nature and thus non-natural. Guyer resolves the second paradox by noting that the meaning of happiness that is part of the highest good as the final end of nature is not one's own happiness, but the happiness of all human beings. Moreover, Kant does not commit himself to the strong claim that nature guarantees our happiness but only to the weaker claim that nature accompanied by the right moral choices enables us to work for as much happiness as possible.

In his paper, "The Role of the Organism in the Transcendental Philosophy of Kant's *Opus Postumum*", Ernst-Otto Onnasch discusses Kant's views on the special status of organisms, especially concerning our grounds for asserting their actuality, given that we cannot intuit them a priori. After connecting Kant's views to Stahl's and showing that Kant wants to reject hylozoism as inconsistent with Newton's law of inertia, Onnasch argues that over time, and most clearly in the *OP*, Kant comes to adopt a model of cognition according to which experience is possible only if it is assumed that it is had by an organism. For to subsume the data given in intuition under concepts, it is necessary that a subject forms an intention or purpose that selects which data are relevant. So Kant is committed to the organism being not the object, but rather the subject of knowledge, and such an organism receives its organization not from without, but rather from its own spontaneous self-affection. The actuality of the organism is thus not determined through intuition (as it is for empirical objects), but rather as a condition of the possibility of experience, a view that becomes, Onnasch argues, more prominent in the *OP*.

2.3 Part Three: Kant's Theory of Biology in the Present Time

Part Three consists of two substantive essays that explore the significance of Kant's views for contemporary biology and philosophy of biology. In "Oughts without Intentions: A Kantian Account of Biological Functions", Hannah Ginsborg articulates and defends an account of biological functions in terms of a notion of normativity that draws on Kant's account of organisms as natural purposes (insofar as his account involves a notion of purposiveness that does not invoke actual objective purposes). According to Ginsborg, in the context of biology, a function of a trait or entity is neither what it was in fact designed, or selected, to do (as it would be on historical or etiological approaches to functions), nor what it contributes to what the organism is designed to do (as it would be on the causal role approach), but rather simply what it should, or ought, to do, which one might think of as a kind of 'natural ought'. Ginsborg's notion of normativity is, however, not the non-normative notion of prediction. It is more fundamental than either the notion of function that it is intended to explain or the notion of an intention in a designer or rational agent, which is often thought to be required for normativity. Ginsborg illustrates the notion by appealing to what she takes to be a primitive awareness of normativity in the workings of our own cognitive faculties, drawing in part on the later Wittgenstein's analysis of pre-conceptual rule following. She clarifies her account of a function further by showing that it is intended not to provide a naturalistic reduction of the notion of function in biology, but rather simply to remove the main conceptual obstacle to making functional ascriptions in biology (namely that it can seem problematic to ascribe oughts without intentions). She also clarifies that her account is not committed to any specific account of the circumstances in which function ascriptions are or are not justified, though she notes that it could be supplemented by a range of such accounts. In this way, Ginsborg draws on some basic features of Kant's account in laying the foundation for a contemporary theory of functions in biology.

In "Kant, Polanyi, and Molecular Biology", Siegfried Roth argues that Kant's anti-reductionist account of organisms is highly relevant to molecular biologists today. Though molecular biology, as currently practiced, seems to be reductionist insofar as, e.g., phenotypic features of organisms are explained on the basis of particular molecular changes, Roth argues that the reductive approach is not, in fact, as extensive as is often maintained. Roth sets the stage for his argument by providing an analysis of Kant's and Polanyi's views. On his interpretation, Kant adopts a generic preformationist view, with a concept-like self-representation of the organism controlling the epigenetic process by which the special kind of complexity distinctive of organisms arises, whereby the representation of the whole exists prior to its realization. Though he employs different terminology, Polanyi similarly maintains that the genetic material within an organism that contains the instructions for self-production is a system with higher and lower levels of control. Roth then turns to the contemporary context and shows that the structure of the sequence-based macromolecules of living cells (DNA, RNA, and proteins) cannot be explained purely on chemical grounds, but requires functional and evolutionary considerations and thus represents structures under dual control. Therefore, modern biologists ought to be sympathetic to Kant's and Polanyi's views and recast Kant's famous remark that there will never be a Newton who could make comprehensible the generation of a blade of grass, such that this is just as true even of the simplest bacterium.

3 Acknowledgements

The idea for a major conference that would bring together researchers in the field of Kant's theory of biology for the first time in the hopes of intensifying and enlivening the discussion (something that was, and continues to be a great need in the field), was born at the end of a research journey to California that Ina Goy made in spring 2009 to work closely with Hannah Ginsborg in Berkeley and with Eric Watkins in San Diego. The conference was supposed to take place at the end of April 2010 but was prevented by the eruption of the Eyjafjallajökull in Iceland (challenging our beliefs in the teleology of nature), since it forced the cancellation of most travel to and within Europe for several days, including that of our conference speakers. Fortunately it was possible to reschedule the conference for December 2010, when it finally took place in the beautiful castle of Hohentübingen surrounded by an exceptionally romantic and snow-covered countryside. We would like to thank the Fritz-Thyssen-Foundation and the Universitätsbund Tübingen for their financial support, especially for their flexibility and generosity regarding the repeated planning of the conference. We also thank all who were involved in the extensive logistical preparations for the project, in particular Christoph Wehle, who helped us to organize the conference twice, and Michael Demo, who helped us to edit the conference volume. Eva Oggionni and Julius Alves also provided support during the conference.

Ina Goy and Eric Watkins   Tübingen and San Diego, August 2013

References

Allison, Henry E. 2003, Kant's Antinomy of Teleological Judgment, in: *Kant's 'Critique of the Power of Judgment': Critical Essays*, ed. by Paul Guyer, New York: Rowman & Littlefield, 219–36.

Ameriks, Karl 2011, Das Schicksal von Kants *Rezensionen* zu Herders *Ideen*, in: *Immanuel Kant. Schriften zur Geschichtsphilosophie*, ed. by Otfried Höffe, Berlin: Akademie Verlag, 119–36.

Breitenbach, Angela 2009, *Die Analogie von Vernunft und Natur. Eine Umweltphilosophie nach Kant*, Berlin/New York: Walter de Gruyter.

Cheung, Tobias 2009, Der Baum im Baum. Modellkörper, reproduktive Systeme und die Differenz zwischen Lebendigem und Unlebendigem bei Kant und Bonnet, in: *Kants Philosophie der Natur. Ihre Entwicklung im Opus Postumum und ihre Wirkung*, ed. by Ernst-Otto Onnasch, Berlin/New York: Walter de Gruyter, 25–49.

Fisher, Mark 2007, Kant's Explanatory Natural History: Generation and Classification of Organisms in Kant's Natural Philosophy, in: *Understanding Purpose. Kant and the*

Philosophy of Biology, ed. by Philippe Huneman, Rochester: Rochester University Press, 101–21.

Fisher, Mark 2008, Organisms and Teleology in Kant's Natural Philosophy, Emory University.

Förster, Eckart 2002a, Die Bedeutung von §§76, 77 der *Kritik der Urteilskraft* für die Entwicklung der nachkantischen Philosophie, *Zeitschrift für philosophische Forschung* 56 (2), 170–90.

Förster, Eckart 2002b, Die Bedeutung der §§76, 77 der *Kritik der Urteilskraft* für die Entwicklung der nachkantischen Philosophie (Teil II), *Zeitschrift für philosophische Forschung* 56 (3), 321–45.

Förster, Eckhart 2008, Von der Eigentümlichkeit unseres Verstandes in Ansehung der Urteilskraft (§§74–78), in: *Immanuel Kant. Kritik der Urteilskraft*, ed. by Otfried Höffe, Berlin: Akademie Verlag, 259–88.

Förster, Eckart 2011, *Die 25 Jahre der Philosophie*, Frankfurt/M.: Klostermann.

Forster, Georg 1786, Noch etwas über die Menschenrassen, *Der Teutsche Merkur* 56, 57–86, 150–66.

Frank, Manfred/Zanetti, Véronique 2001, Dialektik der teleologischen Urteilskraft, in: *Schriften zur Ästhetik und Naturphilosophie*, ed. by Manfred Frank and Véronique Zanetti, Frankfurt/M.: Suhrkamp, vol. 3, 1286–306.

Frigo, Gian Franco 2009, Bildungskraft und Bildungstrieb bei Kant, in: *Kants Philosophie der Natur. Ihre Entwicklung im Opus Postumum und ihre Wirkung*, ed. by Ernst-Otto Onnasch, Berlin/New York: Walter de Gruyter, 9–23.

Ginsborg, Hannah 2001, Kant on Understanding Organisms as Natural Purposes, in: *Kant and the Sciences*, ed. by Eric Watkins, Oxford: Oxford University Press, 231–58.

Ginsborg, Hannah 2006, Kant's Biological Teleology and its Philosophical Significance, in: *A Companion to Kant*, ed. by Graham Bird, Oxford: Blackwell, 455–69.

Goy, Ina (forthcoming), The Antinomy of Teleological Judgment.

Goy, Ina 2012, Kant on Formative Power, *Lebenswelt* 2, 26–49.

Höffe, Ottfried 2008, Der Mensch als Endzweck, in: *Immanuel Kant: Kritik der Urteilskraft*, ed. by Otfried Höffe, Berlin: Akademie Verlag, 289–308.

Lenoir, Timothy 1980, Kant, Blumenbach, and vital Materialism in German Biology, *Isis* 71, 77–108.

Lenoir, Timothy 1981, The Göttingen School and the Development of Transcendental Naturphilosophie in the Romantic Era, *Studies in History of Biology* 5, 111–205.

Lenoir, Timothy 1982, Blumenbach, Kant and the Teleomechanical Approach to Life, in: Timothy Lenoir, *The Strategy of Life. Teleology and Mechanics in Nineteenth-Century German Biology*, Dordrecht/Boston/London, 17–34.

Look, Brandon 2006, Blumenbach and Kant on Mechanism and Teleology in Nature. The Case of the Formative Drive, in: *The Problem of Animal Generation in Early Modern Philosophy*, ed. by Justin E. H. Smith, Cambridge: Cambridge University Press, 355–72.

McLaughlin, Peter 1989a, *Kants Kritik der teleologischen Urteilskraft*, Bonn: Bouvier; transl. 1990, *Kant's Critique of Teleology in Biological Explanation. Antinomy and Teleology*, Lampeter: Edwin Mellon Press.

Trembley, Abraham 1744, *Mémoires pour servir à l'histoire d'un genre d'eau douce, à bras en forme de cornes*, Leiden: Jean & Herman Verbeek; English in: Lenhoff, Sylvia G./Lenhoff, Howard M. 1986, *Hydra and the Birth of Experimental Biology—1744. Abraham Trembley's Memoirs Concerning the Natural History of a Type of Freshwater Polyp with*

Armes Shaped like Horns, with a translation of Trembley's *Memoirs* from the French, Pacific Crove: The Boxwood Press.

Quarfood, Marcel 2004, *Transcendental Idealism and the Organism: Essays on Kant*, Stockholm: Almquist & Wiksell.

Richards, Robert J. 2000, Kant and Blumenbach on the Bildungstrieb: A Historical Misunderstanding, *Studies in the History and Philosophy of Biological and Biomedical Sciences* 31 (1), 11–32.

Richards, Robert J. 2002, Early Theories of Development: Kant und Blumenbach, in: Robert Richards, *The Romantic Conception of Life: Science and Philosophy in the Age of Goethe*, Chicago: University of Chicago Press, 207–37.

van den Berg, Hein 2009, Kant on Vital Forces. Metaphysical Concerns versus Scientific Practice, in: *Kants Philosophie der Natur. Ihre Entwicklung im Opus Postumum und ihre Wirkung*, ed. by Ernst-Otto Onnasch, Berlin/New York: Walter de Gruyter, 115–35.

Watkins, Eric 2008, Die Antinomie der teleologischen Urteilskraft und Kants Ablehnung alternativer Teleologien (§§69–71 und §§72–73), in: *Immanuel Kant. Kritik der Urteilskraft*, ed. by Otfried Höffe, Berlin: Akademie Verlag, 241–58.

Watkins, Eric 2009, The Antinomy of Teleological Judgment, in: *Kant Yearbook 1*, ed. by Dietmar Heidemann, Berlin/New York: Walter de Gruyter, 197–221.

Zammito, John 2007, Kant's Persistent Ambivalence Toward Epigenesis, 1764–90, in: *Understanding Purpose. Kant and the Philosophy of Biology*, ed. by Philippe Huneman, Rochester: University of Rochester Press, 51–74.

Zammito, John 2012, The Lenoir Thesis Revisited: Blumenbach and Kant, *Studies in History and Philosophy of Biological and Biomedical Sciences* 43, 120–32.

Zuckert, Rachel 2007, *Kant on Beauty and Biology. An Interpretation of the* Critique of Judgment, Cambridge: Cambridge University Press.

**Part I. Kant's Theory of Biology and Research on
Nature in the Seventeenth and Eighteenth
Centuries**

Mark Fisher
Metaphysics and Physiology in Kant's Attitude towards Theories of Preformation

The relevance of eighteenth-century biological theory to Kant's philosophy has become increasingly clear in recent years. Disputes between supporters of the theory of preformation and supporters of the theory of epigenesis,[1] which have long been known to historians of biology, have become familiar to historians of philosophy and Kant scholars as well. Kant's use of terms associated with these embryological theories makes it natural to think that a closer look at the state of embryology in the latter part of the eighteenth century could lead to useful insights concerning the development and articulation of his critical philosophy. Whatever this contextual approach might bring to our understanding of Kant's thought, when we look at his *CPJ*, with an eye towards situating his own views on organic generation within these disputes, several puzzling points emerge.

The first is that Kant's use of the terms '*Präformation*' and '*Epigenesis*' suggests that he understands the relation between these theories to be one of partial inclusion, rather than one of opposition between mutually exclusive options. Kant discusses *epigenesis* as a specific kind of preformationist theory that differs from evolutionist versions of *preformation* in important respects (*CPJ* V 376.10). A second point is that Kant's taxonomy treats the distinction between *preformation* and *epigenesis* as being of secondary importance for classifying and evaluating views of organic generation. The first distinction Kant draws is between *occasionalism* and *prestabilism* (*CPJ* V 375.20–35). A third peculiarity is the lack of stress Kant places on the role of particular investigations in providing a motivation for, or a defense of, one's choice of embryological theory. He does claim (*CPJ* V 378.7–9) that empirical grounds favor *epigenesis*; however, he does not follow this claim with any detailed discussion of the grounds he has in mind. Instead, he turns directly to the supposed advantages of the theory from the standpoint of reason (*CPJ* V 378.9–18), which he thinks should be convincing to those who remain uncertain about it on empirical grounds. The result of his explicit discussion of these theories in the *CPJ* appears to be a qualified adherence to *epigen-*

---

1 Hereafter, I will use 'preformation' to refer to the theory of preformation, and 'epigenesis' to refer to the theory of epigenesis.

esis, with the relevant qualification being that the causality responsible for the epigenetic capacities of organic bodies be conceived in preformationist terms.[2]

In the following, I will suggest that these features of the discussion of theories of organic generation in the *CPJ* are best understood by seeing Kant's attitude towards *preformation* as something that develops within his own particular approach to the project of early modern metaphysics. More specifically, his attempt to articulate a conception of a unified *order of nature* that includes both purely corporeal and embodied psychological events leads Kant to a preformationist position concerning the causality responsible for the vital functioning of organized bodies. The relevant contrast in the context of this broader conception of nature, however, is not between *preformation* and *epigenesis* (as it is in disputes concerning plant and animal generation), but between *prestabilism* and *occasionalism*.

Kant believes that the common early modern correlate of this metaphysical *prestabilism*, namely, the theory of preexistence,[3] conflicts with the aims and methods of natural science just as much as would the appeals to divine intervention into the *order of nature* they were developed to avoid. Accordingly, an interest in reconciling an appeal to immaterial or vital principles with the methods of natural science ought to lead one to reject *preexistence*. These same interests should also lead us, in Kant's view, to avoid the conclusion that the causality responsible for organic generation is not preformed, or constrained within determinate natural limits, in any way. Drawing such conclusions concerning the vital powers of organized beings opens us up to precisely the same objections leveled at theories of divine intervention and *preexistence*. None of these allows us to conceive of the causality responsible for organic generation as natural causality.

If Kant is right, the only way to appeal to vital principles in accounting for organic generation, while also maintaining that this generation occurs within the *order of nature*, is to accept a theory of *generic preformation*. Rather than attributing vital powers to matter, and thereby contradicting the commitment to its es-

---

2 Kant claims that epigenesis is rightly thought of as a theory of generic preformation because the "productive capacity of the progenitor is still performed in accordance with the internally purposive predispositions that were imparted to its stock, and thus the specific form was preformed *virtualiter*" (*CPJ* V 423.6 – 8, italics in text).

3 Hereafter, I will use *preexistence* to refer to the theory of preexistence. This view, which Kant refers to as "individual preformation", should be distinguished from the view of preformation that pre-dates the modern development of *epigenesis* as well as from the preformationism of eighteenth-century naturalists such as Haller and Bonnet. Unlike these views of preformation, *preexistence* denies that parent organisms play any causal role in the production of offspring. Instead, parent and offspring alike are produced by the creative act of God. The importance of this distinction in the history of embryology is stressed by Roger (1963/²1973, 325 – 6).

sential inertia, the generative power of organized beings is attributed to the non-empirical ground of both the matter and the specific form of self-generating organic bodies.[4] I will make my reasons for suggesting this understanding of Kant's attitude towards *preformation* clear by starting with a look at the general argument he provides for rejecting *preexistence* (section I). I will then consider what I take to be significant similarities between the argument against *preexistence* offered by the theoretical and experimental naturalist Pierre-Louis Moreau de Maupertuis (section II)[5] and the argument of Kant's own *Argument* essay (section III). This will be followed by a discussion of Kant's reasons for resisting the rejection of *preformation* altogether.

I Metaphysical Reasons for Rejecting *Preexistence*

In the 1763 *Argument* essay, Kant follows Maupertuis in citing metaphysical reasons for preferring the two-seed (epigenetic) approach to the formation of animal bodies taken by Descartes in his *Le Monde ou Traité de la Lumiére* to the one-seed (pre-existence)[6] approach taken by Malebranche in his *De la Rechereche de la vérité*. Unlike Maupertuis, however, Kant does not go on to develop his own account of how the seeds from existing organisms provide the grounds for the production of a genuinely new organism. Kant is content to argue for his cosmological point about the *order of nature*, which would stand regardless of whether a modified one-seed theory (like the preformationist theories of Haller and Bonnet)[7] or a modified two-seed theory (like the epigenetic theories of

---

4 Thus, Kant's view is preformationist in the sense that the active cause of the generation of a specific form in the matter of individual organized beings is that specific form itself as it is expressed in the matter of other individuals of the same kind.
5 I agree with Zammito (2006) about the importance of Maupertuis in this context. My understanding of Maupertuis' rejection of preexistence is informed by Hoffheimer (1982). Pyle (2006) provides a helpful discussion of the empirical and systematic grounds on which Malebranche bases his arguments for preexistence.
6 Preexistence theories are characterized by a commitment to the existence of all organized bodies in seed form from the time of conception. The most common version of this theory in the mid-eighteenth century was ovism, which held that the seeds of all future generations are contained in the first female, but there were also proponents of animalculism or spermism, which identified the first male as the location of these seeds. For discussion of the early modern treatment of animal generation, and of Maupertuis' relation to it, see the introduction to Roe (1981, 1–20).
7 Here I agree with Pyle (2006, 195) that one should be careful to distinguish between *preexistence*, which holds that "the new organism was always present either in the egg (ovism) or the sperm (animalculism)", and *preformation*, which holds that "the new organism is 'elaborated' in

Maupertuis and Buffon)[8] is determined to be the best replacement for *preexistence* in physiology. The same reasons that lead Kant to reject *preexistence* also lead him to favor a modified two-seed theory of the generation of animal bodies, together with a creationist theory of the origin of distinct kinds of animals.

A distinguishing feature of Kant's argument is that it does not have any clearly empirical premises. The argument is based almost entirely on metaphysical claims about the *order of nature*. This would be peculiar if Kant saw himself as entering into contemporary debates in physiology, such as the one between Buffon and Haller or the one about to develop between Haller and Wolff. It would be more understandable, however, if the immediate target of his criticisms were other metaphysical theories concerning the causal basis of the *order of nature*, such as those associated with Descartes, Malebranche, and Leibniz.[9] This leads me to think that these theories are his primary targets, and that the issue of plant and animal generation provides Kant with an illustration of what he takes to be a particularly important advantage of his own approach to the metaphysics of nature. My suggestion receives external support from other texts in which Kant addresses similar issues, e.g., his *New Elucidation* (1755), his *Inaugural Dissertation* (1770),[10] and §81 of the *CPJ* (1790). It becomes even more plausible, I believe, when we look at the text of the *Argument* and

---

the testes of the father, or the ovaries of the mother, in a process governed by the soul of the respective parent". Hoffheimer (1982) appeals to this distinction in his discussion of Maupertuis' rejection of *preexistence*, and Sloan (2002) draws our attention to its importance for understanding Kant's use of the vocabulary of *preformation* and *epigenesis*.

8 Maupertuis does not explicitly defend a two-seed theory in *Venus physique*, but he does state a preference for the view that the fetus arises from the "mixture of the seminal fluids given off by both sexes" (Maupertuis [2]1745 in 1966, 7).

9 Watkins (2005) argues persuasively that metaphysical disputes concerning the causal powers of created substances are directly relevant to the development of Kant's own views on causality. I follow him in rejecting the interpretation of early modern philosophy that views the theories of *occasionalism* and *preestablished harmony* primarily as responses to the particular problem of mind-body interaction. Broader considerations of the causal grounds of the *order of nature* provide the basis for particular approaches to psychophysical causality.

10 On both occasions where Kant presents and defends his own views publicly in front of the philosophy faculty at Königsberg, he situates these views concerning the causal principles responsible for the order of cosmological phenomena in relation to the *occasionalism* of Malebranche and the *preestablished harmony* of Leibniz, see *New Elucidation* (I 415.17–416.4) and *Inaugural Dissertation* (II 409.2–26).

compare Kant's argument there to a similar argument provided by Maupertuis in his *Venus physique* (^{2}1745).[11]

In these texts, both Kant and Maupertuis criticize the hypothetical apparatus through which metaphysicians like Malebranche attempt to save a commitment to understanding and explaining the phenomena of plant and animal generation by appeal to the *order of nature*. Neither is impressed by the particular way in which Malebranche seeks to combine theological, cosmological, and physiological considerations in arguing for *preexistence*.[12] Even if peculiar features of the production of plants and animals should lead a theistic metaphysician to accept that God's role in nature is not limited to the creation of *res extensa* and the legislation of motion, it is not clear that there is any compelling reason to follow Malebranche in asserting that *each individual* in the plant and animal kingdoms is the direct result of God's creative activity.

Postulating that each plant and each animal was actually produced simultaneously, with generation after generation nested within one other, does not make it any easier for us to conceive of how the production or formation of plants and animals is possible. Positing such a hypothesis over other competing hypotheses —maintaining, e. g., that animals are formed successively out of the seminal fluid(s) of the parent(s)—also does not provide any advantages for the practice of natural science. These appear to provide good reasons, independent of any particular empirical considerations, for rejecting *preexistence*.

II Maupertuis' Argument against *Preexistence* in his *Venus Physique*

In order to fill in the details of the argument against *preexistence*, and to see why this metaphysical argument alone does not decide the issue between *preformation* and *epigenesis*, I will turn to Maupertuis' *Venus Physique* before looking at

---

11 *Venus physique* (^{2}1745) is the title of the second edition of Maupertuis' *Dissertation physique à l'occasion du nègre blanc* (1744). It is not clear that Kant knew this particular text, but Maupertuis published his views on generation in several forms during the 1740s and 1750s. Kant owned a copy of *Versuch von der Bildung der Körper* (1761), a German translation of the Latin dissertation *Dissertatio inauguralis metaphysica de universali naturae systemate* (1752) in which Maupertuis presented his views. I refer to *Venus physique* (^{2}1745) because it is the edition most often discussed in the literature on generation. Nothing in my interpretation of Kant's view depends on the claim of direct influence. I am grateful to Jennifer Mensch for her help on this point.
12 Pyle (2006) offers a compelling analysis of the line of thinking that leads Malebranche to accept *preexistence*.

Kant's *Argument*. Although Kant will not follow Maupertuis' particular suggestions for providing an epigenetic account of generation (*Argument* II 115.3–6),
he is sympathetic both to Maupertuis' argument against *preexistence* and to
his commitment to attributing to some natural bodies a capacity to produce
new individuals of their kind.

Maupertuis begins his argument against *preexistence* by providing a reconstruction of its appeal to thinkers in the early modern period:

> Most modern Physicists, guided by the analogy of plant growth, which might seem like the
> production of new parts, but is really the development of parts already present in the seed
> or the bulb, could not understand how an organized body might actually be produced.
> Such scientists wished to reduce all generation to simple development. Consequently
> they believed that all animals of each kind were already completely formed inside either
> the father or the mother, thus obviating the necessity of a new reproduction (Maupertuis
> [2]1745 in 1966, 40).

According to Maupertuis, *preexistence* is presented as a solution to a general
problem faced by modern natural philosophers that gains some measure of plau-
sibility through an analogy between plant physiology and animal physiology.[13]
Dissection of plant bulbs has revealed that parts that become visible only later
are already present, articulated, and folded in on one another in the bulb.
Thus, accepting that animals develop in ways analogous to plants, we might
think that what seems to be the inexplicable production of new animal parts
is really only the development of animal parts that are already present in seed
form. Regardless of how compelling this analogy may be, however, Maupertuis
argues that preexistence is not actually a way of solving the problem of understanding how organized bodies are produced:

> As a fact, when it is most difficult to understand how an organic body, such as an animal,
> can be formed with each new generation, would it be easier to conceive how an infinite suc-
> cession of animals, contained in each other, could have been formed at once? This is self-
> delusion, adopted in the hope of resolving the problem by simply making it more remote.
> The difficulty remains the same unless it becomes an even greater one to conceive all these
> highly organized creatures being formed one in the other and all within the first member of
> the species rather than believing that they are produced successively (Maupertuis [2]1745 in
> 1966, 41).

Maupertuis' interpretation of the move to *preexistence* is not a very charitable
one. If he is right, thinkers like Malebranche have deluded themselves in think-

---

13 For a discussion of the role that a philosophical commitment to modern physics plays in the
rise of *preexistence* in general, and of Malebranche's version in particular, see Pyle (2006).

ing they have provided an explanation of animal generation, when in fact they have only pushed the problem back all the way to creation. Since doing this does not allow us to understand how animal bodies are formed the appeal to *preexistence* does not accomplish what supporters claim it does. It may even make matters worse by requiring us to conceive of the original production in a peculiar sort of way; i.e., as the simultaneous production of each and every individual member of a particular species with each successive generation enveloped within the prior generations.

Given the obviousness of this criticism, one might wonder whether the move to preexistence could be seen in a more favorable light,[14] and if so, whether it would still be open to criticism. A slightly more charitable interpretation would be to see it as a way of saving modern theories from the appearance that animal generation poses any problem specifically for them. The fact that these theories do not allow us to understand how organized bodies are produced is a legitimate criticism, only if there actually is some explanation of animal generation that we *could* understand. Moreover, it would be a legitimate criticism of one's choice to adopt a modern theory over, say, an Aristotelian or Scholastic alternative, only if the latter actually provided such an explanation.

However, if we think that the attempt to understand the production of animals is like the attempt to understand the production of substances, or of a world, then a modern natural philosopher can contend that although *there is* an explanation here, which we can legitimately *refer to* in some contexts, it is not an explanation that we can *understand*. The difficulty we started with remains, of course, for the defender of *preexistence*—we are still left with the problem of animal generation. This difficulty is simply no longer considered to pose a problem specifically for our modern theory, or for our choice to side with the moderns over the ancients or the scholastics. Still, Maupertuis seems right to contend that even if it is *no more* difficult to conceive of how an infinite number of successively smaller animals could be encased within one another than it is to conceive of how a single animal could be produced from others, it is also *no less* difficult. The self-delusion he claims to see here may be a product of thinking that one can save a modern theory in this way, while still criticizing other theories for lacking the resources to render animal generation intelligible to us.

If we allow the defender of *preexistence* to claim that we cannot understand how an animal body can be formed, this claim ought to be open to defenders of

---

14 Pyle (2006) suggests that Malebranche accepts preexistence as the result of an argument from elimination. The three possibilities for accounting for complex functional structures are *epigenesis*, *preformation*, and *preexistence*, and the only option that renders their existence intelligible is *preexistence*.

other theories as well. This means that it no longer counts as a criticism of theories of animal generation that they do not allow us to understand how animal bodies are produced. Of course, this might lead us to wonder what the point of such theories is. If we insist that their real point is to deliver this understanding, we are not likely to be impressed by *preexistence*, or by any other theories that agree with them here. If we remain open to the possibility that our theorizing has other legitimate ends, however, we might be able to accept this consequence of *preexistence* without accepting the theory itself. If other standards guide our theory choice, someone like Maupertuis might even be able to argue that we have principled reasons for not accepting *preexistence* that are largely independent of the recognition that the theory leaves the process of animal generation obscure:

> If metaphysical reasons can count for something in such a case, Descartes' ideas on the creation of the embryo through the mixture of the two seminal elements have something eminently worth considering. We have no reason to suspect that he entertained the idea out of respect for the Ancients or because he could think of nothing better. If we do believe that nature's Creator does not simply leave the creation of animals to laws of motion, and if we do believe that He had first to lend a hand and create all these animals contained in one another, what is gained by thinking He created them all at once? And what has natural science lost by the idea that animals are formed successively? For God, is there any real difference between one moment in time and the next (Maupertuis [2]1745 in 1966, 42)?

What is interesting about these passages is not that Maupertuis takes it that metaphysical considerations might play a role here, but rather that he suggests that the theistic metaphysical principles of those who adopt *preexistence* do not actually provide the support necessary to maintain it as a distinct view. Granting the common objections to Descartes' theory of generation, and ascribing to God a more immediate causal role in the production of animals, should leave it an open question whether or not Descartes was right to think of animals as being generated *out of* a mixture of two seeds, rather than being created *out of* nothing and then developing from a single seed. That animals are produced *by* God, and are produced *in* one another, does not have any necessary consequences concerning *when* they are produced or *what* they are produced out of.

What is more, there seem to be no scientifically or theologically relevant advantages to postulating that they are all produced out of nothing in the beginning. Why not allow that God's atemporal agency is causally responsible for creating and sustaining a world containing series of natural events that include the successive formation of a new animal *out of* the mixture of seminal elements provided by both parents in the womb of the mother? Defenders of *preexistence* have already taken the most obvious objection to this view out of play; i.e., the objection that we cannot understand how this occurs, since it would apply equally to *preexistence*. Perhaps it would be even more correct to say that God *creates* in such a way that at the moment of conception a new animal *is generated*, than it would be to index God's activity itself to a prior point in time and say that God *created* all these animals in the beginning. If, despite our inability to understand how it works, a physiological model that resembles Descartes'[15] could provide us with a more compelling way of saving our natural scientific commitments, then we appear to have good reasons for preferring it. Thus, if I understand Maupertuis' intent in these passages correctly, he is arguing that one ought to reject *preexistence* on those same rational or metaphysical grounds that might appear to support it, and that this rejection is justified independently of whether or not some opposing model of organic generation is deemed to be a plausible replacement.

III Kant's Argument against *Preexistence* in the *Argument*

It is striking how closely Kant's argument in the *Argument* follows the logic of the argument provided by Maupertuis in his *Venus physique*.[16] Kant too accepts the objection to Descartes' theory of generation that is common among defenders of *preexistence*, and denies that *preexistence* follows directly from a rejection of the Cartesian theory:

> Now, it would be absurd to regard the initial generation of a plant or animal as a mechanical effect incidentally arising from the universal laws of nature: nonetheless, there is a twofold question, which has remained unanswered for the reason mentioned. Is each individual member of the plant- and animal-kingdoms directly formed by God, and thus of supernatural origin, with only propagation, that is to say, only the periodic transition for the pur-

---

15 The model that Maupertuis goes on to suggest will differ from Descartes' in a very important respect; namely, Maupertuis will attribute vital powers to the material particles that arrange themselves into complex functional structures. Zammito (2006) is right to see this as part of the objection that Kant has to Maupertuis' model.
16 Zammito (2006) also draws attention to this similarity. While I think he is right about the significance of Maupertuis' work, I read the *Argument* in a way that is importantly different. Where he sees Kant describing *preformation*, I see Kant describing *preexistence* (*Argument* II 114.21–4). Where he sees Kant describing *epigenesis*, I see Kant describing a commitment that is common to *preformation* and *epigenesis* (*Argument* II 114.24–8). My interpretation makes better sense of the passage as a whole and provides an answer to a question Zammito raises about the relation to Haller. What Kant rejects in this context is not Haller's *preformation*, but *preexistence*.

poses of development, being entrusted to a natural law? Or do some individual members of the plant- and animal-kingdoms, although immediately formed by God and thus of divine origin, possess the capacity, which we cannot understand, actually to generate their own kind in accordance with a regular law of nature, and not merely to unfold them (*Argument* II 114.15 – 24)?

We see from this passage that Kant thinks the mere fact that a theory of generation does not allow us to understand how organic bodies are generated does not rule out the choice-worthiness of that theory. On one side of the disjunction we have individual members of the plant and animal kingdoms directly formed by God (*preexistence*). On the other side, we have original members of kinds directly formed by God with the capacity to generate others according to natural laws (*preformation* or *epigenesis*). Neither of these theories provides an explanation of animal generation that satisfies our desire to understand; yet we also do not appear to be in a position to withhold assent altogether here. Like Maupertuis, Kant suggests that, faced with a decision between theories that are equally unable to provide this understanding of organic generation, we appeal to metaphysical considerations in deciding between them:

> There are difficulties on both sides, and it is perhaps impossible to make out which difficulty is the greatest. But our concern here is merely to determine the relative weight of the various reasons, in so far as they are metaphysical in character. For example: in the light of everything we know, it is utterly unintelligible to us that a tree should be able, in virtue of an internal mechanical constitution, to form and process its sap in such a way that there should arise in the bud or the seed something containing a tree like itself in miniature, or something from which such a tree could develop. The internal forms proposed by *Buffon*, and the elements of organic matter which, in the opinion of *Maupertuis*, join together as their memories dictate and in accordance with the laws of desire and aversion, are either as incomprehensible as the thing itself, or they are entirely arbitrary inventions. But, leaving aside such theories, is one obliged for that reason to develop an alternative theory oneself, which is just as arbitrary, the theory, namely, that, since their natural manner of coming to be is unintelligible to us, all these individuals must be of supernatural origin (*Argument* II 114.28 – 115.15)?

It is clearly a form of objection to characterize a theory as either 'as incomprehensible as the thing itself' or 'an arbitrary invention'; however, when the objection of utter unintelligibility is leveled at the alternative, as it is here, we clearly are not taking it as decisive for rejecting one theory and accepting the other. Kant is saying that, even if Buffon or Maupertuis were right about what goes on in the tree when the bud is formed, it would not follow that they have comprehended, understood, or rendered this process intelligible. There is no independent basis of knowledge concerning the explanatory grounds they assume (elements of organic matter, internal molds) that would allow us to grasp the generation possibility of plants and animals *a priori*, or to see that this process was possible, even if we did not already know that it was actual. If we are merely providing more detailed, micro-level descriptions of what is taking place, then as interesting and informative as the accounts may be, we are not actually rendering this process any more comprehensible than it had been before. We have a finer-grained picture of the *explanandum*, but we still have not understood it through a prior grasp of the *explanans*.

If, alternatively, these theorists are postulating entities whose prior existence would make it possible that something like this could take place, the *ad hoc* nature of the introduction of these principles (elements of *organic* matter, *internal* moulds) would undermine their supposed role in rendering this process intelligible to us (even if the existence of such principles would, in fact, render this process intelligible *per se*). Elements of matter may be principles available to us for our explanation of the natural generation of bodies. Insofar as matter is a generic concept under which the concepts of particular kinds of bodies can be subsumed, however, this concept does not refer specifically to whatever real grounds distinguish particular kinds of body from one another. Adding the term '*organic*' does specify this generic term, and it might appear reasonable to think that elements of specifically *organic* matter stand in the same relation to specifically *organic* bodies, as elements of matter stand to bodies. Even if this turned out to be the case, though, this conceptual exercise would not have somehow removed the special difficulties involved in understanding the natural generation of *organic* bodies. Similarly, positing moulds involved in organic generation that are unlike the moulds *into which* we pour metals in that they serve as *internal* constraints on the form a body can take, does not allow us to understand organic forms by reference to the physical mechanisms through which the causality responsible for their formation is exercised. In both of these cases, we would appear to have tried to explain something we do not understand by positing something we understand even less, which we never would have conceived of except as the supposed cause of these particular effects.

On the issue of comprehensibility or intelligibility, Kant sees no real advantage to siding with Maupertuis and Buffon against those who defend a creationist account of all members of the plant and animal kingdoms. The advantage of answering the central two-fold question in the same general way becomes apparent only when we bring metaphysical considerations of the *order of nature* into our evaluation of *preexistence*. Here again, I think we can see Kant adopting the general strategy employed by Maupertuis:

In this case, the origin of all such organic products is regarded as completely supernatural; it is, nonetheless, supposed that the natural philosophers have been left with something when they are permitted to toy with the problem of the manner of gradual propagation. But consider: the supernatural is not thereby diminished; for whether the supernatural generation occurs at the moment of creation, or whether it takes place gradually, at different times, the degree of the supernatural is no greater in the second case than it is in the first. The only difference between them relates not to the degree of the immediate divine action but merely to the *when*. As for the natural order of unfolding mentioned above: it is not a rule of the fruitfulness of nature, but a futile method of evading the issue. For not the least degree of an immediate divine action is thereby spared. Accordingly, the following alternatives seem unavoidable: either the formation of the fruit is to be attributed immediately to a divine action, which is performed at every mating, or, alternatively there must be granted to the initial divine organization of plants and animals a capacity, not merely to develop their kind thereafter in accordance with a natural law, but truly to generate their kind (*Argument* II 115.13 – 23).

Kant's point here is that if the only difference between *preexistence* and what he will later call *occasionalism* (i.e., the view that God creates a new individual *on the occasion* of the mixing of the parent's seeds) is one concerning *when* the divine action occurs, metaphysical considerations of the *order of nature* will lead us to lump these supposedly distinct theories together as denying that the causal agency responsible for organic generation is natural agency. If we are committed to the view that (whether we can understand and explain them in this way or not) all events are to be treated as natural events, and the class of theory that now includes both *preexistence* and *occasionalism* does not allow us to do so with many common events, then we appear to have good reason to reject this class of theory. This reason, however, is not the kind of reason that one would expect to enter directly into disputes between physiologists about the proper interpretation of observations and experimental results. Rather, it is the kind of reason that would appeal to the metaphysician (or transcendental philosopher) who is concerned with identifying and, where possible, understanding the sources and limits of the fundamental presuppositions that guide the practice of natural philosophy.

IV Prestabilism and Epigenesis-as-Generic-Preformation

If I am right, then looking at the above passages from Maupertuis and Kant allows us to identify a common argumentative strategy aimed at undermining the credibility of *preexistence* as a theoretical framework for plant and animal physiology. This strategy does not, however, require that we deny all forms of *preformation*. According to the alternatives Kant discusses, we can choose the occasionalist view of the formation of the fruit or seed, or we can choose the view that the formation of such fruits or seeds is a natural process. The occasionalist view is just as incompatible with the fundamental presuppositions of natural philosophy as is *preexistence*, so the conclusion to be drawn is that we ought to reject it and "concede to the things of nature a possibility, greater than that which is commonly conceded, of producing their effects in accordance with universal laws" (*Argument* II 115.26 – 8). Kant's considered view is that the only consistent way of conceding this possibility to the things of nature is to adopt a form of *preformation*. He is more explicit in the *Argument* essay than he is in his later works about what kind of *preformation* he has in mind,[17] but it is clear in the *CPJ* (§§80 – 1) that Kant thinks of *epigenesis* as a preformationist theory. We might wonder, then, what exactly distinguishes the version of *preformation* he thinks is necessary from *preexistence*, on the one side, and from views that involve a more thoroughgoing rejection of *preformation*, on the other.

In response to the first of these, the version of *preformation* Kant thinks is necessary holds that the individual fruits and individual seeds, some of which eventually develop into mature plants and animals, are themselves products of the natural functioning of other plants and animals of the same kind. This is the rule of the fruitfulness of nature that is lacking in the account of the unfolding of preexisting seeds. The admission that we cannot understand how it is possible for the functioning of these plants and animals to produce such fruits and seeds should not preclude us from presupposing that it, in fact, does. This view is still preformationist in that it presupposes the natural functioning of plants and animals of various kinds as the real grounds of the process through which a new plant or animal of the same kind is generated; i. e., the specific form that the particular matter composing this new body comes to have during the process is the active cause of the process itself.[18] The visible, outward form by

---

17 This seems to me to be an important difference between Kant's treatment in the 1760s and the way he addresses these issues in the 1780s and in the third *Critique*. He appears to be more careful in the critical period to distinguish the claim that the existence of the natural species in some form is an ontological presupposition of the generation of any individual member of that species from the claim that the temporal regress in generations must terminate in an *ungenerated* (but dependent and, therefore, *created*) generator. For more on this point, see Fisher (2008, 235 – 45).

18 If we understand the distinction between *preexistence* and *preformation* along the lines suggested by Pyle (2006), Hoffheimer (1982), and Roger (1963/²1973), the active cause of generation for the preformationist is the soul of the mother or of the father. In my view, Kant attempts to overcome the problem this presents for understanding heredity and the worry about

reference to which we identify *this* particular body as a plant or animal of *this* kind is the temporally posterior effect of the process, but it is the "inner form" in virtue of which it is already a member of this kind before it takes on this outward form that is central for a causal theory of generation.

The outward form can be exhibited by the matter of the embryo in sequential stages, on this view, only if the 'inner form' is already present as the real ground of the unity of this sequence *as* stages within the life cycle of a plant or animal of this kind. An epigenetic account of the generation of individuals out of apparently unformed masses of matter, on Kant's view, presupposes the 'inner form' of the kind of which these individuals are members. Thus, this account cannot also provide us an explanation of the origin of this kind. If there is a genetic or natural historical account of the origin of specific kinds (species of a common genera, varieties or races of a common species) it will have to be an appeal to the confluence between the generative or formative power of some more general kind and other powers at work in nature. The generative or formative power is assumed as providing the real grounds for the development of the individual members, while the other powers of nature act as occasional causes of those characteristics that distinguish these members from other members of the same general kind in specific ways.[19]

Kant denies that the *explanans* in plant and animal physiology has to be an *explanandum* for one of our more basic scientific endeavors, if the practice in which we appeal to it is to be legitimate. What he does not deny is the requirement to conceive of this *explanans* in a way that is generically similar to the way in which we conceive of the *explanans* in other natural scientific endeavors. Thus, even if the specific generative powers or formative drives that we are constrained to appeal to in accounts of animal generation cannot be derived from the motive powers common to all bodies, we are still constrained to think of these specifically different powers in generically similar ways. We treat them as individually necessary conditions for the development of some specific form within a body, which become sufficient for the development of that particular form only in combination with other individually necessary conditions. It means, further, that we conceive of these powers as exercised in such a way that the resulting series of events can be subsumed under natural laws. If we do not conceive of them in this way, as stable or lawful relations between natural substances and their manifold states, we end up sacrificing all the gains we

---

the appeal to the soul by treating the *generative power* of the species (*Races* II 430.2) or the *self-propagating formative power* (*CPJ* V 374.28) as providing for the unity of the species.
19 Kant provides detailed discussions of this model in the 1775 and 1785 essays on races of man (*Races* II 429.6–443.30, *Human Race* VIII 91.1–106.5).

made in rejecting *preexistence* and *occasionalism* as frameworks for plant and animal physiology. Kant sums this view up nicely in his review of Herder's *Ideas:*

> [I]f the cause organizing itself *from within* were limited by its nature only perhaps to a certain number and degree of differences in the formation of a creature […] then one could call this natural vocation of the forming nature also 'germs' or 'original dispositions'. [One could do so] without thereby regarding the [germs] as primordially implanted machines and buds that unfold themselves only when occasioned (as in the system of evolution), but merely as limitations, not further explicable, of a self-forming faculty, which latter we can just as little explain or make comprehensible (*Review of Herder* VIII 62.20 – 63.2).

This is one of the central points to be cognizant of in appreciating Kant's attitude towards *preformation* and the role it plays in his theory of biology. If we grant that animals are not machines, but are, rather, bodies animated by some principle distinct from matter and the motive forces of bodies, we have to be careful about how we conceive of these principles and how far we extend their supposed influence. In avoiding the false dilemma between the equally unsatisfactory options of a strictly mechanical account of generation and *preexistence*, we also need to be careful not to attribute characteristics to these animating principles that would render their effects scientifically inscrutable. Kant thinks that *preformation* is necessary to steer this course in a way that is consistent both with seemingly obvious facts about nature and with the legitimate demands that are made on our theorizing about the causes responsible for these facts.

V Conclusion

If the preceding interpretation is correct, understanding Kant's somewhat complex attitude towards *preformation* requires us to be sensitive to the different perspectives from which one can address the problem of organic generation. According to Kant, the physiologist is warranted in providing an epigenetic account of generation, but is also constrained by the need to appeal to the vital functioning of organic bodies in such an account. The metaphysician or the transcendental philosopher is warranted in appealing to immaterial principles as grounds of this possibility, but is also constrained by the requirement to do so in a way that will not undermine a natural-causal account of the production of these bodies from other bodies of the same kind. Malebranche and other adherents to *preexistence* fail to recognize this latter constraint, and Kant agrees with Maupertuis that this failure provides us with non-empirical reasons for rejecting *preexistence*. The commitment to a fundamental generative or formative power, by contrast, allows Kant to offer a model of organic generation that recognizes both of these constraints. That model involves adherence to a version of *epigenesis* that can also be described as *generic preformation*.[20]

References

Churchill, Frederick B. 1970, The History of Embryology as Intellectual History, *Journal of the History of Biology* 3, 155 – 81.

Descartes, René 1964 – 76, *Œvres de Descartes*, ed. by Charles Adam and Paul Tannery, Paris: Vrin.

Detlefsen, Karen 2006, Explanation and Demonstration in the Haller-Wolff Debate, in: *The Problem of Animal Generation in Early Modern Philosophy*, ed. by Justin E. H. Smith, Cambridge: Cambridge University Press, 235 – 61.

Fisher, Mark 2007, Kant's Explanatory Natural History: Generation and Classification of Organisms in Kant's Natural Philosophy, in: *Understanding Purpose. Kant and the Philosophy of Biology*, ed. by Philippe Huneman, Rochester: Rochester University Press, 101 – 21.

Fisher, Mark 2008, Organisms and Teleology in Kant's Natural Philosophy, Emory University.

Greene, Marjorie/Depew, David 2004, *The Philosophy of Biology: An Episodic History*, Cambridge: Cambridge University Press.

Guyer, Paul 2001b, Organisms and the Unity of Science, in: *Kant and the Sciences*, ed. by Eric Watkins, Oxford: Oxford University Press, 259 – 81.

Hoffheimer, Michael 1982, Maupertuis and the Eighteenth-Century Critique of Preexistence, *Journal of the History of Biology* 15, 119 – 44.

Huneman, Philippe 2007b, Reflexive Judgment and Embryology: Kant's Shift Between the First and the Third Critique, in: *Understanding Purpose: Kant and the Philosophy of Biology*, ed. by Philippe Huneman, Rochester: University of Rochester Press, 75 – 100.

Malebranche, Nicolas 1674, Recherche de la vérité, in: Nicolas Malebranche, *Œvres complètes*, ed. by Geneviève Rodis-Lewis, Paris: Vrin 1965, 20 vols., vol. 1.

Maupertuis, Pierre-Louis Moreau de 1744/²1745, *The Earthly Venus*, trans. by Simon Brangier Boas, New York & London: Johnson Reprint Corporation 1966.

Pyle, Andrew 2006, Malebranche on Animal Generation: Preexistence and the Microscope, in: *The Problem of Animal Generation in Early Modern Philosophy*, ed. by Justin E. H. Smith, Cambridge: Cambridge University Press, 194 – 214.

Roe, Shirley 1981, *Matter, Life, and Generation: Eighteenth-Century Embryology and the Haller-Wolff Debate*, Cambridge: Cambridge University Press.

Roger, Jaques 1963/²1971, *Les sciences de la vie dans la pensée fraçaise au XVIIIe siècle*, Paris: A. Colin.

Sloan, Phillip R. 2002, Preforming the Categories: Eighteenth-Century Generation Theory and the Biological Roots of Kant's A Priori, *Journal of the History of Philosophy* 40 (2), 229 – 53.

---

20 I would like to thank Peter McLaughlin, Ina Goy, and Eric Watkins for their written comments on earlier versions of this paper.

Watkins, Eric 2003, Forces and Causes in Kant's Early Pre-Critical Writings, *Studies in History and Philosophy of Science* 34, 5–27.

Watkins, Eric 2005, *Kant and the Metaphysics of Causality*, Cambridge/New York: Cambridge University Press.

Zammito, John 2006a, Kant's Early Views on Epigenesis: The Role of Maupertuis, in: *The Problem of Animal Generation in Early Modern Philosophy*, ed. by Justin E. H. Smith, Cambridge: Cambridge University Press, 317–54.

Zammito, John 2007, Kant's Persistent Ambivalence Toward Epigenesis, 1764–90, in: *Understanding Purpose. Kant and the Philosophy of Biology*, ed. by Philippe Huneman, Rochester: University of Rochester Press, 51–74.

Ina Goy
Epigenetic Theories: Caspar Friedrich Wolff and Immanuel Kant

Around the middle of the eighteenth century, a new interpretation of the origin of organic life arose and began to replace preformation theories: the doctrine of epigenesis. Defenders of preformation theories claimed that the origin of an organism is explained by a divine preformed germ, which—like a russian doll—contains in miniature all features of the prospective living being. Different preformation theorists held different views on the nature of the divine preformed germ: ovists believed the female egg to be the germ; animalculists, in contrast, the male sperm. Advocates of ovistic theories in the sixteenth and seventeenth centuries were William Harvey (1578–1657)—who at the same time was an eclectic Aristotelian and early defender of epigenesis—Marcellus Malpighi (1628–1694), and Jan Swammerdam (1637–1680). In the eighteenth century Albrecht von Haller (1708–1777), Charles Bonnet (1720–1793), and Abbé Lazzaro Spallanzani (1729–1799) advocated ovistic preformation. The most important animalculist theories were developed from the second half of the seventeenth until the beginning of the eighteenth centuries. Prominent animalculists include Antoni van Leeuwenhoek (1632–1723), Nicolaas Hartsoeker (1656–1725), and Gottfried Wilhelm Leibniz (1646–1716).

The theory of epigenesis belongs to the tradition of Aristotelian biology. William Harvey transferred it into early modern science. According to epigenetic theories, organic life begins with a self-organizing natural power that inheres in unstructured matter. Due to different views on the nature of organic matter and on the formative powers, a variety of forms of epigenetic accounts appeared. The earliest representatives of epigenesis in the seventeenth and eighteenth centuries, like Pierre-Louis Moreau de Maupertuis (1698–1759), Georges-Louis Leclerc de Buffon (1707–1788), and John Turberville Needham (1713–1781), understood epigenetic powers mechanically. Later defenders of this doctrine established increasingly vitalistic accounts of the formative power. Caspar Friedrich Wolff (1734–1794) marked the transition from mechanistic to vitalistic accounts of epigenetic powers. A version of the latter was advocated, for instance, by Johann Friedrich Blumenbach (1752–1840). Whereas most of the mechanical interpretations of epigenesis explained the self-organizing processes of organisms in Newtonian terms of attraction and repulsion, vitalistic accounts ascribed entirely new capabilities to the epigenetic powers, such as sensitivity, irritability, intelligibility, and spontaneity.

In this paper, I investigate the relation of Immanuel Kant's (1724–1804) theory of biology to epigenetic accounts of organic generation and development. In the literature, a dispute about similarities between Blumenbach's epigenetic account and Kant dominated the debate for many years (see Lenoir 1980, 1981, and 1982, 17–34, Richards 2000; 2002, 207–37; Look 2006, and van den Berg 2009). Some more recent interpreters claim that Wolff's, more than Blumenbach's account plays the pivotal role in the development of a vitalistic conception of epigenesis in Kant (see Dupont 2007 and Huneman 2007).

Although I myself hold the view that Kant's position contains preformistic and epigenetic characteristics, in the current paper I focus solely on an investigation of epigenetic elements in Kant's account and compare them to the corresponding epigenetic elements in Wolff's theory. Section I of the paper is devoted to an analysis of Wolff's most important epigenetic theorems: the notion of the essential power (*vis essentialis*) and the conception of the part-whole composition of organized matter. Although Wolff describes the essential power vitalisticly, as a principle of life, he understands it as the cause of mechanical motions explaining the generation, nourishment, and the growth of an organism. Wolff's model of the part-whole composition of organic matter is subtle, but committed to fundamental mechanistic assumptions, such as that the organism as a whole is composed of inorganic parts. In section II, I analyze the corresponding elements in Kant's theory: the notion of the formative power and the conception of the whole-part composition of organized materials. Kant describes the formative power as a principle that causes the purposive form of an organized being such that matter and mechanism are the means to the purpose of the being as their end. The purpose of the whole is a functional unit which is in principle superior to the form and matter of the subordinate parts. The parts are combined into such a whole in being mutually cause and effect of each other and in being related to the superior whole. In section III, I respond to the debate in the literature. Against Dupont (2007) and Huneman (2007) I argue that, according to Wolff, the *vis essentialis* accounts for mechanic effects in matter, whereas, according to Kant, the formative power explains the intentional order (form, end, purpose) of an individual organized being, its parts, and its species. Since this view is closer to Blumenbach than to Wolff, the ongoing comparison between Kant and Blumenbach in the literature is justified. However, the emphasis on the specific part-whole composition that Kant considers to be the determining feature of an organized being can be found only in Wolff and not in Blumenbach— though Wolff and Kant describe it in opposing ways. This increases the systematic importance of Wolff for Kant. Thus, a fresh look on the historical debate is required.[1]

---

1 A few historical remarks in advance: within the whole Kantian œuvre not a single passage refers to Caspar Friedrich Wolff (including the reference to a "vis essentialis" in *Lect. Met. Herder* XXVIII/1 49.18). However, Kant was indirectly aware of Wolff at least from the description of his account in Johann Friedrich Blumenbach's dissertation *Über den Bildungstrieb und das Zeu-*

1 Epigenetic Elements in Caspar Friedrich Wolff's Account of the Organism[2]

1.1 Wolff's Conception of the Essential Power (*vis essentialis*)

Organic bodies, according to Wolff, are "formed during the process of genera-tion". Therefore a theory of epigenesis must investigate those "powers" that are the cause of this formation (Wolff 1764, 61–3). Wolff claims that living bodies have a certain power that "distributes the nutritive fluids through the parts of the

---

gungsgeschäfte, which Kant verifiably read (see *Teleological Principles* VIII 180.31–5; *CPJ* V 424.19–34; *Correspondence* XI 184.29–185.25, 211.1–23; see also Löw 1980, 175–80). But Blu-menbach's writing (1781, 14, 17–8) contains only a few sentences on Wolff, in two short passages. And it is not clear what to make of those remarks: in the first passage, Blumenbach attributes a mechanistic account of the *vis essentialis* to Wolff, while in the second passage he attributes one that is vitalistic. In the first passage, he warns the reader not to "mingle th[e] [formative] drive with the vis plastica, with the vis essentialis, with chemical fermentation and blind expansion, or with other merely mechanical powers". He explicitly states that the *nisus formativus* cannot be reduced to mechanical powers, and names the *vis essentialis* among these mechanical powers. In the second passage, Blumenbach argues that the "distinction between the nisus formativus and the so-called vis essentialis is easier to overlook". He recommends to compare "the definition of the vis essentialis that its famous inventor [*H. Casp. Friedr. Wolf *Theorie von der Generation*, p. 160] introduced" with his own definition of the formative drive, and quotes Wolff's definition as follows: the *vis essentialis* "is that power through which, in vegetable bodies, all features are initiated that cause us to ascribe life to them. It is on these grounds that I have named this power a vis essentialis of these bodies; namely, because a plant would cease to be a plant if this force were removed from it. It is found in animals just as much as in plants, and everything that animals and plants have in common depends on this power alone". With regard to the second passage, it is not easy to grasp why Blumenbach did *not* feel close to Wolff's definition of the *vis essentialis* (since it is similar to his own description of the formative drive).— Beside Blumenbach's dissertation, between 1777 and 1779, Kant also intensively studied Johann Nicolas Tetens' *Philosophische Versuche über die menschliche Natur* (1776/77), which outline the debate on embryology between the advocates of preformation and epigenesis, including Wolff's account of the *vis essentialis*. However, both sources convey only Wolff's conception of organic powers but not of his conception of part and whole in organized matter (which Wolff himself considered to be of great importance). Two letters by Johann Georg Hamann (1777 and 1779 in 1957, 337 and 1959, 81) to Johann Gottfried Herder from October 15, 1777 and May 17, 1779 testify to Kant's Tetens studies.

2 In this paper, I argue on the basis of Wolff's Latin dissertation, *Theoria generationis* (1759), and the extended German version of the dissertation *Theorie von der Generation*, published in 1764. In the years after 1766, Wolff also published the writing *De formatione intestinorum* in two volumes of the Proceedings of the Academy of the Sciences at Petersburg. His final work on embryology was the essay *Von der eigenthümlichen und wesentlichen Kraft* (1789). All English translations of Wolff's Latin and German writings in this paper are mine.

body" and thereby triggers the "formation" of the whole (ibid., 37). The initiating power of all processes of generation in organic bodies is the so-called essential power (*vis essentialis*). This force is the "first principle of generation" in nature ("[p]rimum [...] generationis principium"; Wolff 1759 [1999], §233) and is considered to be "that specific power which in vegetable bodies initiates all features that cause us to ascribe life to organic beings" (Wolff 1764, 160). This determination is important because it explicitly entails a reference to the "life" of organisms. It is therefore one of the most vitalistic functional definitions of the essential power. Its presence in Wolff's account could strengthen the consensus among recent researchers that Wolff—with his conception of the specific power in organisms—intends to make a turn to a vitalistic position (see Huneman 2007 and Dupont 2007). Thus let us consider Wolff's description of the essential power in closer detail.

The essential power is essential because "a plant" would not continue to "be a plant" without this force (Wolff 1764, 160). It is a *"sufficient principle for the generation of both plants and animals"* (Wolff 1759 [1999], §242). It effects a "precisely determined distribution of the fluids" (Wolff 1764, 163), *"accumulating the fluids from the surrounding earth, coercing them to enter the roots, distributing them through the plant, partly saving them at different places, partly excreting them again"*. Whereas the "ingestion", "distribution", and "evaporation" of "fluids" serve as an impetus to plant growth (Wolff 1759 [1999], §1), the *"diminution of the quantity"* of the nutritive fluids inhibits it (ibid., §95).

Variation in the nutritive fluids caused by the essential power leads to the formation of the substance of a plant. The essential power causes variation in the absorption (ibid., §§1, 3), and attraction (ibid., §§2, 4), distribution (ibid., §§1, 5, 7, 22–3.), penetration (ibid., §§7, 22, 60), and transition (ibid., §§5, 7, 22), repletion (ibid., §81), expansion, augmentation (ibid., §§4, 21–2, 25), deposition (ibid., §§22–3), and finally the excretion and exhalation of nutritive fluids (ibid., §§1–2, 26–7). Reflection on these different effects leads to the conclusion that the essential force, although described as a principle of vegetable life, is the embodiment of physical determinations; it is the cause of all the different mechanical motions of a plant.

Wolff also discusses a faculty of "solidification", which inhibits growth through a kind of coagulation and cohesion of the vegetable substance (ibid., §§242–3). Roe (1979) and Dupont (2007) take the faculty of solidification to be a second principle that is distinct from the essential power. So Roe (1979, 5–6) says that in his dissertation "Wolff proposed a model for development in plants and animals based on two factors: the ability of plant and animal fluids to solidify, and a force, which he named the *vis essentialis* (essential force)". Similarly Dupont (2007, 40) claims vegetable or animal "development" to be "based on two factors: the essential force and the tendency of plant and animal fluids to solidify" (see also Duchesneau 2006, 173, 177).

However, it is not clear whether Wolff does indeed distinguish the faculty of solidification from the essential power: in some passages consolidification and deposition seem to belong to the effects of the essential power (Wolff 1759 [1999], §187; see also §§22–3, 26–8, 61). Furthermore, he says that *"the essential force together with the faculty of solidification"* is "one sufficient principle *both of the development of plants and animals"* (ibid., §242, my italics). However, in other passages Wolff argues for a functional distinction between both faculties and designates them as two principles whose interplay effects the "order of all parts of a plant and their specific composition" (ibid., §93). Regardless of which position one takes in this dispute, it is of *systematic* importance that the faculty of solidification, like the essential power, plays a role in a physical-mechanical process.

The formation of a plant begins from a specific vegetation point. It is located at the end of the caulis or the stem. From this point, the development of the leaves, blossoms, and fruits proceeds by emission and excretion. It is an initial area that continuously generates vegetable meristem. The kind of structure that is produced depends on the amount of nutritive fluid that reaches the vegetation point (ibid., §§43–4). Since Wolff assumes that generation is similar in plants and animals (see Wolff 1764, 164–5, 203), he argues that animals must have something like the vegetation point found in plants. In animals it is located in the embryonic disk and the surrounding area of the umbilicalis. It is at this point that arteries, veins, and the heart first arise from the unstructured matter of the yolk (Wolff 1759 [1999], §§173–81). Just as in plants, in animals, the *"essential power"* determines the transport of *"nourishing parts"* from the yolk to the evolving living being and from the egg to the embryo (ibid., §§168–9). Wolff claims that *"forwarding the materials from the fetus"* can be *"caused only by the same essential power"* that causes their *"separation"* (ibid., §187).

As we have seen, Wolff defines (ibid., §§187, 233) the essential power by reference to a multiplicity of functions and effects. Thus, it seems all the more astonishing that in several passages of the text he claims that he is unable (and unwilling) to describe the essential power more precisely (ibid., §4)—a statement which seems founded in scepticism concerning the sufficiency of the conceptual means chosen to describe the nature of this force. Whereas he outlines the basic function of the essential power as a vitalistic one (it is the principle of life in plants and animals), the conception of the force that generates this life remains mechanical.[3] Therefore Wolff's account contains an internal discrepancy, which might have been the reason for Wolff's discontent.

1.2 Wolff's Conception of the Formation of Organic Matter

In the third part of his Latin dissertation and also in the German publication *Theorie von der Generation*, Wolff describes the formation of organic substance or matter using a specific part-whole relation. The most important feature of this relation is that the whole arises from the composition of the parts. It always occurs later than the parts. A researcher of nature "who is not able to talk about the structure of the parts and the composition of the body, and who cannot indicate the principles for the parts and their composition, and demonstrate how the parts and the composition are determined by these principles cannot explain generation either" (Wolff 1764, 13).

Wolff claims that the "formation" of an organic body occurs "little by little by the addition of matter or by the congregation of parts" (Wolff 1759 [1999], §235). This is so because "without the composition of the parts", the "transfer of nutritive substances" that is fundamental for the formation of an organism cannot proceed (ibid., §238). Wolff gives two reasons for his claim that the parts of an organism form a whole: first, "*the parts cannot exist alone without each other*"; second, all of the individual parts "*receive some of their nourishment*" from other parts of the body (ibid., §236). The degree of "*organization de-*

---

3 Breidbach (1999, xxii–iii, my translations) interprets Wolff as offering an entirely mechanical account, especially in the *Theoria generationis*. He claims that the "basic power postulated by Wolff" is "understood as a mechanism" that cannot be interpreted "in a vitalistic manner". The "shape of an organism" not only with regard to its "function" but also "with regard to its generation is mechanical". It is the "result of a mechanism" and the "product of a process that can be explained in an entirely naturalistic fashion". But Breidbach's interpretation misses Wolff's determination of the *vis essentialis* as a principle of life which is not mechanistic even though Wolff's terminology remains mechanical. Duchesneau also emphasizes the mechanical nature of Wolff's *vis essentialis*, but he more cautiously reconstructs the intricate structures of this mechanism which at some points seem to transcend its mechanical nature: the *vis essentialis* is a "material force", that "selects among the material elements for the sake of organic structuring, but this "for the sake of" is only a metaphorical formula, and the discriminating function of this force should be compared with "chemical affinities" and with mechanical phaenomena dependent on attraction/repulsion" (Duchesneau 2006, 172). The *vis essentialis* "seems to foreshadow the notion of a *vital principle*. But, looking more closely, it appears that those bodies on which vegetation acts are inorganic in their ingredients; and the end products of this process are devices of a complex mechanical type on which the function of the resulting organism depend" (ibid., 177).

pends on the amount of the composed parts" dedicated to its nourishment if the "*common source of nourishment for all parts remains the same*". The organization "*dwindles if the amount of sources of the nourishment increases*", and organization "*completely disappears if the body is dissolved into inorganic parts*" (ibid., §237).

In the *Theoria generationis* (1759) Wolff distinguishes three types of parts that are formed by the entering fluids: separate, distinct, and imaginary parts. Separate parts are formed "*by the excretion from that part of the stem which they rest on*"; distinct parts are formed "*by the sedimentation from that part of the stem by which they are enwrapped*"; and imaginary parts are formed "*neither by excretion nor by sedimentation but rather by the mere extension of the substance in which they occur*" (ibid., §239).

The organization of an organism follows a composition of the parts of the whole which is yet inorganic, and which becomes organized afterwards. The preliminary production of an inorganic body occurs according to the internal dependency of the parts. The ontologically superior part is generated first. Only once all of the parts necessary to the structure of the organism have been formed is the body considered organized and organic. Only once all of the inferior parts —which remain inorganic until their production is complete—are structured into their respective wholes do they become organized and organic. For Wolff, the organic whole is never prior to the parts, and the organization, as a structuring principle, never precedes the structure.

In the *Theorie von der Generation* (1764), Wolff modifies his description of the three kinds of part. The first class consists of simple and ultimate parts, out of which all other parts are composed. The second class consists in composition of simple and ultimate parts that cannot exist independently, and that themselves are parts of other parts. And the third class of parts consists of compositions of simple and ultimate parts from the first class and of compositions of parts from the first and the second class (Wolff 1764, 145–6). As in the *Theoria generationis* (1759), Wolff describes the relationship between the essential power and the formation of an organism from its parts as a process of *inorganic* production preceding a process of the *organization* of matter. He tells us that it could be considered as a "general law of the formation of natural bodies" that "every organic body or part of an organic body" is first produced "without any organic structure" and only afterwards "is rendered organic" (Wolff 1764, 163). However, it is also difficult in *Theorie von der Generation* to pinpoint the transition from inorganic to organic based on Wolff's account.

First, a preliminary production merely produces the outer outline of the parts of the whole organism. The fluids penetrate the young, still inorganic parts (which lack all vessels) and are distributed equally through the young part. Locations growing up equally necessarily receive the same amount of nourishing fluids; however, locations that expand faster and stronger, necessarily receive a greater amount of nourishing fluids. This first step of inorganic production generates all inorganic parts (ibid.), followed by a second process of inner organization. This further step consists in a differentiation and consolidation of structures and results in the actual vitalization and organization of the parts: the nourishing fluids distributed by the essential power now produce the vessels and vesicles.

Just as the vitalizing effects of the essential power (the grounding of organic life) are to be produced by mechanical causes (by powers of penetration, attraction, expansion, excretion), Wolff claims that the generation of organic matter is to proceed in a mechanical fashion first by an aggregation and composition of parts that form an inorganic whole and afterwards by an internal structuring that adds as much complexity as necessary for the body to be an organic being.

2 Epigenetic Elements in Immanuel Kant's Account of Organized Beings: Formative Power and the Formation of Organized Materials

The precise formulation 'formative power [bildende(n) Kraft]' appears in fourteen passages within the whole Kantian œuvre.[4] Two of those passages—*CPJ* V 374.21–6 in §65 and *CPJ* V 423.12–424.6 in §81—belong to Kant's published writings; though only the passage *CPJ* V 374.21–6 in §65 refers to Kant's own account. All other appearances occur in lectures, notes, reflections, and fragments—texts which Kant himself did not authorize for publication. The term 'formative power' is a rare term. Nevertheless, placed at the center of §65, the formative power might be an indispensable part of Kant's account of biological causation.

---

4 There are, of course, more passages in which Kant discusses epigenetic conceptions of powers under varying names, for instance a "capacity for [...] formation [Bildungsvermögen]" (*CPJ* V 371.25) in §64 of the *CPJ*, and a "generative power [Zeugungskraft, zeugende Kraft]" in his two early writings on races (*Races* II 435.1–436.8, *Human Race* VIII 98.11–99.12). In §81 of the *CPJ*, he mentions Blumenbach's "*formative drive* [Bildungstrieb]" (*CPJ* V 424.34) in a review of epigenetic positions. Furthermore in §58 he talks about a chemical version of "formation [Bildung]" (*CPJ* V 348.11, 21, 25; 349.1; 350.1). The limitation but also the value of the following investigation is its concentration on those selected passages where Kant precisely uses the term 'formative power'.

Reading all fourteen passages results in a surprisingly clear picture of two different treatments of the term 'formative power' in Kant's writings. The earlier meaning appears in eight passages and belongs to *epistemology*. All passages stem from the 1770s. In this early (pre-critical) view, Kant treats formative power as a source of the spontaneous production of mental representations in the human (seven passages) and in the animal's mind (one passage). The term 'formative power' designates a productive force of the consciousness to spontaneously generate representations on both the sensual and the conceptual level. In its most elaborate version, Kant distinguishes six kinds of spontaneously generated *sensual* representations: "re-formations [Abbildungen]", "post-formations [Nachbildungen]", and "pre-formations [Vorbildungen]", "in-formations [Einbildungen]", "anti-formations [Gegenbildungen]", and "ex-formations [Ausbildungen]" (*Lect. Met. L$_1$* XXVIII/1 235.24–237.28). In addition, he identifies two kinds of spontaneously generated *conceptual* representations: concepts (categories) and laws of the understanding.

The later meaning of the term 'formative power' appears in six passages, one of them stems from the early 1780s, five from the 1790s onwards until Kant's latest notes. All six passages belong to the field of *biology*; however, their particular contents and backgrounds are so diverse that they cannot be used to interpret each other. Thus, for the purposes of the current investigation, I only consider the allegedly well-known passage in §65 of the *CPJ*, since this is the most pertinent source for our understanding of Kant's notion of a formative power.[5] In this passage, Kant treats the formative power as a natural force that is responsible not for creating or generating organized matter, but for establishing and sustaining the organized teleological *order* or *form* of organized beings.[6]

---

5 For more detailed discussions see my paper "Kant on Formative Power" (2012).
6 Here is a list of these passages:

Lect. Met. L$_1$
(1) XXVIII/1 230 – 240   mid 1770s   epistemology
(2) XXVIII/1 276   mid 1770s   epistemology
Notes and Fragments on metaphysics
(3) XVII 736, refl. 4811   phase τ 1775 – 6?, μ 1770 – 1?   epistemology
Notes and Fragments on anthropology
(4) XV/1 95, refl. 251   phase ν^1 1771?, ρ^1 1773 – 5?, φ^1 1776 – 8, χ^1 1778 – 9 epistemology
(5) XV/1 127, refl. 321   phase λ 1769 – 70?, ξ 1772?   epistemology
(6) XV/1 383, refl. 872   phase υ 1776 – 8   epistemology
(7) XV/2 699, refl. 1484   phase σ 1775 – 7   epistemology
Lect. Moral Phil. Mrong.
(8) XXVII/2.2 1498   mid 1770s   epistemology
Notes and Fragments on metaphysics

The crucial passage in §65 consists of only one intricate sentence:

[a] An organized being is thus not a mere machine, for that has only a *motive* power, [b] while the organized being possesses in itself a *formative* power, [c] and indeed one that it communicates to the matter, which does not have it ([d] it organizes the latter): [e] thus it has a self-propagating formative power, which cannot be explained through the capacity for movement alone (that is, mechanism). [Ein organisirtes Wesen ist also nicht bloß Maschine: denn die hat lediglich *bewegende* Kraft; sondern es besitzt in sich *bildende* Kraft und zwar eine solche, die es den Materien mittheilt, welche sie nicht haben (sie organisirt): also eine sich fortpflanzende bildende Kraft, welche durch das Bewegungsvermögen allein (den Mechanism) nicht erklärt werden kann] (*CPJ* V 374.21–6).

The sentence contains the following five claims: a) the formative power distinguishes organized beings from machines with which they share motive powers; b) the formative power belongs to the organized being in itself; c) the formative power is communicated by the (organized) being to materials, materials do not have formative power; d) when communicated to materials the formative power organizes a being; e) the formative power is a self-propagating formative power.

What do these claims mean? a) An organized being is partly identical with a machine, namely insofar as it possesses motive power. But it differs from a machine insofar as it possesses a formative power that cannot be identified with the capacity of motion alone. The formative power can involve but cannot be reduced to the mechanisms of motion. b) The formative power is an *intrinsic* power in the organized being. It does not externally cause the organized being (as for instance the formative power of an artisan that produces the artificial object). The formative power is a natural capacity of and is effective in the organized being.

---

(1) XVIII 574, refl. 6302	phase ψ^2 1783–4		biology
CPJ			
(2) §65, V 374	1790		biology
(3) §81, V 423–4	1790		biology
Lect. Met. K$_2$			
(4) XXVIII/2.1 761	early 1790s		biology
OP			
(5) XXI 475	1786–98		biology
(6) XXI 630	1798–9		biology

The dating of the passages follows the editors of the "Academy Edition" and the editors of "The Cambridge Edition of the Works of Immanuel Kant". The reflections on metaphysics and anthropology are dated by Adickes, see the editorial remarks in *Notes and Fragments* XIV xxv–liv. For dates of the lectures on metaphysics, see the editorial remarks by Ameriks and Naragon in the "Cambridge Edition" of the *Lectures on Metaphysics* (1997, xxii).

Section c) is a difficult, ambiguous part of the sentence, which also has bearing on the different meanings of section d). The "Cambridge Edition" translations of c) and d) fail to convey an important aspect of the German text. The original Kantian text says in c) and d) that an organized being has a formative power "und zwar eine solche, die es den Materien mittheilt, welche sie nicht haben". Using 'den Materien' in c), and correspondingly 'welche', and 'haben' in d), Kant indicates plural, i.e., he does not suggest that the formative power acts upon matter, but upon several materials. The "Cambridge Edition" translators write: "the matter, which does not have it". Using 'matter' in c), and correspondingly 'does' in d), they—at first glance—indicate a singular, although 'matter', as the word 'Materie' in German, does not exclusively designate a singular.

One possible reading of Kant's claim that an organized being communicates formative power to materials, which do not have formative power is: cα) The organized being communicates the formative power to materials (reading "den Materien" as 'allen Materien'), none of which have formative power. This non-restrictive reading suggests that in an organized being the formative power acts upon all kinds of matter, and that no kind of matter itself has formative power. The consequence of this reading is that the formative power is itself not material, for otherwise it would be part of matter. The formative power then is an immaterial power. In addition, an organized being that contains formative power "in itself" cannot be an entirely material being, for at least its formative power is an immaterial element "in" the organized being. In line with this reading, White (1997, 134) stresses that the formative power "is in some essential sense distinguishable from the matter" it determines.

cβ) Emphasizing "*den* Materien" in the sense of 'only those', an alternative reading is that the organized being communicates the formative power only to those kinds of materials that do not have it. In this restrictive reading it is possible to interpret the formative power itself as part of matter. It could be a material power that occupies some parts of matter (organized materials), whereas it is communicated to all other raw matter (unorganized materials) that do not have formative power and that will be formed by the formative power. The distinction between cα) and cβ) is that the formative power in cα) is immaterial whereas in cβ) it is material. The ambiguity of the passage allows both readings. Frigo (2009, 13, 15) describing "matter as formative power [Materie als Bildungskraft]" and "matter as formative drive [Materie als Bildungstrieb]" seems to hold cβ).

Look's (2006, 372) proposal, in contrast, provides indirect support for cα). He argues that precisely since Blumenbach identifies the formative power as a part of matter, Kant thought that he had to depart from Blumenbach. Kant criticized Blumenbach for determining the formative drive as "a feature of *all* matter". Furthermore, a defender of cα) could stress that Kant's text suggests at several places that the formative power as the cause of the purposive form of nature is analogous to the human will and the "practical faculty of reason" (*CPJ* V 375.24–5) as the cause of the purposive form of our human actions. The human will and its faculty of reason is an immaterial power for Kant. However, the analogy between the formative power and the human will does not imply that Kant ascribes reason to nature, since, in the Kantian sense, nature does not have practical reason. Thus, some of Müller-Sievers' (2000, 61) remarks go too far in saying that the "formative drive" is "the expression of a will for self-organization in nature". But even if the formative power is not identical with practical reason, it can be an immaterial power. I am inclined to say that cα) has more support in the text.

d) The formative power acts upon matter and thereby organizes the materials. In d) Kant describes the effect of the formative power: it organizes matter. The meaning of 'organization' is explained in a brief footnote where Kant says that an organized being is a "whole" in which each part is "not merely a means, but at the same time also an end, and, insofar as it contributes to the possibility of the whole, its position and function should also be determined by the idea of the whole" (*CPJ* V 375.34–7). In an organized being, whole and part stand in specific relations to each other. For a body,

> therefore, which is to be judged as a natural end in itself and in accordance with its internal possibility, it is required that its parts reciprocally produce each other, *as far as both their form and their combination is concerned*, and thus produce a whole out of their own causality, the concept of which, conversely, is in turn the cause [...] of it in accordance with a principle; consequently the *connection* of *efficient causes* could at the same time be judged as an *effect through final causes* (*CPJ* V 373.26–34, first two italics are mine).

It is at this point that Kant's view on the formative power meets his view on the relation between part and whole in an organism. Like the formative power that involves but cannot be reduced to moving power, the composition of the organized being that results from the formative power involves but cannot be reduced to a mechanical composition of the being. Organization consists in at least four different types of causal processes between whole and part: 1) Parts of type A mechanically cause parts of type B. Or more precisely: the materials of parts of type A in an organized being have an effect on the materials of other parts based on mechanical laws and motive powers. 2) Material parts cause the composition of the whole as an aggregate of all parts based on mechanical laws and moving powers. 3) The purpose of a part of type A teleologically causes parts of type B. The purposive form or function of one part has a teleological effect on other parts based on the formative power: parts B serve to support the purpose of a part A. Parts exist *"for the sake of the other[]"* parts (*CPJ* V 373.35). And 4), the purpose of the organized whole teleologically causes the parts. The purpose of the organized whole has a teleological effect on all parts, since all parts are thought to stand in supportive relations to the purpose of the whole. This support is caused by the formative power, which directs the mechanical motions of a being towards the purpose of the being. The former two part-whole relations are mechanical, the latter two are teleological.

The most astonishing claim is e): the formative power is "a *self*-propagating formative power [eine *sich* fortpflanzende bildende Kraft]" (my italics). Kant does not say 'a propagating power [eine fortpflanzende Kraft]'; i.e., he does not claim that the formative power causes the process of the impregnation and generation of organized beings, at least not on a material level. Instead, he says "a *self*-propagating formative power [eine *sich* fortpflanzende bildende Kraft]" (my italics). The word "self" might be read in two ways:

eα) In German 'to propagate [sich fortpflanzen]' is used as a metaphor to say that something spreads out or extends itself. If we say that a wave, caused by a tsunami, spreads out in the ocean and along the coast, we could say: 'Die Welle pflanzt sich im Meer und an der Küste fort'. The domino effect of an economic crisis in one country, which causes a crisis in the neighboring countries, would be another example for 'sich fortpflanzen'. For we could say: 'Die Krise pflanzt sich in den benachbarten Ländern fort'. This meaning does not necessarily describe a new generation of something, but only an extension of something (a form or order or even disorder) in something else, without the new generation of this something else. The 'formative power' in this sense would be an immaterial power that is transferred to and spread out in something else: namely matter, without generating matter. It only generates a new form of matter—its organization. The cited sentence would say that in the organized being an immaterial formative power is transferred to and spread out in matter, which does not have a formative power originally. It thereby generates a new organization in this matter. It self-organizes matter.

eβ) In German 'to propagate [sich fortpflanzen]' is used literally with regard to plants, animals, and humans. However, Kant claims the self-propagating capacity not with regard to plants, animals, and humans but with regard to a power. In this sense, the "self-propagating formative power [eine sich fortpflanzende bildende Kraft]" can have a self-reflexive meaning, namely 'a formative power that propagates itself [eine sich selbst fortpflanzende bildende Kraft]'. The formative power would then be a power that generates and/or preserves *itself*. How can we make sense of such a claim without making it sound mystical? A possible self-reflexive reading would be to say that a formative power is a self-explanatory and self-evident basic power. In his writing *Teleological Principles*, written two years before the *CPJ* in 1788, Kant describes such a basic power as follows:

> [A] basic power that is effectuated through an organization has to be thought as a cause effective according to *ends*, and this in such a manner that these ends have to be presupposed for the possibility of the effect. But we know such powers, *in terms of their ground of determination* only in *ourselves*, namely in our understanding and will [...]. In us understanding and will are basic powers, of which the latter, insofar as it is determined by the former, is a faculty to produce something *according to an idea* which is called an end (*Teleological Principles* VIII 180.18 – 181.14).

The formative power then would be a final and fundamental purpose (or end) setting force of nature, which cannot lead back to another principle. Equivalent to the human understanding and will as inner capacities (causes), it brings about an end as its effect and generates order among the means to achieve this end. Given d) and e), it is likely that the formative power itself is not a power of generation. Although Kant calls this power 'fortpflanzend' it does not necessarily function as seminal fluid. The immaterial, natural formative power is a basic, ordering and *form*[7] giving principle which is directed towards an end or purpose, and spreads out its organizing and ordering capacity in matter. But it does not necessarily bring matter into existence.

3 A Response to the Kant-Wolff-Debate in the Literature

Now I am in the position to respond to contemporary scholars who argue that Wolff's account seems to be an important predecessor of Kant's theory of the organism. Dupont (2007, 37 – 8), for instance, writes: "even though it is to Blumenbach and not to Wolff that Kant refers [...] in the third *Critique*, Wolff's embryological works do represent a condition of realizability of the Kantian project for the biology". Similarly, Huneman (2007, 75) argues: "the Wolffian embryology, exposed in the *Theorie von der Generation* (1764) [...] enabled Kant to resolve the philosophical problem of natural generation, and subsequently to determine what is proper to the explanation of living processes".

---

7 The majority of passages throughout the second half of the third *Critique* support a reading according to which the formative power is responsible for the form of the being; see for instance *CPJ* V 369.33 – 370.15, 373.4 – 34, 377.1 – 23, 378.12 – 379.9, 407.13 – 409.22, 410.16 – 411.29.

Huneman's claim is based on a comparison of the conception of epigenetic powers, but not on the part-whole composition of organisms in Wolff and Kant. He interprets Wolff's conception of an essential power as a precursor of Kant's notion of a formative power, serving the epistemological function of guaranteeing the systematicity of an immanent, natural order of the organic. For Wolff, he claims, two events that causally and temporally follow each other do not belong to the same series because they presuppose the same, initially tiny, invisible, and later more and more visible *form* as preformism argued. Rather, these two events belong to the same series because they presuppose the continuity of the same *force* (ibid., 83–4). The essential power serves as a reason for the causal connection between unstructured matter and structured parts whose form and shape do not obviously follow from the unstructured matter.

According to Huneman (ibid., 78), Wolff's theory of epigenesis anticipates the solution of the "generation dilemma", i. e., "to provide any intelligible account of generative mechanisms". For generation is a process that occurs in discontinuous phases. It is precisely this discontinuity of generation that requires a *"principle of continuity"* (ibid.), as a guarantee that it is the same generation that constantly occurs. The essential power brings continuity and temporal order into the discontinuity of generation: it *"sets* the discontinuous phases seen by the observer *into series* and *order"* (ibid., 84). Therefore, it cannot be explained by mathematical or mechanical powers; which means it cannot be a Newtonian force that causes regular effects in correspondence to mathematical laws.

First, there is a general argument against Huneman's emphasis on Wolff's importance for Kant, since the transition from *form* to the unity of *force* as a justifying reason for the generation of an organism is not merely part of Wolff's account, but a general feature of all epigenetic accounts. Therefore the presence of this transition in Wolff tells us nothing specific about Wolff's role in the genesis of Kant's ideas. Second, Huneman's thought is plausible only if one follows his exegesis claiming Wolff transcended the mechanistic way of thinking (ibid., 82–3). But, as demonstrated above, in his *Theoria generationis* and the *Theorie von der Generation*, Wolff characterizes the essential power as an aggregation of mechanical determinations. Moreover, he argues explicitly that "all appearances occurring in the world" can be "produced and originated by *physical* causes" alone (Wolff 1764, 51, 57, my italics). Wolff's model of generation and growth in which "nutrition provides the key for analyzing generative processes" (Duchesneau 2006, 174) is to the greatest possible extent based on mathematical and mechanical considerations, where the amount of nourishing fluids is proportionate to the degree of growth and development—even though in some passages Wolff transcends mechanical descriptions of the essential force. Wolff's position is not able to resolve the "generation dilemma" of a mechanical account. Kant's notion of a formative power, however, is part of a finalistic account of causality, which includes but cannot be reduced to mechanical powers of motion.

Whereas the systematic proximity between Wolff and Kant regarding their conceptions of epigenetic powers is questionable, it is important that Wolff, before Kant, uses a part-whole model to explain the generation of an organism, and in this way, anticipates one of the most central ideas in Kant's theory of biology. However, Wolff's and Kant's positions are also opposed to each other on this front. Kant does not divide the process of production into an inorganic and an organic phase. For Kant production is *always* an organizing formation of the parts, for it occurs with regard to the whole. Furthermore, according to Kant, the organized whole or part is not only caused by a mechanical aggregation of materials, but by the purpose of the organized whole and also by the more specific purposes of parts that are causally prior to the materials and mechanisms of parts.

In contrast, on Wolff's account the materials and mechanisms of the parts precede the whole. The organic whole is the final result of an inorganic organization followed by an organic formation of all parts. Even in passages that seem to suggest that a superior whole is structured prior to the parts—for instance if first of all the outer shape of a blossom is produced without any internal organization—it is not true that the organic whole, the blossom, precedes its parts. Based on the systematic distinction between an inorganic production and an organic formation, Wolff claims that the superior whole is formed inorganically and never precedes its parts as an organic but only as an inorganic whole.

On the other hand, the appearance of a part-whole model in Wolff's account before Kant's emphasis of the whole-part relation as an essential feature of organized beings increases the systematic importance of Wolff's view with regard to Kant and relativizes the dominance of the Blumenbach-Kant-debate as *the* central historical target in the literature, since Blumenbach misses one of the most central concerns in Kant's account of biology: the part-whole relation as a significant feature of organized beings.[8]

---

8 I would like to thank the German Research Foundation for a four-year research followship that enabled me to conduct the studies for this essay.

References

Blumenbach, Johann Friedrich 1781, *Über den Bildungstrieb und das Zeugungsgeschäfte*, Göttingen: Johann Christian Dieterich.

Breidbach, Olaf 1999, Einleitung. Zur Mechanik der Ontogenese, in: Caspar Friedrich Wolff, *Theoria Generationis. Ueber die Entwicklung der Pflanzen und Thiere. I., II. und III. Theil (1759)*, Thun/Frankfurt/M.: Harry Deutsch, i–xxxiv.

Duchesneau, François 2006, Essential Force and Formative Force: Models for Epigenesis in the 18th Century, in: *Self-Organization and Emergence in Life Sciences*, ed. by Bernard Feltz, Marc Crommelinck, and Philippe Goujon, Dordrecht: Springer, 171–86.

Dupont, Jean-Claude 2007, Pre-Kantian Revival of Epigenesis: Caspar Friedrich Wolff's *De formatione intestinorum* (1768–69), in: *Understanding Purpose: Collected Essays on Kant and the Philosophy of Biology*, ed. by Philippe Huneman, Rochester: University of Rochester Press, 37–49.

Frigo, Gian Franco 2009, Bildungskraft und Bildungstrieb bei Kant, in: *Kants Philosophie der Natur. Ihre Entwicklung im* Opus Postumum *und ihre Wirkung*, ed. by Ernst-Otto Onnasch, Berlin/New York: Walter de Gruyter, 9–23.

Huneman, Philippe 2007b, Reflexive Judgment and Embryology: Kant's Shift Between the First and the Third Critique, in: *Understanding Purpose: Kant and the Philosophy of Biology*, ed. by Philippe Huneman, Rochester: University of Rochester Press, 75–100.

Johann Georg Hamann. Briefwechsel, ed. by Walther Ziesemer and Arthur Henkel (vols. 1–3) and ed. by Arthur Henkel (vols. 4–7), Wiesbaden: Insel 1955–1979.

Lenoir, Timothy 1980, Kant, Blumenbach, and Vital Materialism in German Biology, *Isis* 71, 77–108.

Lenoir, Timothy 1981, The Göttingen School and the Development of Transcendental Naturphilosophie in the Romantic Era, *Studies in History of Biology* 5, 111–205.

Lenoir, Timothy 1982a, Blumenbach, Kant, and the Teleomechanical Approach to Life, in: Timothy Lenoir, *The Strategy of Life. Teleology and Mechanics in Nineteenth-Century German Biology*, Dordrecht/Boston/London: Springer, 17–34.

Löw, Reinhard 1980, *Die Philosophie des Lebendigen*, Frankfurt/M.: Suhrkamp.

Look, Brandon 2006, Blumenbach and Kant on Mechanism and Teleology in Nature. The Case of the Formative Drive, in: *The Problem of Animal Generation in Early Modern Philosophy*, ed. by Justin E. H. Smith, Cambridge: Cambridge University Press, 355–72.

Müller-Sievers, Helmut 2000, From Preformation to Epigenesis/Self-Generation in Philosophy: Kant, in: Helmut Müller-Sievers, *Self-generation: Biology, Philosophy, and Literature around 1800*, Stanford: Stanford University Press, 26–64.

Richards, Robert J. 2000, Kant and Blumenbach on the Bildungstrieb: A Historical Misunderstanding, *Studies in the History and Philosophy of Biological and Biomedical Sciences* 31 (1), 11–32.

Richards, Robert J. 2002, Early Theories of Development: Kant und Blumenbach, in: Robert J. Richards, *The Romantic Conception of Life: Science and Philosophy in the Age of Goethe*, Chicago: University of Chicago Press, 207–37.

Roe, Shirley A. 1979, Rationalism and Embryology: Caspar Friedrich Wolff's Theory of Epigenesis, *Journal of the History of Biology* 12, 1–43.

Roe, Shirley A. 1981, *Matter, Life, and Generation. Eighteenth-Century Embryology and the Haller-Wolff Debate*, Cambridge: Cambridge University Press.

Tetens, Johann Nicolaus 1776/7, Von der Entwicklung des menschlichen Körpers, in: Johann Nicolaus Tetens, *Philosophische Versuche über die menschliche Natur*, Hildesheim/New York: Georg Olms 1979, vol. 2, 459–64.

van den Berg, Hein 2009, Kant on Vital Forces. Metaphysical Concerns versus Scientific Practice, in: *Kants Philosophie der Natur. Ihre Entwicklung im Opus Postumum und ihre Wirkung*, ed. by Ernst-Otto Onnasch, Berlin/New York: Walter de Gruyter, 115–35.

White, David A. 1997, Kant's Notion of a Purpose, in: Special Issue: Final Causality in Nature and Human Affairs, ed. by Richard F. Hassing, *Studies in Philosophy and the History of Philosophy* 30, 125–50.

Wolff, Caspar Friedrich 1759, *Theoria generationis*, Halle: Litteris Hendelianis; reprinted in: Wolff, Caspar Friedrich, *Theorie von der Generation in zwei Abhandlungen erklärt und bewiesen/Theoria generationis*, Hildesheim: Georg Olms 1966.

Wolff, Caspar Friedrich 1764, *Die Theorie der Generationen, in zwo Abhandlungen erklärt und bewiesen*, Berlin: Friedrich Wilhelm Birnstiel; reprinted in: Wolff, Caspar Friedrich, *Theorie von der Generation in zwei Abhandlungen erklärt und bewiesen/Theoria generationis*, Hildesheim: Georg Olms 1966.

Wolff, Caspar Friedrich 1768/9, De formatione intestinorum praecipue, tum et de amnio spurio aliisque partibus embryonis gallinacei, nondum visis, observationes, in ovis incubatis institutae, in: *Novi Commentarii Academiae Scientiarum Imperialis Petropolitanae*, Sankt Petersburg: Petropoli Typis Academiae Scientiarum, Tom. XII (1768), 403–507 [§§1–119 [sic!] = Pars I/II]; Tom. XIII (1769), 478–530 [§§119 [sic!]–155 = Pars III].

Wolff, Caspar Friedrich 1789, *Von der eigenthümlichen und wesentlichen Kraft der vegetabilischen, sowohl als auch der animalischen Substanz*, St. Petersburg: Imperial Academy of Sciences.

Rachel Zuckert

Organisms and Metaphysics: Kant's First Herder Review

In January 1785, Kant reviewed the first volume of Johann Gottfried Herder's *Ideas towards a Philosophy of the History of Man* for the *Allgemeine Literaturzeitung*.[1] In this volume, Herder presents the foundation for his treatment of human history in the work as a whole: an account of the nature of organic life, and of human beings as part of the organic world. Kant's review is quite critical: though he makes praising gestures, he takes Herder to task for lacking philosophical precision, and criticizes his claims as "metaphysics, indeed [...] very dogmatic" metaphysics (*Review of Herder* VIII 54.7). Scholars have taken Kant's criticism here to anticipate his later positions concerning the judgment of organisms in the *CPJ*. In particular, Zammito (1992, 180–213 and 2007) has argued that Kant's criticism of Herder (and hence his later position) is motivated by Kant's own metaphysical commitments concerning the distinctiveness of human beings as opposed to the rest of organic nature, and of organic nature as opposed to lifeless matter. For such reasons, Zammito contends, Kant was opposed to any suggestion, like Herder's, that organisms have a self-organizing force, which is both the cause of human capacities and a basic force of matter.

As I shall discuss, Kant's review provides some support for Zammito's contention, and it is not unlikely that, in a review written in the same year in which he published the *Groundwork* and composed the *Metaphysical Foundations*, Kant is preoccupied with morality and reason as distinctively characterizing human beings among organic beings, and with matter as fundamentally characterized by physical laws (not organic force). But—as Zammito intends to suggest—such metaphysical grounds for criticism would seem not to justify Kant's accusation that Herder's position is problematically metaphysical, but rather to render it hypocritical. In this paper, I shall suggest, then, an alternative reading of Kant's criticism of Herder as a dogmatic metaphysician, arguing that it is ground-

---

1 Kant also reviewed the second volume in November 1785; in March 1785, he also replied to a criticism of his original review (written by Carl Leonhard Reinhold), both in the same journal. I concentrate on the first review. Herder was wounded by these reviews, particularly the first (see Haym 1954, 279–83), and paid back Kant's criticism with interest in his later bitter anti-Kantian polemics, *Metakritik zur Kritik der reinen Vernunft* and *Kalligone* (though he also later writes much-quoted praise of Kant in his *Letters for the Promotion of Humanity*). Kant's review is thus the first salvo in the complicated, antagonistic relationship that developed between him and his former student.

ed not in Kant's metaphysical commitments, but in epistemological concerns articulated in the *CPR*, i.e., in Kant's predominant critical treatment of metaphysics. As I shall also suggest, Kant's arguments in the review are perhaps not quite representative of his position in the *CPJ* either, but rather represent a transitional position in his thinking concerning the explanation of organisms.[2]

I shall begin by considering Zammito's line of interpretation in more detail (section 1), before turning to investigate other aspects of Kant's review that support an alternative interpretation. Here I discuss first Kant's criticism of the most obviously metaphysical of Herder's arguments (left unmentioned by Zammito), namely his attempt to prove the immortality of the soul (section 2). Though this argument surely does in part prompt Kant's criticism, it does not suffice to explain why the entirety of Herder's view, not solely this argument, might be characterized as "dogmatic metaphysics". In the subsequent sections (3 and 4), therefore, I reconstruct a more comprehensive Kantian epistemological argument that, in promulgating his view concerning the one force in nature, Herder oversteps the limits on human, discursive understanding and therefore engages in dogmatic metaphysics. I conclude with some remarks concerning the relationship between this argument and Kant's later discussion of organic life in the third *Critique* (section 5).

1 Kant's Criticisms: Based on Metaphysical Commitments?

Kant's discussion of Herder's doctrines is compressed, and ranges over a number of topics. Nonetheless the focal point of his criticism is clear: Herder's claim that there is a "unity of [...] organic force", which is "self-forming in regard to the manifoldness of all organic creatures" (*Review of Herder* VIII 54.25 – 7). Kant appears to have in mind here Herder's presentation of organic nature as comprising a progression of living things from less to more complex, ultimately culminating in human beings (*Review of Herder* VIII 52.25 – 8), which progression expresses, arises from the workings of a basic, natural, self-organizing and animating force.

As noted above, Zammito argues that Kant's opposition to this proposal—his relegation of it to "dogmatic metaphysics"—arises from his own metaphysical commitments. Kant cannot accept Herderian natural history or a single organic force because he is committed to a strong metaphysical distinction between

---

2 Here I concur with Ameriks (2009, 54 – 6) that in the period of these reviews, Kant holds an intermediate, transitional position, though Ameriks focuses on Kant's concept of freedom.

human beings (as rational and free) and other organisms, and between these and matter as such (as essentially non-living) (Zammito 1992, 189 – 99, 206). Moments in the review do support this claim. Kant quotes passages in which Herder suggests that an organizing, living force may characterize inorganic matter; though he does not comment upon this claim, he may mean to flag it as problematic (as it is a view of which he is critical throughout his philosophical career). Kant also refers in passing to metaphysical conclusions (about human souls) that might be drawn on the basis of the "thinking principle" in human beings (*Review of Herder* VIII 53.27) or on moral grounds (*Review of Herder* VIII 53.21). And in a famous passage Kant writes that Herder's proposal of a

> *relationship* among [species], where either one species would have arisen from the other and all from a single original species or [...] a single procreative maternal womb, would lead to *ideas* so monstrous that reason recoils before them (*Review of Herder* VIII 54.18 – 22).[3]

Kant does not state *which* ideas are "so monstrous". But one might think (as Zammito contends) that this horror reflects metaphysical orthodoxy[4], specifically a resistance to understanding human beings to have arisen out of, to be on a metaphysical continuum with, primordial slime. Kant seems too to refer to such orthodoxy in writing that one might have "reservations" about attributing to Herder the view that human souls are not "particular substance[s]" but only temporary configurations of the one organic force (*Review of Herder* VIII 53.30 – 3; see VIII 54.20).

Perhaps Kant means only to save Herder from the opprobrium of their metaphysically orthodox contemporaries. But these remarks are connected to doctrines Kant himself holds dear. For, though Kant argues that we cannot prove that the soul is a substance, he also holds that we cannot disprove this claim, which may be crucial for morality. Herder's view does deny a firm distinction between human souls and animals (or even rocks), which may threaten to undermine claims concerning human freedom. Thus Kant may well be concerned, as Zammito argues, by the way in which Herder's view challenges human distinc-

---

3 I have modified the Cambridge translation, replacing "*affinity*" with "*relationship*" to translate "*Verwandtschaft*".

4 One might class here as well Kant's criticisms (e. g., *Review of Herder* VIII 54.32 – 55.2) of Herder's proposal that the emergence of human rationality may be explained by human erect posture, though (as Kant makes clear at *Review of Herder* VIII 57.4 – 9) this criticism too appears ultimately to be epistemological: empirical evidence alone cannot establish that rationality can be instantiated only in an erect form (a claim to metaphysical necessity).

tiveness, and Kant's commitment to the latter (as well as, perhaps, to "lifeless" matter) might explain his vehemence.

There are, however, also reasons to resist this interpretation of Kant's claims and their motivations. First, its textual support is relatively scanty: as noted, Kant only mentions (but does not explicitly criticize) the Herderian thesis that matter has living force, and discusses the distinctiveness of human beings (as rational or free) only in passing. Correspondingly, this interpretation ignores much of Kant's substantive discussion in the review (which will be discussed below), relying heavily instead on the passage (quoted above) concerning the "monstrous" ideas to which Herder's view might give rise. This passage is, however, somewhat ambiguous. It supports Zammito's interpretation only if it is quite clear that Kant's commitment to human freedom—the most important source of human distinctiveness on his view—must prompt him to reject the proposal that human beings arose from animals. But it is not entirely clear that it must. If the possibility of free will is not threatened by the fact that all human actions and states of mind are causally determined (as Kant argues in the "Third Antinomy"), it is not clear that it should be threatened by claims concerning the origin of human beings as a natural species either.[5] Finally, and most importantly, this interpretation does not explain Kant's criticism of Herder. For on this interpretation, Kant's claim that Herder's views are problematic *because* they are distinctively, dogmatically metaphysical is unexplained, indeed appears unwarranted and hypocritical (as Zammito intends to suggest): Kant is taken to be motivated simply by his own opposing metaphysical views in making this charge, and to provide no argument as to why Herder's claims are more dogmatically metaphysical than his own. For these reasons, I suggest, it is worth taking another look at the review, to see if Kant might offer reasons there for his charge that Herder's claims are "dogmatic metaphysics", ill-grounded or illegitimately formulated in some way. It is to this I now turn.

---

[5] It is also not clear that Kant may use claims concerning human freedom to weigh for or against any theoretical claim, given the structure of his systematic thought: practical claims have their own (practical-rational) source of legitimacy, but they may not conflict with theoretically established claims; hence indeed Kant's claim that transcendental idealism—in establishing an arena in which we cannot know anything—makes room for faith in freedom, immortality, and God.

2 The Immortality of the Soul

We may begin, as Kant begins the critical portion of his review, with the most obviously metaphysical aspect of Herder's project: Herder's argument for the immortality of the soul, on the basis of his view of progressively self-organizing nature. The progression in nature, ending in human beings, shows that nature (the one organic force) is always aiming to perfect itself, to produce new, more complex kinds, Herder contends. In nature, there appears to be no further step in this progression beyond human beings. But by the "analogy of nature", Herder argues, we may expect that there is nonetheless further progression; the progressive force cannot simply disappear on the death of human beings. So we may believe that the human soul (this formation of the organic force) persists after death, at which point it rises to a new, higher level of complexity (Herder, 1785 in 1887, 167–80).[6]

Kant objects that even if one grants Herder his hypothesis concerning the progressiveness of organic force *and* that one may extend this progression beyond the nature we experience, this would still not establish the immortality of human souls—but only that there is a higher species than human beings. For this would be the true "analogy" to the natural progression that Herder proposes (*Review of Herder* VIII 53.1–5). And, in writing that Herder sees the human soul "not as a particular substance", but as the "effect on matter of an invisible" force that "works within it and animates it" (*Review of Herder* VIII 53.30–2), Kant also suggests, though less explicitly, that Herder's argument from the conservation of force would establish only that the *force* is preserved, not that the particular configuration *of* that force, i.e., this soul, is conserved.

Herder's claim here does seem straightforwardly metaphysical: the immortality of the soul is a traditional topic of metaphysics; this is a question that concerns the ultimate being of human beings, as transcending the experienced, physical world. As Kant writes, "nature lets us see nothing other than that she abandons individuals to complete destruction and preserves only the kind" (*Review of Herder* VIII 53.10–2). Kant's criticism of it seems fair as well. Herder's argument seems bad, for precisely the reasons stated. Because Herder comes to unjustified conclusions concerning metaphysical matters, via arguments in which he uncritically applies concepts used to explain natural events to supra-natural entities, his view could, then, be called dogmatic.

This objection does not establish, however, that Herder's claim concerning the one organic force is itself metaphysical in any problematic sense. The failure

---

6 Kant summarizes this argument at *Review of Herder* (VIII 52.18–33).

of this argument would trouble Herder, given his aspirations to propose a view of the natural world consonant with empirical observation *and* (more or less) traditional religious views. Showing this failure is also important for Kant, since he aims to limit the pretensions of human reason to establish claims about (e.g.) the immortality of the soul. But as Kant's objections suggest, Herder's claim about the one force does not seem to commit him to asserting the immortality of the soul—quite the contrary (as it does not entail this conclusion). Thus, though Kant seems warranted in suggesting that the claims concerning the immortality of the soul are both metaphysical and dogmatic, other reasons are needed for suggesting that Herder's central proposal is as well; I now turn to Kant's more specific criticisms of that proposal.

3 The One Force

As one might predict (given his emphasis on the *one* organic force), Kant objects that Herder's hypothesis fails to explain *differences* among organic beings. Kant claims, first, that Herder might think that it is reasonable to propose that there is a single organic force because of an illusion:

> The smallness of the distinctions, if one places the species one after another in accordance with their *similarities*, is, given so huge a manifoldness, a necessary consequence of this very manifoldness (*Review of Herder* VIII 54.13–5; see *CPR* A 668/B 696).

Kant here suggests that by arranging organic kinds in a spectrum, from less to more complex, Herder produces an impression of greater similarity among organic kinds than there is in fact. Those kinds placed close to one another on the spectrum will appear more strongly similar to one another than they might otherwise—for they are placed within a context in which their similarities are emphasized (by contrast to the "farther" species in the spectrum). Moreover, because of these local similarities, because it appears that one may move gradually from one kind to the next along the spectrum, one might be tempted to think that there could be one force explaining the behavior of all. But this move occludes the differences among the kinds that first lent the impression that the kinds "closer" to one another are strongly similar. It is, Kant suggests, only such an illusory impression that makes prima facie plausible the proposal that diverse organic kinds and behaviors are produced by a single force.

This passage immediately precedes Kant's exclamation concerning the "monstrous" ideas that arise from the proposal that all genera arise out of a common "procreative womb" (*Review of Herder* VIII 54.15–9). Thus Kant may mean to suggest (though he does not explicitly argue) that Herder not only exaggerates the similarity among organic kinds, but also wrongly infers from that purported similarity to a single causal origin (the organic force). Kant might, that is, be anticipating the distinction made in his later essay, *Teleological Principles* (1788), between "natural description" and "natural history", i.e., between classificatory claims concerning similarities of shape, behavior, etc. of species (as in Linnaean biology[7]), and causal claims concerning the origin of such species (*Teleological Principles* VIII 161–2). Classificatory similarity (on a spectrum) does not entail a common cause. Kant argues, moreover, that Herder's organic force conceived

> as self-forming in regard to the manifoldness of all organic creatures, and later in accordance with the difference of these organs working through them in different ways, is supposed to constitute the entire distinctiveness of its many genera and species (*Review of Herder* VIII 54.26–9).

This passage seems somewhat opposed to the previous passage, for here Kant suggests that the organic force is supposed to be used to explain "manifoldness" (not, now, the purported similarity among all kinds). Kant is indeed turning his attention to the one force understood as driving the natural-historical progression of species: as such, this force is supposed to underlie all the different behaviors of all organisms. Herder mentions nutrition, sexual reproduction, irritability, and more as different characteristic behaviors of organisms, all of which are to be understood as operations of the same force.

Moreover, the very fact *that* there are such differences—which comprise the level of complexity of the organism (plants have nutrition and reproduction only, while animals also have irritability, humans have reason, etc.)—is also supposed to be traced to the operations of this force. For its "self-forming" drive is to explain the progression of species toward greater complexity, i.e., to explain not only wherein the differences among species consists (more or less complexity), but also why there are different species at all. Indeed, Herder uses his organic force to characterize at once embryonic development and natural reproduction —i.e., to explain how individuals come to have the character of their kind, and to produce other individuals of the same kind—*and* the transformation of species.[8] Thus this force is supposed not only to explain a vast variety of effects,

---

7 As purists will note, 'biology' is not a term developed by Kant's time; because it conveniently refers to the study of organisms in general, I use it despite this anachronism.
8 See, e.g., Herder (1785 in 1887, 173, 179–80). Herder speaks here of "forces" (plural), rather than one organic force. But after presenting the progression of living things as increasing in organization, as having different specific powers, he claims that all are manifestations of the

but also to explain both why effects (activities of the same kind of objects) are the same, *and* why effects (of objects of different species) are different.

Kant raises legitimate questions here about the theoretical utility and justification of Herder's proposal. As Kant writes, Herder seems to be proposing "to explain *what one does not comprehend* in terms of *what one comprehends even less*" (*Review of Herder* VIII 53.35 – 7). We do not understand why organisms have the particular capabilities and behaviors they do (by contrast either to one another or to inorganic objects). A force that is taken to be responsible for all of these behaviors—even for opposed effects (producing other objects that are the same, producing other objects that are different)—seems, however, of little use for understanding those behaviors and to be at least equally inexplicable (how can one force produce all such effects?). But we may still ask what Kant means in claiming that, because it is plagued by such problems, Herder's proposal is "dogmatic metaphysics". Why is it not just an unsatisfying scientific (or "natural philosophical") proposal?

4 The Metaphysics of the One Force

Kant claims that Herder's proposal is metaphysical because the one force "lies wholly outside the field of the observational doctrine of nature and belongs merely to speculative philosophy" (*Review of Herder* VIII 54.29 – 30). This claim is, I suggest first, directed ad hominem against Herder and, correspondingly, serves as an indirect defense of the project of the *CPR*. Kant points out that Herder claims that he is not doing metaphysics, but basing his claims upon empirical observation alone (*Review of Herder* VIII 54.7 – 8).[9] But, Kant objects, we do not ever empirically *observe* the single force (only its purported manifold ef-

---

"one organic principle of nature, that we here call plastic, there call impulsive, there sensitive, there artificing" (Herder 1785 in 1887, 102; my translation).

As Roth pointed out (private conversation), Herder's view on this point may be more defensible than the Kantian position I sketch suggests, for it is in fact akin to proposals (in biology at his time, and, mutatis mutandis, in contemporary biology) that ontogeny recapitulates phylogeny. More generally, I should note that I aim here to reconstruct Kant's concerns as based on legitimate epistemological (rather than metaphysical) grounds but not to suggest that they are therefore correct, or borne out by the best biological practice then or now—as is also relevant, e. g., in relation to Kant's commitment to the fixity of species.

9 This seems accurate. Though Herder refers to God, and seems to have a Spinozistic view of the identity of nature and the divine, he claims that his doctrines are based on empirical observation: he is "leaving aside metaphysics", to base his claims on "physiology and experience" or on the "analogies of nature" (Herder 1785 in 1887, 110, 177; my translation).

fects). Claims about force are then, at least in part, a priori. So too Herder's stated principle concerning when and why one posits forces: one does so in order to explain effects—for every effect, there is a force from which it arises (Herder 1785 in 1887, 84)—which (on Kant's view) is a version or corollary of the synthetic a priori causal principle. Thus such claims are, in that Kantian sense, metaphysical, and they require transcendental or other justification to explain their applicability to objects of experience, in order to avoid dogmatism.

Insofar as Herder does not recognize the need to justify such claims, he may be said to engage in "dogmatic metaphysics". Apart from this ad hominem point —and defense of the *CPR* project—though, Kant's accusation that Herder's proposal is (problematically) metaphysical *because* it posits something unobservable seems dubious. Kant after all believes that something like Herder's principle *is* justified, and himself posits forces in the *Metaphysical Foundations* (published a year after this review). In his essays on race and history published both before and after this review, likewise, Kant posits "*Naturanlagen*" (e. g., *Races* II 434 – 6, *Universal History* VIII 18.20 – 32, *Human Race* VIII 98.7)—dispositions of a species —to explain embryonic development, growth, racial differentiation, and even the shape of human history. But neither the physical forces nor the *Anlagen* are observable.[10]

Perhaps we ought to read Kant's claim differently, however: it is not that Herder's force is unobservable, but that the claims concerning it are not based on *scientific* observation, which aims at formulating *laws*. We may not be able to know the ultimate causes of organic behaviors, Kant writes, but at least we can come to know laws governing them through experience (*Review of Herder* VIII 53.37– 54.1). The forces of Kant's dynamics and the organic *Anlagen* are unobservable indeed, but they correlate to laws governing behaviors of determinate, observable sorts; they are taken to underlie (or be the ultimate subject matters of) laws of a form, "If z happens to objects of x kind, those objects will do y". These laws, in turn, may be seen as universalizations of inductive inferences

---

10 As Hegel (1807 in 1970, 107– 36) suggests in the "Force and Understanding" chapter of the *Phenomenology of Spirit*, positing forces may itself be ineliminably metaphysical (in a pejorative sense) in that force is always posited as some sort of "entity" that at once is supposed to explain some effects and yet is known *only* through such effects (or is itself "inexplicable"); thus force is always a misleading or self-deceptive attempt at explanation. At *Review of Herder* (VIII 62.37– 63.2), Kant too claims that fundamental organic forces are ultimately inexplicable, and acknowledges that we cannot "comprehend the possibility of the fundamental [physical] forces" at *Metaphysical Foundations* (IV 524.39 – 40). Unlike Hegel, however, Kant does not appear to hold that this ultimate inexplicability renders it illegitimate to use forces in explanation, and thus he cannot consistently adduce such difficulties with the concept of force to criticize Herder's invocation of the one organic force.

concerning objects (i.e., re-identifiable, repeated/able events), and thus bear a close relationship to empirically observed correlations.[11] Though Kant contrasts such laws with Herder's "conjectured" laws (*Review of Herder* VIII 55.24), moreover, Kant's worries might be more accurately glossed by saying that it is not clear that the workings of Herder's force would be characterized by *any* laws: it is supposed to be responsible for so many different behaviors, it is not clear how its activities might be described by laws.

By suggesting, too, that this force is responsible for the differences among species, Herder undercuts the most obvious, experientially-offered basic terms of such laws in biology, namely organic kinds: things of *this sort* will do y (when z happens); for precisely what makes them do this (on Herder's view) *also* might lead them to do something else, or to be of a different kind altogether.[12] Thus in his later review of Herder's second volume of the *Ideas*, Kant endorses Herder's claim that organisms develop by virtue of some internal force, but requires the added qualification that this force be "limited by its nature" only "to a certain number and degree of differences in the formation of a creature" (*Review of Herder* VIII 62.29–31), that is (I suggest), that it may be characterized by laws governing a particular kind of thing; therefore, presumably, the forces governing the behaviors of different species would be different, not one.

Similarly, this Herderian proposal would make it impossible to do natural history in the only way Kant deems legitimate. Kant argues that because (or insofar as) we do not have evidence concerning very early history, our only guide in speculative reconstructions of such history is to take things to have behaved then according to the laws that govern them now (*Teleological Principles* VIII 161.35–162.3; see *Conjectural Beginning* VIII 109.12–6). Thus a proposal that events or progressions in earlier history operate independently of laws governing kinds, indeed contravene kind-distinctions, will seem, for Kant, unavoidably speculative. This worry might in fact explain his claim that reason would "recoil" from such ideas—and even perhaps why these ideas are "monstrous" (*Review of Herder* VIII 54.19) (insofar as "monsters" are animals that violate species categories)—for reason (or, more properly, the understanding) must proceed by using universal concepts, here of organic kinds.

---

11 See *Metaphysical Foundations* (IV 502, 534); *Teleological Principles* (VIII 180.27–30) and following note; Kant here emphasizes also that these forces may be taken to be governed by *mathematical* laws (though this would not be true, at least as yet, of the *Anlagen*).

12 Herder claims that the transformation or the production of new species has now stopped, but he gives no reason for this change in the operations of nature (Herder 1785 in 1887, 177, 180). (Kant himself faces a similar explanatory burden in his theory of race differentiation via adaptation to climate, as Forster objects; see *Teleological Principles* [VIII 173–4]).

The fact that Herder's force cannot be characterized in terms of laws may, I suggest then, allow us to understand Kant's accusation (of "dogmatic metaphysics") here. Kant's own proposals (of forces or *Anlagen*) may be less "metaphysical" in that his forces are not posited as entities that are ontologically separate from sensible things, which use or act through those things. Kant's forces are, rather, taken just to characterize the behavior of those very sensible objects, to describe the way in which these objects are governed by, act according to, laws. By contrast, Herder (1785 in 1887, 172, 174–5) suggests that (on his view) the one force is something over and above particular sensible entities or behaviors, especially in his descriptions of it as "invisible" or "spiritual", and (as I have been suggesting) it cannot be taken to be a correlate to scientific laws as Kant understands them.[13]

Kant may be taken here, moreover, to be employing his diagnosis of (problematic) metaphysical thinking in the "Dialectic" of the first *Critique*. As is well known, Kant there argues that reason naturally engages in metaphysics, makes a priori claims about entities inaccessible in experience, because it seeks the unconditioned, that which explains but requires no further explanation itself. As Grier (2001, 117–30) has argued, metaphysical thinking arises (specifically) because reason not only takes it that it should *seek* the unconditioned, but assumes that it has *found* it, as somehow intrinsic to, analyzable out of, the conditioned of which we are aware. Kant suggests that this sought-after, purportedly-found unconditioned can take two forms: it may either be one fundamental unconditioned item—paradigmatically God as the first cause—from which all conditioned things descend, or it may be a whole—paradigmatically the infinite series of causes—on which each of the conditioned items depends, as part of that whole.

In concluding the review, Kant echoes the "Dialectic" in writing that Herder's attempt is a "bold" and "natural" one, given the "drive for inquiry" of human reason (*Review of Herder* VIII 55.17–9). Herder's proposal of the one organic force may, moreover, be understood to be a claim to have "found" the unconditioned in both of the Kantian senses: this force is meant to be the explanans for all organic form and behavior and its autonomous creativity is also supposed to account for the whole of organic life, the place of every organism in the progression of species. This proposal aspires, in other words, to the totalizing, final vision characteristic of metaphysical reason, its claim to grasp ulti-

---

13 Herder is, thus, not an orthodox Spinozist, as he appears to believe that nature (as divine) is something (such force[s]) over and above individual sensible things (in which the force is expressed).

mate and global truths that are beyond the reach of human, limited intellect and experience. Thus Herder suggests that he can find, has found, the unconditioned within the conditioned, the ultimate cause as immediately, transparently active within the given, empirical particulars.

This dogmatic metaphysical claim is, finally, intimately connected to the disjunction between Herderian force and the articulation of laws discussed above. It is not just that the operations of such a force cannot be articulated in the form of laws (and thus is not in concert with the aims of scientific observation on Kant's view), but also that claims about such a force are, therefore and correspondingly, not based upon, built up from, observed empirical correlations or (resultant) hypothesized empirical laws either. Such correlations and laws are, however, the characteristic forms of cognition for human, limited intellects, the characteristic forms (specifically) of *conditioned* knowledge: knowledge reliant upon inductive investigation of the empirically given, and knowledge had by human, discursive intellects, who can understand objects and their qualities as governed by lawful relations, indeed, but do not have insight into the essence of things independently of those relations, as ultimately grounding such laws and relations, as they are in themselves. In invoking the one force, independently of any knowledge of laws, then, Herder lays claim to some sort of intellectual intuition, an immediate grasp of the inner nature of things, and therefore—in the words of the "Appendix" to the "Dialectic"

> presumes to dispense with all the natural investigation of cause[s] [...] by, as it were, *passing over* the immanent sources of cognition in experience through an edict [...] of transcendent reason (*CPR* A 690/B 718).

5 Relation to the *CPJ*

Kant's review may thus be understood as expressing a similar view to that in the first *Critique:* it is beyond human reason to attain knowledge of the unconditioned, but Herder claims to do just this in proposing his one organic force. We ought, instead, to confine ourselves to attempting to establish laws linking similar behaviors of objects of particular kinds, and to accept that we do not have, and may never have, a unified explanation of all organic behaviors. The best we can do, as Kant discusses at length in the "Appendix" to the "Dialectic", is to attempt to formulate empirical laws, and then to "reduce" such laws (or the forces correlative to them) to more "fundamental" laws, more "fundamental" forces. By contrast, when Herder proposes the one organic force, he not only claims to grasp an overarching principle of explanation that is not itself a law or describable in accord with laws, but also, in so doing, "leaps over" empirical observation and formulation of specific laws, claiming to grasp that one fundamental explanans of all particulars immediately, simply within (intuitive?) experience of them. Contra Zammito, then, it is not Kant's metaphysical commitments to lifeless matter or human freedom, but his epistemological commitments to scientific observation and explanation in terms of laws (or a system of laws) that drive his criticism of Herder as dogmatic metaphysician.

In conclusion, I wish to suggest that this line of argument may not be quite representative of the position in the *CPJ* either. Kant of course does retain there many views that ground his criticism of Herder: the critical account of dogmatic metaphysics, the view that the articulation of laws is the aim of scientific observation, the view that the formulation of laws and not intellectual intuition, is the appropriate, indeed the only possible, form of knowledge for human, discursive intellects. But there are two significant differences between the position of the review and the *CPJ*—both more specifically about biological investigation or laws.

First, there is no mention in the review of the dominant concern of the *CPJ*, namely teleology or purpose. Herder does attribute purposes to nature; e.g., his one organic force *aims at* the creation of human beings as the most complex form. In the terms of the *CPJ*, such claims would be inappropriate determinative teleological judgments of nature or a "realism about purposes in nature" (*CPJ* V 392.6–10) that is the target of critique, but such issues are not themes in the review.

Second, Kant apparently changes his mind concerning the Herderian proposal of the progression of species. In the *CPJ*, Kant no longer deems this proposal to generate "monstrous" ideas (*Review of Herder* VIII 54.18–9) from which reason recoils, but suggests, rather, that this approach to organic life (including, explicitly, human beings [V 419.2]) "allows the mind [...] a weak ray of hope that something may be accomplished here [i.e., in the explanation of organic form] with the principle of the mechanism of nature, without which there can be no natural science at all" (*CPJ* V 418.30–3).[14] Kant registers some doubts that such a "daring adventure of reason" (*CPJ* V 419.26–7) will find empirical support. But—contra the Herder review—he explicitly claims that it may not be ruled out a priori (*CPJ* V 419.26–38), i.e., deemed to be part of a dogmatic metaphysics.

---

14 The discussion at *CPJ* V 418–20 is more complicated than I can address here. For example, as Richards (2000, 30) points out, Kant holds that biological investigation must still respect the distinction between organized (living) matter and the inorganic and falsely praises Blumenbach for holding to this distinction.

These two differences are, I think, related. They both reflect Kant's sharpened critical position concerning teleological judgment in the *CPJ*: that natural teleology is not only a regulative principle, but is one that may not be used to form or (directly) to ground determinative judgments, may not be a principle of explanation proper; teleological judgment is always merely reflective judgment. By contrast, I suggest, the *Review of Herder* belongs to a transitional period (1784–8, roughly) in Kant's thinking concerning the status of teleological judgment, a period in which he seems to hold that there can be legitimate determinative teleological judgment of organisms.

That is: as is well known, Kant argues in the "Appendix" to the "Dialectic" that the idea of God's design, or of a purposive natural order, is a regulative principle for the investigation of nature; it is not a constitutive principle governing the nature of objects as such, but it may guide investigation, help us to find and formulate laws governing natural objects. Kant appears, moreover, to suggest that this principle could be useful in investigating all manner of objects (including, e. g., the shape of the earth), indeed *recommends* using this principle as widely as possible in investigating nature in order to avoid imposing our ideas about particular divine purposes onto particular objects, thus failing to engage in actual empirical investigation of the course of natural causes (*CPR* A 687/B 715–A 693/B 721).

In the "Appendix", then, Kant does not single out organic behavior as particularly requiring teleological explanation; he remains silent, too, concerning whether this principle may help us to find teleological *laws* governing objects (whether, that is, teleology might be determinatively true of objects, even if it is not a priori constitutive of all objects as such). Both of these positions appear to change in the works of (what I am calling) the "transitional" period, perhaps because here Kant devotes more focused attention to the organic world and to promulgating explanations thereof. Here Kant not only employs a principle of natural purposiveness as a regulative principle—a proceeding defended again in *Teleological Principles* (1788)—but also seems to propose teleological laws concerning organic form and behavior *in particular*.[15] Our *Anlagen* suit human be-

---

15 The claim that there may be determinative teleological laws governing organic behavior is not inconsistent with the "Appendix" position, though it does represent a step beyond it (to closer attention to biological laws in particular). For in the "Appendix"—unlike in the *CPJ*—Kant does not suggest that teleological laws of nature are themselves non-scientific, in conflict with mechanical laws, which are the only true explanation of nature as we can know it. Rather, he suggests only that we *may* so consider nature, which may lead us to be able to formulate laws (possibly including teleological laws, and not necessarily about any particular sort of object). Kant is most concerned to argue in the "Appendix" that we should not take such laws or

ings, Kant argues, to flourish in particular climates (in the race essays). Human reason must have some end (which is not happiness) for it is an "instrument" in an organized being, all of which are adapted for an end (*GMM* IV 395.4–8). And the first thesis of *Universal History* reads that "all natural predispositions of a creature are determined [...] to develop themselves completely and purposively" (*Universal History* VIII 18.20–1). Kant claims, moreover, that this last principle is confirmed by "observation" of all animals (*Universal History* VIII 18.21–2; see *Teleological Principles* VIII 159.6–7, 13–4).[16] Thus in these essays, Kant appears to think that a regulative principle of purposiveness not only guides investigation, but may guide us, specifically, to find (at least potentially determinative) teleological laws governing organic behavior. When Kant refers in the review to laws gleaned from observation, I suggest that he is taking such laws to *include* laws invoking purposes.

In the *CPJ*, Kant continues to hold that teleology is specially relevant to judgment of organic behavior, but he of course denies that there may be determinative teleological judgments concerning organisms. This difference makes a difference concerning Kant's criticism of Herder. For that criticism seems implicitly to rely on the assumption that it is scientifically legitimate to invoke purposes, specifically purposively-defined kinds, in proposing biological laws. This assumption seems to facilitate Kant's claim that scientists are able, based on observation, to formulate biological laws in the form he is demanding (against Herder), namely the correlation of like causes to like effects concerning objects of a kind, or (correspondingly) to avoid associating too many diverse effects with the same force or cause. For, as Kant argues in the *CPJ*, it is characteristic of organisms that they combine many diverse activities, which may be understood as unified only if understood as directed toward a common purpose. In the transitional period, the purposively-defined kind, as the subject matter or basic term for biological laws, seems to do the "work" of unifying organic behaviors, such that they may be taken to be sufficiently related to one another, so as to be characterized by laws.

---

purposes to be evidence for claims concerning God's existence, and that we should not allow specific ideas of God's purposes to intrude into natural scientific investigation; he clearly continues to hold these latter positions throughout the critical period.

16 Kant's essay, *Teleological Principles*, marks the end of this transitional period, however, since Kant asserts (in the passage cited) that we can learn about natural purposes from "physical" observation (not metaphysics), but also that teleological principles may be invoked only when "[legitimate scientific?] theory forsakes us" (*Teleological Principles* VIII 159.17). It is not coincidental, I suggest, that Kant first begins here as well to articulate his *CPJ* view of natural purposiveness (*Teleological Principles* VIII 181–2).

Kant's claim in the *CPJ*, by contrast, that properly scientific laws must be mechanical, not teleological, undermines this assumption. This change in Kant's position may, moreover, explain his change of mind concerning the Herderian proposal of the progression of species. Because Kant has come to hold that properly scientific laws may not employ purposively-defined kinds to secure their unity or lawfulness, suggestions concerning the transformation of species no longer constitute a proposal about which *"nothing at all can be thought"* (*Review of Herder* VIII 57.27). Rather, as suggested in the above-quoted passage, this proposal is consonant with the project of mechanical explanation of organic behavior (so far as possible) endorsed in the *CPJ*. For mechanical laws may consider organisms in a more fragmented manner, not (that is) considering behavior in the context of the organism as a purposive whole, or as behavior of a purposively-defined kind; such potentially kind-crossing (or kind-ignoring) laws are not unthinkable, but rather the only laws (Kant now holds) that we may legitimately formulate concerning organic behavior.

This is not to say that the Kant of the *CPJ* is more sympathetic to Herder's organic force. In the *CPJ*, far more explicitly than in the review, Kant still objects to "hylozoism", the view (endorsed by Herder among others) that matter as such is characterized by living force: though the proposal concerning transformation of species now appears unproblematic, even promising, to Kant, he emphasizes that one must still uphold a strong distinction between the inorganic ("lifeless" matter) and the organic (*in toto*). Kant's famous denial that there could be a "Newton who could make comprehensible even the generation of a blade of grass" may, too, express his continuing rejection of an approach that invokes a single organic force, as Newton invoked attraction (*CPJ* V 400.18–9). But it is to say that this rejection of Herder's position needs to rest on somewhat different grounds from those adduced in the review. By the time of the *CPJ*, Kant himself admits that in order to grasp the unity of diversity in organic behavior—by contrast to, indeed defining the contrast to, inorganic causal relations—we must (albeit reflectively) invoke a metaphysical concept: in his case, the concept of an intelligent designer, the (as if) origin of natural purposes. Thus the true burden for Kant's mature opposition to a Herderian view lies, I suggest, in the question of *which* metaphysical, problematic, concept—the one organic force or purposiveness—may serve better as a regulative principle, as a guide to (proper) scientific investigation, which (on Kant's later view) may employ neither of these as a determinative principle. But a consideration of this question must wait for another occasion.[17]

17 I am grateful to Les Harris and participants at the international symposion on Kant's theory

References

Ameriks, Karl 2009, The Purposive Development of Human Capacities, in: *Kant's Idea for a Universal History with a Cosmopolitan Intent*, ed. by Amélie Oskenberg Rorty and James Schmidt, Cambridge: Cambridge University Press, 46–67.

Grier, Michelle 2001, *Kant's Doctrine of Transcendental Illusion*, Cambridge: Cambridge University Press.

Haym, Rudolf 1954, *Herder*, Berlin: Aufbau Verlag, vol. 2.

Hegel, Georg Wilhelm Friedrich 1807, Phänomenologie des Geistes, in: Georg Wilhelm Friedrich Hegel, *Werke*, Frankfurt/M.: Suhrkamp 1970, vol. 3.

Herder, Johann Gottfried 1785, Ideen zur Philosophie der Geschichte der Menschheit. Erster und Zweiter Teil, in: Johann Gottfried Herder, *Sämmtliche Werke*, ed. by Bernhard Suphan, Berlin: Weidmannsche Buchhandlung 1887, vol. 13, 1–484.

Herder, Johann Gottfried 1797, Briefe zu Beförderung der Humanität, in: Johann Gottfried Herder, *Sämmtliche Werke*, ed. by Bernhard Suphan, Berlin: Weidmannsche Buchhandlung 1883, vol. 18, 1–302.

Herder, Johann Gottfried 1799, Eine Metakritik zur Kritik der reinen Vernunft, in: Johann Gottfried Herder, *Sämmtliche Werke*, ed. by Bernhard Suphan, Berlin: Weidmannsche Buchhandlung 1881, vol. 21.

Herder, Johann Gottfried 1800, Kalligone, in: Johann Gottfried Herder, *Sämmtliche Werke*, ed. by Bernhard Suphan, Berlin: Weidmannsche Buchhandlung 1887, vol. 22, 1–332.

Richards, Robert J. 2000, Kant and Blumenbach on the *Bildungstrieb*: A Historical Misunderstanding, *Studies in the History and Philosophy of Biological and Biomedical Sciences* 31 (1), 11–32.

Zammito, John 1992, *The Genesis of Kant's* Critique of Judgment, Chicago: University of Chicago Press.

Zammito, John 2007, Kant's Persistent Ambivalence Toward Epigenesis, 1764–90, in: *Understanding Purpose. Kant and the Philosophy of Biology*, ed. by Philippe Huneman, Rochester: University of Rochester Press, 51–74.

---

of biology held in Tübingen in December 2010, particularly Eric Watkins and Angela Breitenbach, for comments on previous drafts of this paper.

Luca Illetterati
Teleological Judgment: Between Technique and Nature

1 Introduction

In Kant's philosophy, teleology is an essential topic and plays a fundamental role in the different areas of his thought. In this paper, I will only deal with the role teleology plays with respect to theoretical-epistemological problems. More specifically, I want to consider the role teleological judgment plays in the scientific explanation of the natural world. On the one hand, according to Kant, the teleological conceptual framework is necessary in order to understand some kinds of natural objects, namely living beings. On the other hand, this very approach is problematic insofar as it takes our comprehension of living beings beyond the scope of scientific discourse.

I will start with a brief analysis of the modern criticism of teleology (section 2) and will show how Kant's discussion of teleology and the contemporary epistemological debate concerning the possibility of a teleological or functionalistic approach in natural science are focused on the same set of problems. Kant's aim is to find a coherent justification of a teleological explanation of organic life, namely an explanation free from the contradiction implied by the application of a teleological approach to living beings (section 3). This concern leads Kant to draw the distinction between internal and external purposiveness (section 4). The aim of the paper is to focus on some tensions involved in Kant's conception of living beings, and more precisely on the problematic relation between epistemological and ontological commitment (how can we explain the way of being of organisms? and what is the constitution of organisms?) within this conception. This tension depends on the impossibility of ascribing a constitutive value to teleological judgment with respect to natural objects. This impossibility seems to presuppose technical-practical finality as the model at the basis of the Kantian notion of internal purposiveness. In section 5, I highlight the problems entailed by the assumption of the conceptual framework of technical-practical behavior as an explicative pattern for the comprehension of the natural world. Finally, my aim in the concluding remarks is to consider the implications of this critical analysis with regard to the possibility of a foundation of a science of living beings.

2 Spinoza's (and the Modern) Criticism of Teleology

The development of modern natural science has followed the path of a sort of liberation from any reference to extra-natural, transcendent elements in the explanation of the natural world. Ends and purposes, and the reference to the teleological approach, seem to require a reference to a transcendent element of this sort. This is why modern natural science refuses to adopt models based on final causality and admits efficient causes as the only kind of causality capable of explaining natural phenomena.

Generally, anti-teleological approaches consider teleology and more precisely the reference to final causes as an illegitimate application of models explicative of human behavior as something purposively determined to non-human nature. In this sense, Spinoza's criticism of our prejudice in favor of final causes is paradigmatic[1]. According to Spinoza, if we want to grasp the truth of things, we need to remove any prejudice in favor of final causes. This misleading approach is rooted in the human mind and consists in the idea that not only artifacts, but also natural objects, are determined with respect to a purposively determined end. In the *Ethics* (1677 in 2000, 108), Spinoza tries to show "that all final causes are nothing but human inventions". This prejudice has its origin in the fact that humans usually think of objects in terms of utility, and consequently they tend to do the same with natural objects as well. In the case of artifacts, utility is the aim of their production. However, humans do not produce natural objects. Therefore, they presuppose the existence of someone who has thought of and produced these objects in accord with their utility for humans. Since the teleological approach only works for artifacts, Spinoza cannot accept it as a conceptual framework applicable in our understanding of nature. Therefore, the assumption of purposes as the real causes of these objects is not possible; natural purposiveness is rather to be seen as an anthropomorphic ascription of ends to something that is commensurable with them.

The problematic issue in natural teleology is the necessary connection of purposiveness with the intention of a producer. The realm of artifacts is conceived by Kant as belonging to the technical-practical sphere. The thought of such kinds of objects as purposively determined is legitimated by the fact that their purpose is connected to the intention of their designer, who can be considered the efficient cause of the object itself. Recognizing the intention that deter-

---

1 On Spinoza's position on teleology, see Garrett (2003). Here I just refer to what Spinoza defines as "finalistic prejudice".

mines the object allows the avoidance of the logical *aporia* of a sort of *backward causation* (something coming later—the function—causing something which comes earlier—the functionally characterized object). The logical incoherence implied by the *aporia* is removed by anticipating the purpose in the intention of a designer. The removal of the *aporia* implies the removal of the final cause itself, which leaves only efficient causes behind. Consequently, if we take Spinoza's critique seriously, in order to think of a natural object as purposively structured, we have two alternatives, both of which are impossible: either we fall back into the *aporia* of backward causation, or we connect the object with the intention of a designer as in the case of artifacts.

The cogency of a criticism such as Spinoza's (a criticism that I assume as typical of the modern approach to natural teleology[2]) is not sufficient for the elimination of the reference to the teleological approach in natural science. The discussion of the legitimacy of this approach has continued in twentieth-century epistemology. The debate is now particularly focused on the notion of function, which seems to be fundamental in biology. Although the prevailing philosophical position on teleology in the twentieth-century debate is reductionism or even eliminativism, the use of teleological notions is quite common in the language of natural science. When we say: "the heart is the organ whose function is to pump blood", we identify an object with a notion (its function) that is *prima facie* a teleological one. Therefore, when we try to explain a natural object like the heart, the strategy we adopt is not so different from the one we use when we speak about a telephone, a pen, a microphone, etc.: in all of these cases, we identify objects on the basis of their functions or purposes. This is why we are warranted in speaking of an *artifact talk in biology* or, more radically, in conceiving of biology as a science that assumes an *artifact model of nature*[3].

To summarize, the main issues concerning the epistemological debate on teleology are: (a) the *necessity* of concepts and terms that deal with the teleological approach within scientific discourse, and (b) the epistemological justification of the use of teleological notions within scientific discourse. These issues are examined by Kant in his discussion of teleological judgment in the second part of the *CPJ*.

---

2 I use the expression 'modern approach' because in Greek thought, and specifically in Aristotle, the possibility of final causes with respect to nature is not connected with the intentions of a designer.

3 See Lewens (2004, 1–4).

3 The Kantian Project of Natural Teleology

Kant's theory of teleological judgment with respect to natural organisms can be considered as a strategy to justify the use of teleological notions in scientific explanations of natural objects. This theory is rooted in two apparently incompatible assumptions. The first is that the constitution of organisms and their parts can be recognized, identified, and understood only by considering the ends that they manifest. The second is that, within a modern paradigm, the acknowledgement of a teleological structure of living beings implies a form of theological-metaphysical commitment that is not admitted in natural science. This kind of commitment depends on the connection of purposes with the idea of a designer. Kant's aim is to develop a theory that accepts the first assumption without implying the second.

The starting point of Kant's argument is consistent with modern science's assumptions. As Kant claims at the beginning of the "Critique of the Teleological Power of Judgment", looking for a special kind of causality, namely final causality, within nature as "the sum of the objects of the senses" (*CPJ* V 359.17) does not have any ground. Actually, with regard to the existence of natural purposes, Kant is even more emphatic and affirms that "even experience cannot prove the reality of this to us" (*CPJ* V 359.25–6). Our experience of ends in nature—natural objects that appear to be structured as if they should perform a certain function—does not allow us to explain their constitution on the basis of final causality (even if this causal model seems to be the very one that is effective in explaining them). Actually, this would imply the presupposition of an intelligence as providing the origin and justification of purposes. According to Kant, this presupposition could imply one of the two commitments: (a) nature as an intelligent being or (b) an intelligence beyond nature as an intelligent designer ordering and organizing it. Nevertheless, in the first case the idea of an intelligent being transcends the concept of matter, while the second case involves a pretense that transcends the limits of our knowledge.

Nonetheless, the impossibility of presupposing an intelligence as the origin of ends in nature does not prevent us from using a teleological approach with respect to living beings. In other words, even if one affirms that our tendency of ascribing final causality to nature is unjustified, this does not mean that we may not refer to this kind of causality in order to explain natural items. However, when we adopt the teleological approach, we have to be conscious that the claim that there is final causality in nature is legitimate only by *analogy* with the only coherent way we can think of such a kind of causality; that is, in terms of technical-practical purposiveness. Within the sphere of human action and the arts, this kind of causality is not problematic, because in this case ends always presuppose a will as their origin.

Acknowledging a function or a purpose in an artifact means attributing a causal role to the concept of this function or purpose with respect to the artifact itself:

> If one would define what an end is in accordance with its transcendental determination [...] then an end is the object of a concept insofar as the latter is regarded as the cause of the former (the real ground of its possibility); and the causality of a *concept* with regard to its *object* is purposiveness (*forma finalis*) (*CPJ* V 219.30 – 220.4).

Therefore, we need to be aware of what is implied by attributing a teleological principle to nature. In making such an attribution, we conceive of a natural object's concept as characterized by a causality of the object itself, as if this concept—to use Kant's words—were to be found in nature and was not simply a concept for us thinking about nature: "For we adduce a teleological ground when we ascribe causality in regard to an object to a concept of the object as if it were to be found in nature (not in us)" (*CPJ* V 360.29 – 31).

Thinking of nature as actually characterized by final causality means presupposing that nature itself has some sort of intentional capacity. Through this capacity, it is possible for a concept to perform a causal role with respect to an object. In other words, introducing a teleological principle into nature means to "conceive of nature as *technical* through its own capacity" (*CPJ* V 360.34), or to conceive of nature as a technique without a designer, namely as a technical/poietical work, whose process does not correspond to an intentional project and therefore does not imply a designer's intentions.

Whereas in the technical model the contradiction implied in this kind of *backward causation* can be avoided by positing the purpose within the agent's intention, the analogical application of this teleological model in nature allows us to use a teleological approach only problematically, namely without claiming that it can actually grasp the way of being of natural objects. The conceptual framework explicative of living beings is not supposed to say anything about their ontological status. Rather, it simply sheds light on the way we think of them:

> Nevertheless, teleological judging is rightly drawn into our research into nature, at least problematically, but only in order to bring it under principles of observation and research in *analogy* with causality according to ends, without presuming thereby to *explain* it (*CPJ* V 360.21 – 5).

Thinking of this causality not analogically, but rather as constitutive of nature, implies the introduction of a new kind of causality into natural science, namely a causality that we "merely borrow from ourselves and ascribe to other beings, yet without wanting to think of them as similar to ourselves" (*CPJ* V 361.10 – 1). This analogical use entails that teleological judgment is only reflective and cannot shed light on the ontological status of living beings. Otherwise the notion of purpose would "no longer belong to the reflecting, but to the determining power of judgment" (*CPJ* V 361.5 – 6). Therefore it is a concept employed "for guiding research into objects of this kind" (*CPJ* V 375.20 – 1). The constitution of objects can be grasped only by the objectively valid categories of understanding, i.e., only through determining judgments.

4 External and Internal Purposiveness

The analogical strategy which borrows from the technical/practical sphere in order to understand the natural one is not the only move that Kant makes in order to legitimize the teleological approach with respect to nature. Kant also distinguishes between two different models of purposiveness: external and internal purposiveness. Only the second one has an epistemic value with respect to our knowledge of the natural world. External or relative purposiveness provides information about the general order of the natural world, but only internal purposiveness highlights the way of being of a specific natural object. By contrast, external purposiveness has to be excluded from any scientific consideration of nature. *External purposiveness* takes place when an entity or a natural event appears to be oriented towards something else's utility[4]: "By external purposiveness I mean that in which one thing in nature serves another as the means to an end" (*CPJ* V 425.4 – 5). Purposiveness, in this case, is "contingent in the thing itself to which it is ascribed" (*CPJ* V 368.12). Internal purposiveness—which in a certain way is reminiscent of some features of Blumenbach's *Bildungstrieb*—takes place when a single thing is simultaneously a "cause and effect of itself" (*CPJ* V 370.36 – 7), that is, when the end of an object is the realization of the object itself.

This distinction is essential for Kant. On the basis of this distinction he can rehabilitate the notion of purposiveness and save it from the criticism of teleol-

---

4 More precisely, according to Kant, relative purposiveness can be called "usefulness" ("Nutzbarkeit") when referred to human beings, and "advantageousness" ("Zuträglichkeit") when referred to any other creature (see *CPJ* V 367.8).

ogy which characterized the birth of modern science. According to this criticism humans usually interpret the natural world in the same way they interpret artifacts. This criticism seems to be effective against the teleological approach built on the model of external purposiveness.

External purposiveness can be described in two ways. In §63 of *CPJ*, Kant characterizes it as a *relative* purposiveness, as an organizational model of nature, as a whole in which every natural object is an end in relation to another one. External purposiveness is also the model of purposiveness at the basis of artifacts, because it depends on an entity different from the artifact itself. These two descriptions of external purposiveness have the same logical structure: an object can be an end for something else only if its purposiveness finds its origin in another item external to it that plays a causal role with respect to the object itself. In this sense, external purposiveness implies the presupposition of an intelligence as the origin and justification of purposes.

On the one hand, the clarification of this concept and its distinction from internal purposiveness allows Kant to part ways with the anthropocentric teleology criticized by Spinoza or from what we call cosmic teleology, which involves the kind of progression towards perfection described by Mayr (1982, 50). On the other hand, Kant can attribute an explicative capacity to internal purposiveness with respect to living beings (albeit only at the level of reflecting judgment, and not through determining judgment), because this kind of purposiveness is not targeted by the modern criticism of teleology, insofar as it has nothing to do explicitly with intentions.

Natural objects comprehensible through the notion of internal purposiveness are, in Kant's words, natural ends (*Naturzwecke*). A natural item is conceivable as an end (and then the reference to a teleological judgment is justified) only if it resists mechanical explanation. However, in order to say that something exists as a natural end, something more is needed. Since it is not possible to refer to a will, we can claim that something exists as a natural end only if it is the *cause and effect of itself*, even if this is acknowledged by Kant as "a somewhat improper and indeterminate expression" (*CPJ* V 372.17).

The products of nature which satisfy these conditions, and which thus can be considered as natural ends, are living beings. According to Kant, a living being can be both a cause and effect of itself in at least three senses. Firstly, with respect to the species, in the sense that an organism, by producing another, "continuously preserves itself, as species" (*CPJ* V 371.11–2). It is, therefore, both a cause and an effect of the survival of the species. Secondly, with respect to the *individual*, in the sense of growth, which "is to be taken in such a way that it is entirely distinct from any other increase in magnitude in accordance with mechanical law" (*CPJ* V 371.15–6). Instead, it is a form of generative production or generation (*Zeugung*) necessary to the development of all organisms (being thus the *cause* of itself) "by means of material which, as far as its composition is concerned, is its own product" (*CPJ* V 371.20–1) and therefore, an *effect* of itself. Thirdly, in the sense that the *preservation* of each part "is reciprocally dependent on the preservation of the others" (*CPJ* V 371.31–2), e.g., leaves "are certainly products of the tree" (therefore, its effects), "yet they preserve it in turn" (*CPJ* V 372.2) and are therefore causes.[5]

According to Kant, such features—the ways in which the being of organisms manifest an autopoietic structure in which they are both the cause and effect of themselves—permit us to distinguish the ontological structure of naturally organized products from artifacts. Artifacts differ from organisms because they are products whose ends are always external to themselves.

In fact, following Kant's argument, a machine (a) cannot produce another machine via the self-organization of its matter, (b) cannot, *by itself*, replace its own parts or modify its arrangement spontaneously, and (c) can be the instrument for the movement of other bodies, similar to machines, though it can never be the efficient cause of their production.[6] Thus, the machine is an organized product, but its organization is the result of a 'rational causality' that is different from the matter which composes it. The principle of its organization is a rational causality external to it. Therefore, the distinctive feature of living beings, in contrast to machines, is not their being organized. If we consider them as beings whose parts are what they are only in relation to the other parts and to the whole, then living beings are organized products, like machines. The essential relation of parts and whole is the common feature of all organized products, both artifacts and natural ones. In the same way, both in living organisms and in machines, the parts, isolated from their whole, are not what they were anymore (i.e., they cannot perform the function they are meant to perform within the totality they belong to).

---

5 This is a very interesting point within Kant's argument: the capacity of being both cause and effect of itself in the last sense is at the basis of the extraordinary capacity (possessed only by living beings, and distinguishing them from even the most complex artefacts) of fixing possible deficiencies (self-help) via a transformation of the functions of single parts in order to preserve the whole organism. This capacity can lead to the development of completely novel forms of life, and also to quite odd creatures (see *CPJ* V 372).

6 Each part of an organism has to be thought of as "as an organ that *produces* the other parts" (*CPJ* V 374.3–4), so that each part produces the others reciprocally. In other words, whereas in a technical product "one part is certainly present for the sake of the other but not because of it" (*CPJ* V 374.11), in an organism, "as an *organized* and *self-organized* being" (*CPJ* V 374.6–7), each part can be considered "only *through* all the others" and "*for the sake of the others* and on account of the whole" (*CPJ* V 373.36–7).

Thus, a living organism is not distinguished from a machine by being something organized. Rather, both a living organism and a machine are different from simple aggregates. Referring to a classic Aristotelian distinction (*Metaphysics* 1024a1–4), organized beings are *wholes within which the position of the parts makes a difference*, whereas aggregates are objects describable as *totals within which the position of parts does not make any difference.* This is why *organization* is a feature that characterizes both natural living organisms and machines. Moreover, if we look at the modern origin of the concept of organism, the complexity of the issue becomes explicit.[7]

In early modern thought, the notions of organism and machine often overlap. They both refer to an organization that unifies a plurality of parts interacting with one another. The concept of an organism establishes itself within scientific language, especially in the polemics of Stahl and Leibniz against Cartesian mechanics.[8] Nonetheless, its origin within a context dominated by the idea of the machine is undeniable. In Leibniz, *machina* does not exclusively denote an artificial product, but rather refers to an organized whole of different parts, each of which is an instrument (*organon*) related to the others. In this sense, the notion of *machina* is not set against the idea of organism, but opposed to the idea of a confused and non-ordered aggregate (like a heap of stone or a stack of wood). [9]

---

7 See Cheung (2006). Cheung shows that the first occurrence of the term 'organismus' in a scientific context, even though it appears in some medieval documents, can be found in the works on medicine and physiology by Stahl. However, Cheung stresses that, during the eighteenth century, the word denoted not so much a particular kind of object, but a way of organization of objects. Only at the end of eighteenth century and at the beginning of the nineteenth century did this word start to denote living beings, and progressively became a technical expression in biology. For a general picture of this issue, see Duchesneau (1998).

8 This topic is developed in Nunziante (2004).

9 "*Corpus viventis est Machina sese sustentans et sibi similem producen*" (Leibniz, 1683–1685(?), 568) Leibniz does not overlook the difference between organized structures whose principle of organization is external—namely structures requiring the reference to a producer who realises the project—and structures whose organization is internal and spontaneous. Nevertheless, according to Leibniz, this difference is not the basis of the distinction between *machina* and *organism*. It is rather the criterion differentiating the *artificial* from the *natural*. This is confirmed by the fact that Leibniz refers to structures organized by an external subject with the expression *organica artificialia*, whereas the expression *machina naturalis* refers to self-organized, self-maintaining, and self-reproducing structures. In a certain way, this reference to Leibniz could seem to confirm Heidegger's (1976, 255) idea that the concept of an organism is specifically modern, rooted in a mechanical-technical pattern, a concept through which living beings are conceived as artifacts that produce themselves. According to Heidegger, a radically critical analysis of the concept of an organism seems to be even more necessary if the emphasis on the notion of organism and related expressions is meant to take us beyond the mechanistic inter-

Therefore, in the idea of internal purposiveness, Kant does not merely mean to focus on the organized relation between the whole and the parts in living beings, which characterizes machines as well. In the kind of beings which can be defined as natural ends, the parts are possible only in their relation to the whole, in terms of both their existence and their form. The specificity of those beings that are natural ends, provided they cannot be considered as products of a rational causality external to them, is the relation between their parts and the whole as if *each one was mutually cause and effect of the existence and the form of all the others.*

Since internal purposiveness is not reducible to external purposiveness, and since this implies that there is a kind of purposiveness which cannot be reduced to an intention, the application of a teleological model to the natural world is possible only in relation with the first one. The application of the external purposiveness model (which grounds the structure of artifacts) to natural beings implies unsustainable assumptions, falling under Spinoza's critical argument. In this sense, the distinction between internal and external purposiveness satisfies an epistemological and an ontological demand. This difference shows from an epistemological point of view the futility of scientific accounts of nature using teleological principles based upon external purposiveness (as in the various forms of anthropocentric and cosmic teleology) and the possibility of a teleology based upon *intrinsic* purposiveness in the consideration of organized products of nature, albeit mainly with a regulative and heuristic function. The distinction between internal and external purposiveness also highlights an ontological difference between natural and technical products:

> An organized being is thus not a mere machine, because that has solely a *motive* power [*bewegende* Kraft]; while an organized being possesses in itself a *formative* power, and indeed one that it communicates to the matter [Materien], which does not have it (it organizes the latter): thus it has a self-propagating formative power, which cannot be explained through the capacity for movement alone (that is, mechanism) (*CPJ* V 374.21–6).

Whereas the structure of natural beings is *self*-organized and *self*-organizing (and thus comprehensible through the concept of internal purposiveness), the organizing principle of artifacts is always external to the products themselves (and thus comprehensible through the concept of external purposiveness). Similarly, whereas the living product of nature is characterized by a self-realizing activity (which is why we can speak of *internal* purposiveness), artifacts always

---

pretation of living beings and assume a normative function as an 'alternative' paradigm to the mechanistic one.

point at something external by finding their target in something different from themselves; thus they are characterized by an *external* purposiveness.[10]

5 Functions as "Fictions"? The Analogy between Technical, Practical, and Natural Teleology

The distinction between internal and external purposiveness appears to be very important for Kant. Nevertheless, he does not overcome the classic concept of purpose as something that presupposes the thought of an intentional act. Kant's perspective on the problem of purposiveness accepts Spinoza's presupposition concerning the necessary connection between the idea of purpose and the idea of a designer that conceives of it. On the one hand, Kant refuses to use the model of external purposiveness within the scientific account of nature, and considers the principle of internal purposiveness essential in order to understand the way of being of a living being. On the other hand, according to Kant, the principle of internal purposiveness is not constitutive of living beings themselves: internal purposiveness is *a regulative concept for a reflecting power of judgment*. A teleological judgment about living beings can have only a regulative value because the attribution of a constitutive value would lead *necessarily* to the presupposition of an intention behind things.

From this perspective, Kant does not radically call into question the notion of purpose involved in Spinoza's argument. If we could say that the slogan summing up Spinoza's conception of purposiveness is "Nature's Functions are Human Fictions", Kant's proposal does not appear to challenge the fictional character of the notion of natural end. Rather, he would say the following. First, there are *unjustified* teleological notions that we have to exclude from our scientific explanations of nature—for instance, the representation of the natural world based on the principle of external purposiveness. Second, there are *justified* teleological notions that are not only useful, but also necessary for understanding the natural world—for instance, internal purposiveness with respect to organisms, and in particular with respect to the relation between the whole organism and its parts. But these kinds of notions, even if necessary, seem to maintain a sort of fictional character too: indeed, they have no justification in things themselves, but neither do they have their origin in mere human invention. They rather have their justification in the way subjects necessarily understand living beings.

---

10 On this issue, see Illetterati (2008) and Toepfer (2008).

By restricting the explanatory value of teleological judgment with respect to living beings, Kant's position risks weakening the power of the distinction between internal and external purposiveness. If only a purposiveness supported by an intention can really perform a causal function, internal purposiveness seems to be an external purposiveness without intention. This is why internal purposiveness can be expressed only by a reflective judgment and not by a determining one. This distinction seems to preclude the possibility of answering the ontological demand from which it arises, that is, to provide a criterion to distinguish living beings from artifacts.

Purposiveness can have a constitutive value with respect to an object only if it is somehow explicable according to uni-directional efficient causality, namely only if it is an external purposiveness. The value of the teleological model is only really constitutive of artifacts because it is reducible to the efficient causality model via the recognition of the designer's intention. Ascribing a constitutive role to purposiveness in nature *necessarily* leads to the idea of a designer, and the presence of a designer implies a reference to something external to nature, namely a transcendent principle. Kant claims that purposiveness cannot be considered as constitutive of nature, because, apparently, he cannot think of purposiveness as independent from the intention of a designer.[11] Namely, he cannot think of a final causality different from the technical-practical one, which is the only admissible model of causality according to ends. The point is made via the claim that such an attribution would somehow entail the admission of an architect, or a producer, as the only way to make sense of a constitutive natural purposiveness. Nevertheless, this way of considering the natural ends runs the risk of getting entangled in a contradiction. On the one hand, in order to be natural, the end cannot refer to any intention grounding it. On the other hand, in order to be an end, it necessarily implies a reference to an intention, because only this reference can justify its being an end.[12] Therefore, even if technical-practical purposiveness is considered by Kant only as analogous with purposive-

---

11 This argument is developed by Chiereghin (1990).

12 According to Ginsborg, the regulative character of teleological judgement is not sufficient to avoid the intrinsic *aporia* of the notion of a natural end. Ascribing a regulative value to the notion of a natural end puts the conflict between the notion of "end" (which is linked to the notion of design) and the notion of "natural" (which is independent from the notion of design) on another level: "Otherwise there is no such thing as regarding something *as if* designed, without regarding it as *in fact* designed, in which case we again seem committed to regarding the object as having two contradictory features, and hence adopting an attitude towards it which is apparently incoherent" (Ginsborg 2006, 459). The topic is also developed in Ginsborg (2001).

ness in nature, it is still the model he uses to analyze the internal purposiveness of self-organized structures.

It is very likely that the notion of *purposiveness in nature* is connected in various ways, in Kant, with that of the *technique of nature*. The *technique of nature* surfaces at various points; generally speaking, we may claim that this concept points to the teleological procedure of nature—its proceeding according to purposes, so that, for instance, "we would call the procedure (the causality) of nature a technique, on account of the similarity to ends that we find in its products" (*CPJ* V 390.33–5). Therefore, the expression *technique of nature* refers to the perspective "where objects of nature are sometimes judged merely as if their possibility were grounded in art" (*First Introduction* XX 200.8–9).[13] Insofar as Kant conceptualizes the functioning of organized natural products as a *technique of nature*, he seems to think of them in terms of *production*, namely that kind of action at the basis of the way of being of a machine, or of a technical product—something to which, following Kant, living beings cannot be reduced.[14] Even if Kant meant to underline the irreducibility of living beings to machines through the notion of purposiveness, when he needs to justify their causal model as acting in terms of internal purposiveness, he is compelled to refer to the causal model at the basis of the purposiveness characteristic of the technical-practical field, namely external purposiveness.

This kind of perspective implies a kind of further weakening of the notion of internal purposiveness. Such a weakening justifies those interpretations that think of Kant's resort to purposiveness not as an attempt to provide an autonomous scientific foundation for the science of living beings, but rather as a proof —to use Zammito's (2006, 749) expression—of a sort of "epistemological 'deflation'" with respect to biological and biomedical sciences.[15] From this perspective, the necessity of a reference to purposiveness would suggest that these "sciences" are not "real sciences". And they cannot be real sciences, because they cannot determine the objects they study.

Therefore, Kant's position is characterized by a particular tension. On the one hand, through the distinction between internal and external purposiveness, Kant seems to think of living beings as not reducible to the manner of being of

---

13 Natural beings can be described *as if* their possibility were based upon art, and therefore technique.

14 The expression 'technique of nature' plays a fundamental role in the *First Introduction*; its role is drastically reduced in the *Introduction* Kant published. This might indicate Kant's awareness of the problems faced by the corresponding concept. However, the expression remains in the published version and the plane to which it refers is also manifestly preserved.

15 An interpretation of *CPJ* as a sort of foundation of life-sciences is argued by Ungerer (1922).

artifacts. On the other hand, by ascribing only a regulative and heuristic value to the principle of internal purposiveness, Kant makes explicit that the only effective explanatory purposiveness is the external one. This kind of purposiveness is reducible to a linear causal model, and thus allows us to think of final causality while avoiding the logical problem of backward causation. This tension arises plainly when Kant approaches the relation between causality according to ends, typical of the practical world, and the notion of a *purposiveness of nature*. In effect, in this context Kant himself tends to call into question the analogical perspective that, at the beginning of his path, in the "Analytic of Teleological Judgement", seemed to be the only way of justifying the reference to a teleological model in natural research.

In the "Introduction" to the *CPJ*, Kant clearly claims that "this concept [*end*] is also entirely distinct from that of practical purposiveness (of human art as well as of morals), although it is certainly conceived of in terms of an analogy with that" (*CPJ* V 181.8 – 11). Practical purposiveness presupposes the idea of a will. Since this presupposition is not justified within the field of nature, natural purposiveness appears to be necessarily different from practical purposiveness. At the same time, natural purposiveness is necessarily conceived analogously to practical purposiveness, because this second kind of purposiveness is the only legitimate model of causality according to ends that is coherently conceivable.

However, Kant still underlines the difference between natural purposiveness and technical-practical purposiveness in §65. There, he explicitly discusses natural beings as natural ends, and highlights the difficulties of an interpretation of living beings that refers to artifacts, even if this reference is only analogical: "One says far too little about nature and its capacity in organized products if one calls this an *analogue of art* [*Analogon der Kunst*]" (*CPJ* V 374.27 – 9). According to Kant, when we compare the way of being of the organized products of nature with arts and techniques, we have already transformed living beings from self-organizing products into entities related to a rational being external to them. The thought of the organized products of nature as analogous to artifacts would imply a misunderstanding of their specific ontological status, that is, what characterizes them as self-organizing beings that are both cause and effect of themselves, and consequently contain their own end.

We may approach an understanding of this autopoietic capacity of natural beings (which remains largely unknowable) by calling it, as Kant does, an *analogue of life*. However, this analogy is full of problematic metaphysical consequences: "one must either endow matter as mere matter with a property (hylozoism) that contradicts its essence, or else associate with it an alien principle *standing in communion* with it (a soul)" (*CPJ* V 374.35 – 375.1). Both ways lead to a dead end: in the first case, we presuppose what we aim to explain, that is, organized matter (*and this is the contradiction inherent in any form of vitalism*); in the second case, we take the soul to be the artist of such a construction, thereby subtracting it from nature (*and this is inherent in any form of animism*).

Strictly speaking, Kant claims: "the organization of nature is therefore not analogous with any causality that we know" (*CPJ* V 375.5–7). This is a crucial point: the self-producing structure of living beings—if thoroughly conceptualized—is irreducible to any kind of causality known to us. This irreducibility seems to allow the notion of internal purposiveness to be the one that is able to grasp the ontological structure of the organized products of nature. But, according to Kant, it is impossible to think of purposiveness without at the same time thinking of an intention as its justification. This consideration forces Kant to assume the notion of natural purposiveness as merely regulative, never constitutive.

6 Conclusions

Despite Kant's acknowledgement of the inadequacy of the analogy with the technical model of purposiveness for understanding nature, this model operates implicitly within his theory of teleological judgment with respect to living beings. In this sense, even if Kant explicitly denies the possibility of the analogy, he implicitly assumes an artifactual model of nature. If the assumption of this model represents an effective strategy from an epistemological point of view, Kant does not want to derive any kind of ontological commitment from it. Nevertheless, the assumption of the artifact model as the model of natural inquiry is not neutral with respect to the way of being of nature itself. Therefore, either we think that assuming the hermeneutics of artifacts as the paradigmatic way to understand nature somehow reflects the way of being of nature itself or, if we want to avoid reaching such a conclusion, we have to admit that the results conveyed by the assumption of such a model are nothing but mere subjective constructions having nothing to do with the way of being of the reality we are investigating. On the one hand, the analogical strategy is invoked to justify the reference to teleological judgment in our comprehension of the natural world. On the other hand, this strategy is inadequate if it is meant to grasp the way of being of living organisms and in particular their self-organizing principle (therefore the principle of internal purposiveness).

In this sense, Kant's argument seems to mirror and justify the specific tensions and the intimately problematic structure of the science of living beings of his own time. As Kant claims in the *Metaphysical Foundations*, if "in any special doctrine of nature there can be only as much proper science as there is *math-*

ematics therein" (*Metaphysical Foundations* IV 470.13 – 5); research into living beings cannot be thought as a real science. Nevertheless, in more general terms, Kant's insistence on the merely regulative value of the teleological principle (that is, the principle providing the possibility to define the object of biology) prevents Kant's philosophy of biology from being a transcendental foundation of the science of living beings.

In the *CPJ*, Kant, rather than grounding the possibility of a science of living being, denies the very possibility of it. The acknowledgement of the impossibility of a "Newton of a blade of grass" (*CPJ* V 400.16 – 20) and the acknowledgement of the impossibility of knowing the organizing principle of natural objects—which remains, according to Kant, an *"inscrutable principle"* (*CPJ* V 424.29)—implies the impossibility of knowing the essential structure of living beings and consequently of a scientific biology. Equally, if we can come close to the way of being of a living being only through the analogical strategy, but at the same time we cannot ascribe any kind of determining value to it, the strategy itself is different from a kind of foundation of the possibility of a science.

Nonetheless, Kant's proposal shows its appeal with respect to this point. For it shows how knowledge of life in general cannot be considered a "science" in the same sense as those scientific disciplines whose structure is reducible to mathematics and physics. The merely regulative value of the teleological judgment with respect to natural beings, and the acknowledgement of the impossibility of the knowledge of their organizing principle, have a critical function towards both any kind of reductionism that argues for the possibility of explaining *the organization* of living beings starting from the physical and chemical analysis of their parts, and any kind of vitalism. In effect, according to Kant, antireductionism does not imply the necessity of referring to extra-natural principles within in a scientific context. Nevertheless, vitalism assumes an antireductionist approach to justify a discourse transcending the limits of scientific knowledge. As we have seen, Kant sustains an antireductionist position (organisms are not reducible to a mechanistic explanation) while avoiding any vitalistic thesis (every explanation exceeding the mechanistic one cannot be considered as scientific). In this sense, the Kantian perspective overcomes the dichotomy between mechanism and vitalism.

Therefore, Kant's approach understands living beings on two levels. On the one hand, organic processes, if considered separately from each other can be explained in physical-chemical terms. On the other hand, the physical-chemical model turns out to be inadequate if applied to the higher levels of organization, especially with respect to the connection of different processes within a whole such as the organism. In this sense, in order to provide an account of those characteristics of an organized system that today we define as *emerging features*, we need a principle different from the mechanistic one, namely the principle of natural teleology that Kant tries to justify by starting from the structure of our cognitive abilities and thus without referring to extra-natural elements.

References

Cheung, Tobias 2006, From the Organism of a Body to the Body of an Organism: Occurrence and Meaning of the Word 'Organism' from the Seventeenth to the Nineteenth Centuries, *The British Journal for the History of Science* 39, 319–39.

Chiereghin, Franco 1990, Finalità e idea della vita. La recezione hegeliana della teleologia di Kant, *Verifiche* 19, 127–229.

Duchesneau, François 1998, *Les modèles du vivant de Descartes à Leibniz*, Paris: Vrin.

Garrett, Don 2003, Teleology in Spinoza and Early Modern Rationalism, in: *New Essays on the Rationalists*, ed. by Rocco Di Gennaro and Charles Huenemann, Oxford: Oxford University Press, 310–36.

Ginsborg, Hannah 2001, Kant on Understanding Organisms as Natural Purposes, in: *Kant and the Sciences*, ed. by Eric Watkins, Oxford: Oxford University Press, 231–58.

Ginsborg, Hannah 2006, Kant's Biological Teleology and its Philosophical Significance, in: *A Companion to Kant*, ed. by Graham Bird, Oxford: Blackwell, 455–69.

Heidegger, Martin 1976, Vom Wesen und Begriff der Physis. Aristoteles, Physik B, 1, in: Wegmarken, in: Martin Heidegger, *Gesamtausgabe*, ed. by Friedrich-Wilhelm von Hermann, Frankfurt/M.: Vittorio Klostermann 1975–, vol. 9, 239–301.

Illetterati, Luca/Michelini, Francesca (eds.) 2008, *Purposiveness. Teleology Between Nature and Mind*, Frankfurt/M.: Ontos.

Illetterati, Luca 2008, Being-for: Purposes and Functions in Artefacts and Living Beings, in: *Purposiveness: Teleology Between Nature and Mind*, ed. by Luca Illetterati and Francesca Michelini, Frankfurt/M.: Ontos, 135–62.

Leibniz, Gottfried Wilhelm von 1683–1685 [?], Genera terminorum substantiae, in: Gottfried Wilhelm von Leibniz, *Sämtliche Schriften und Briefe*, ed. by The Berlin-Brandenburgian Academy of Sciences and the Academy of Sciences Göttingen, Berlin: Akademie Verlag 1999, series 6, vol. 4, part A, 566–9.

Lewens, Tim 2004, *Organisms and Artefacts. Design in Nature and Elsewhere*, Cambridge (MA): MIT Press.

Mayr, Ernst 1982, *The Growth of Biological Thought: Diversity, Evolution, and Inheritance*, Cambridge (MA): The Belknap Press of Harvard University Press.

Nunziante, Antonio 2004, "Corpus vivens est automaton sui perpetuativum ex naturae istituto". Some Remarks on Leibniz's Distinction between "Machina naturalis" and "Organica artificialia", *Studia Leibnitiana*, Sonderheft 32, 203–16.

Spinoza, Benedictus de 1677, *Ethics*, ed. and trans. by George Henry Radcliffe Parkinson, Oxford: Oxford University Press 2000.

Toepfer, Georg 2008, Teleology in Natural Organised Systems and Artefacts: Interdependence of Processes versus External Design, in: *Purposiveness: Teleology Between Nature and Mind*, ed. by Luca Illetterati and Francesca Michelini, Frankfurt/M.: Ontos, 163–81.

Ungerer, Emil 1922, *Die Teleologie Kants und ihre Bedeutung für die Logik der Biologie*, Berlin: Bornträger.

Zammito, John 2006b, Teleology Then and Now: The Question of Kant's Relevance for Contemporary Controversies over Function in Biology, in: *Studies in History and Philosophy of Biological and Biomedical Sciences* 37, 748–70.

Predrag Šustar
Kant's Account of Biological Causation

1 Introduction

Although there are several separate issues flowing through §§64–6 of the *CPJ*, in particular in §§64–5, the notion of *biological causation* is both the focus and the most fundamental issue. Accordingly, I will argue in this paper that even the notion of a natural end, which comes to the fore in §64 of the third *Critique*, is ultimately based on a specific notion of causation that holds for relata in the biological world. As a result of the growing interest in reconstructing Kant's overall theory of the biological sciences,[1] it is useful to ascertain a decisively causally-oriented line of thought in the account sketched at the beginning of the "Critique of the Teleological Power of Judgment".

This paper aims at (i) a detailed reconstruction of Kant's account of biological causation, as the account in question is directly associated with his classification of individual objects in the biological world in terms of certain forms of objective, material purposiveness, and (ii) bringing out the most probable reasons for the difficulties in Kant's account of biological causation. With regard to the above two aims, I will try to find support for the following two interpretative hypotheses:

(H1) Kant is advocating a *dispositionalist* view of biological causation.[2]

(H2) Kant is advocating a *deflationary* view about the role that any specific metaphysics should play in the biological sciences.

As to (H1), the following similarities connect Kant's account of biological causation and contemporary dispositionalist theories of causation, both in general and in the biological sciences more specifically:

---

1 See, more recently, Zuckert (2007), especially the first part of the book, Steigerwald (2006), and Ginsborg (2001).

2 With this classification of Kant's view, I simply point out interesting similarities between his findings about how causation should be conceived in the physiological disciplines at the time and contemporary philosophical theories of causation in genetics, such as, most notably, the dispositionalist theory of biological causation, worked out recently in Mumford/Anjum (2011). I will shortly list those similarities, which will be exemplified in sections 2 and 3 of the paper.

1) They both introduce, in particular within the biological subject area, an ontology of *powers, capacities,* or *dispositions*.[3]
2) The ontology mentioned in 1) departs from Humean ontology of "loose and separate", distinct existences (Mumford 2009, 265).
3) According to 2), causation is not a contingent and external relation between loose and separate, distinct existences, but an *internal* relation between a power and its *manifestation.*
4) *Circumstances* enable powers to be manifested, which in this way trace a line between causal powers and their manifestations as distinct existences.
5) Causal efficacy obtains in virtue of *properties*, which are, basically, *dispositional* or *powers*. For example, as it is the case with the causal efficacy of aspirin to relieve headaches (Cartwright 1999).

The above list of general features that distinguish the dispositionalist theory of causation from other current philosophical theories in this area gains major support from our understanding of the main biological processes, especially those in genetics.[4] Thus, for instance, genes, proteins or other biological objects, at higher organizational levels, such as embryos, are characterized in their causal efficacy by distinctive *dispositional modality*, i.e., biological objects merely dispose towards, and do not necessitate certain outcomes. Finally, we frequently work out biological explanations by invoking dispositions or causal powers pertaining to biological objects under consideration.

Now, in showing how (H1), understood by the features given above, relates to Kant's view in §§64–6, I will proceed in the following way: In section 2, Kant's account of biological causation will be reconstructed in detail. In section 3, I will assess the validity and shortcomings of Kant's account by referring to an example from recent biological practice, which shares relevant features with Kant's own example of the tree in §64 of the third *Critique*. In brief, by giving a molecular twist to Kant's physiologically-laden account, we will be in a better position to see: (i) how Kant's dispositionalist account, as reconstructed in section 2, be-

---

3 For an overview of the debate on a possible difference between these three notions, see Mumford (2009), in which, however, according to the line of thought defended by Mumford, the basic ontology is an ontology of powers. In brief, he tries to downplay the importance of that differentiation. Kant's vocabulary contains all three notions, as will be shown in the next section. Nevertheless, it is committed to the same shift toward an ontology of biological powers, as argued in section 3 of this paper. For a new and comprehensive reading of Kant's general account of causation, see Watkins (2005).

4 See Mumford/Anjum (2011), in particular, chapter 10 "A Biologically Disposed Theory of Causation", more specifically, section 10.3 "The Central Features of Causal Dispositionalism".

haves in analogous, but also new scientific situations; (ii) the specific role that the notion of "formative power" (see, e. g., *CPJ* V 374.23) has within Kant's account of biological causation in general; and (iii) the account's main difficulty, which follows from the analysis of point (i) mentioned above. Finally, in section 4 of this paper, I will complete the analysis of Kant's account by seeing how the claims posited by the second interpretative hypothesis (H2) are related to the main difficulty facing biological causal dispositionalism.

2 Kant's Account of Biological Causation: A Reconstruction

2.1 Classifying the Products in the World

Kant's line of argument in §§64–6 in the "Critique of the Teleological Power of Judgment" opens with an attempt to identify the main types of products that we can encounter in the world. According to §64, these types of products (TP) are the following:

(TP.1) mere natural products.

An example that adequately illustrates the above type of product is a grain of sand that can be found on a beach, as reported by Kant's own example. The grain is a product of different activities that involve the natural object in question: the physical force of the waves, the impact of other pebbles and grains, some chemical factors and so on. It should be immediately noted that Kant insists—already in this opening move—on a certain *productivity* or, more specifically, *causal efficacy* feature that distinguishes relationships among the above-mentioned properties of the objects that are involved in the examined token of (TP.1).

Kant's classification proceeds by referring to the type of products that will be the focus of our attention in this paper:

(TP.2) natural products such as natural ends.

Consider here Kant's example of a tree, which, as we will see later on, he uses throughout §64 to illustrate his most basic insight about causation operating in the living world. Now, think of a tree token of (TP.2) in its various stages of individual development: it could be in an embryonic phase, an adult phase ready to reproduce or, even, in a certain regenerative phase, e. g., one concerned with getting new foliage. Again, as in the case of (TP.1), there is an even greater emphasis on the distinctive feature of the causal efficacy of the properties of the tree as a whole towards certain outcomes in changing environmental conditions and, in addition, the causal efficacy that holds reciprocally between properties of biological component-parts at different organizational levels. As will be shown in a moment, Kant's well-known and widely disputed definition of the special character of natural ends arises from this particular causal characterization.

Finally, there is the third type of products that we can encounter in the world:

(TP.3) products of art.

§64 gives a good example that illustrates these types of products, which, according to Kant, are ultimately human artifacts: "a regular hexagon, drawn in the sand" (*CPJ* V 370.19). This particular artifact token is a product of human intentional activity that presupposes a knowledge of determinate rules, which in a given situation results in the material representation of a regular hexagon. As already pointed out in the case of the other two types of products, even human artifacts are brought about by causal efficacy of something that, given certain circumstances, should dispose towards certain outcome, should that be a procedure for constructing 'a regular hexagon' or, more colloquially, a recipe for wedding cake. I will now focus on (TP.2), namely, its causal features.

2.2 Defining the "Natural End" Products

Taking a step forward in his overall account of biological causation, Kant launches the following famous definition of the special character of natural ends, which were classified earlier as one of the three main products that exist in the world: "I would say provisionally that a thing exists as a natural end *if it is cause and effect of itself*" (*CPJ* V 370.31–371.1).

Some points in the above preliminary definition should be clarified before we move to a detailed analysis of biological causation. (i) Although notably cryptic, Kant's brief causal formula *"cause and effect of itself"* (*CPJ* V 371.1) is more comprehensible in the context of the other two types of products in the world, i. e., the kinds of causal relationship that distinguish them. Roughly speaking, in the case of (TP.1), mere natural products, we have a form of unintentional causal relationship between certain properties of the examined grain of sand and corresponding causal powers, e. g., the impact of water, whereas in the case of (TP.3), products of art, a form of intentional causal efficacy responsible for their production as distinct existences. In the case of (TP.2), tokens of individual organized beings, classified by us as natural end products, instantiate a kind of causation that is characteristically *internal*. This feature, on which the dispositionalist theory of causation insists, especially in the biological sciences, gets its fuller meaning if compared to (TP.1) and (TP.3) cases. (ii) I will refer to the definition of the special character of natural ends, through the corresponding kind of causation, the *causation-based definition* (CBD). However, as we will see shortly, this is, in fact, only a 'provisional' step, as claimed by Kant in his above definition, which is followed by, according to the reading worked out here, two other main steps: the *elucidation* of CBD and its "derivation from a determinate concept" (*CPJ* V 372.20 – 1). (iii) In my view, both steps will also determine a barely identifiable reference of the locution "although in a twofold sense" (*CPJ* V 371.1), added to CBD in the second edition of the *CPJ*. At this point, we can move closer to the central part of Kant's account of causation in biology.

2.3 Elucidation of CBD: The Example of a Tree

The elucidation in question opens with one of the three most important causal processes in physiology, i.e., the process of *generation* or *reproduction*. More specifically, the process through which an individual organized being, which in Kant's example is an individual tree, belonging to a corresponding biological species, which "generates another tree in accordance with a known natural law" (*CPJ* V 371.7– 8). It is distinct as an individual organized being, but also the same at the biological organizational level of a species. Now, this simple situation taken from physiology, where the main focus of concern is the distinction between the two organizational levels in the biological world—the tree as an *individual* natural object and the tree considered as a certain natural *kind* or a *species*—is offered by Kant in order to elucidate previously mentioned causal formula.[5]

---

5 With the above characterizations, I am not implying that Kant is endorsing any variant of more recent accounts in the debate on the ontological status of biological species, such as the account of *species as natural kinds* in a stronger sense or the account of *species as individuals*. Nor am I implying that Kant is steering a middle-ground philosophical account in this regard, namely, that he is endorsing a variety of the so-called "Homeostatic Property Clusters" account of the ontological nature of biological species. For an overview of this debate in the history of philosophy of biology, and further application of the debate to other related issues, see Wilson (2005). In this paper, I will stay focused strictly on Kant's account of causation as this issue appears in the biological sciences that he examines mostly in the "Analytic" of the "Critique of the Teleological Power of Judgment".

Let us ignore here the extent to which Kant succeeds in illustrating the preliminary CBD of natural ends with regard to the physiological process of reproduction, and single out, instead, certain points that are more directly intertwined with the issue of biological causation. In the causal process in which an individual natural object, such as a tree belonging to some taxonomic group, produces another individual natural object of the same kind, the following two points are especially interesting:

(i) The process in question occurs, as claimed by §64, "in accordance with a known natural *law*" (*CPJ* V 371.7–8, my italics). More specifically, when an individual tree, through its life-cycle, produces another tree, this causal process is *law-governed*, even though it is not entirely clear throughout the third *Critique* what the characteristics of these particular natural laws are.[6] Thus, putting the point even more schematically, Kant apparently suggests that we have a causal process in physiology in which two individual natural objects are linked by a corresponding natural law. However, by looking more closely at the provisional formula given by CBD of natural end products, that is, "a thing exists as a natural end *if it is cause and effect of itself*" (*CPJ* V 370.31–371.1), it is more correct to say that an already existing individual tree ought to dispose, according to the specific, dispositional modality of biological causal processes, towards a new individual tree as its final manifestation. This understanding of the first part of Kant's elucidation of CBD also helps in finding the right position for his general view on causation, i.e., how and in virtue of what the cause is related to the effect. This basic issue brings us to the second point that should be stressed.

(ii) Taking into consideration the above shift in understanding the causal process of reproduction in plant physiology and Kant's characterization that the tree "unceasingly *produces* itself" (*CPJ* V 371.10), the general nature of causal relation is best understood as one of *production*.

The next process examined by Kant to further elucidate his provisional causal formula is "growth", which is defined as an effect of the causal activity of the tree considered as an *individual* natural end product (see *CPJ* V 371.13–29). This causal process should be distinguished from processes such as those involved in purely chemical crystallization. According to Kant, there is only an "*increase* in magnitude in accordance with mechanical laws" occurring in that class of processes (*CPJ* V 371.15–6, my italics). So, what is so specific about the physiological process of growth, and how does it elucidate Kant's provisional causal formula of natural end products? A metabolic pathway in plant physiology, taken in its

---

6 For more recent attempts to tackle this issue within Kant's philosophy of biology, see esp. Teufel (2011), also Ginsborg (2001) and Šustar (2013).

most general sense, helps here. Consider, for example, a generic process in which a green plant takes in simple chemical compounds such as water and carbon dioxide from its environment and, then, somehow produces from them new, much more complex compounds that are used for growth and many other biological processes.[7] In distinguishing the two types of causal processes, Kant insists on the fact that the "composition" of the newly synthesized compounds in a green plant, e.g., the tree mentioned in the elucidation, is "its own product" (*CPJ* V 371.21). In other words, the individual green plant in question, firstly, breaks down in a precise way the ingested raw material from its environment and, then secondly, builds up, again in a precise way, the new, much more complex, organic compounds.[8]

Several points should be drawn out from this second part in the overall elucidation of the provisional causal formula that distinguishes the natural end products in the world: (i) Although the tree that shows the causal efficacy—as provisionally claimed by CBD—in the physiological process of growth may appear as an "educt", i.e., "as far as the components that it receives from nature outside of itself" (*CPJ* V 371.21–2), it should be regarded as a product. The above *educt-product* distinction is meant to draw, once again, a previously introduced distinction between causal efficacy occurring in the processes such as crystallization and those that are properly physiological. Thus, in contrast to merely chemical processes, the causal processes in, e.g., plant physiology involve, as stated throughout Kant's works in the critical period,[9] the "capacity of separation" and that of "new composition" or the "capacity for formation" (*CPJ* V 371.25).

---

[7] In section 3 of this paper, I will examine more in detail the physiological process of photosynthesis, especially from a molecular point of view. At this point, I am just trying to more vividly illustrate Kant's second part of his elucidation of the causal formula under consideration in §64 of the third *Critique*. For an informative overview of Kant's philosophy of science with regard to the main chemical processes, which bears some relevance also for the biological processes exemplified above, and, more generally, with regard to the changes that affected chemistry at the time of its revolutionary transition from Stahl's to Lavoisier's paradigm, see Carrier (2001) and Friedman (1992, esp. 264–90).

8 Since the newly synthesized compounds are *species-specific* or, in Kant's terminology, of a "quality peculiar to its species" (*CPJ* V 371.18–9), an objection can be raised about Kant's sharp distinction between the individual level of biological organization and the species level. As I will try to show in section 4, the "Critique of the Teleological Power of Judgment" is pushed towards views like this, in particular towards a deflationary view about metaphysical research programs in biology, because of a clear theoretical inadequacy in the traditional "metaphysical systems" (*CPJ* V 392.21), e.g., hylozoism and animism, that accompany the scientific practice of the biological disciplines at the time.

9 For a reference to the most relevant works in question, see Šustar (2008).

(ii) Both capacities bring more information into Kant's overall account of biological causation, in particular, with regard to interpretative hypothesis (H1) according to which Kant endorses a kind of *causal dispositionalism*. For instance, a certain property related to the embryonic form of a tree 'produces' or 'brings about' itself as an adult individual in a range of determined environmental conditions. In other words, we are getting causation out of the above capacities, which, in the biological domain specifically, ought to dispose towards expected outcomes in certain circumstances. (iii) The previous point further confirms the basic *production* approach to causation, i. e., when we say that a cause—whatever that cause is taken to be from the ontological point of view—*causes* the corresponding effect, we say that something *in* the cause 'produces' or 'brings about' the effect.

The last of the three main causal processes in physiology through which Kant elucidates his provisional causal formula is the process of *regeneration* (E3).[10] The novelty, which is most clearly exemplified and introduced particularly in the third part of the elucidation, is related to the issue of biological or organic *component-parts-whole* relationship. For example, as stated at the end of §64, the relationship between leaves and the tree as a whole illustrates the issue in question and the difficulty of characterizing this type of relationship as a causal one, as we will see in detail in the derivation part of Kant's account.

2.4 The "Derivation" of the Notion of Natural End Products

In accordance with the reading advanced here, the derivation part aims at a more conceptually oriented philosophical analysis, once the elucidation has done its

---

10 The corresponding part of §64 in the *CPJ* is less straightforward with regard to this physiological process. Namely, although Kant is more strictly speaking focused on regeneration, understood basically as the biological process by which some organisms replace or restore lost or damaged body parts, there is also —in this part of the elucidation—a brief reference to some other types of processes in physiology, such as the processes of *teratogenesis*, e. g., "miscarriages or malformations in growth" (*CPJ* V 372.6–7). Thus, most probably for the above composite character of the final part of Kant's elucidation, the literature, when referring to these distinct types of physiological processes, emphasizes the notion of preservation or self-maintenance, as a possible common thread in that respect; for a comparison, see Ginsborg (2004). Here, I will put the above subtleties to one side, and instead move on to an important novelty that all these physiological processes puts forward, which also announces more directly the derivation part in Kant's overall account of biological causation. However, all the points that have been mentioned above, which are concerned with interpretative hypothesis (H1) apply in the same way to the causal process of regeneration.

job in illustrating the provisional causal formula. The upshot of this analysis is the characterization of natural end products through a "determinate concept" (*CPJ* V 372.18), in this case, a concept of causation that applies specifically to the physiological area of the biological sciences.

As a first step in Kant's derivation, we may single out the distinction between the following two basic kinds of connection that something called a cause yields with some other thing that we call its effect:

(i) "the connection of *real* causes"; and
(ii) the connection of "*ideal*" causes (*CPJ* V 373.1–3; my italics).[11]

Kant proceeds in bringing to an end his initial classificatory task, i.e., marking the distinction between natural end products and other two types of product, by deepening the analysis of provisional causal formula. In this regard, three distinct causal requirements are set out. I will quote and examine them in the order in which they appear in the "Analytic of the Teleological Power of Judgment".

(R1) "Now for a thing as a natural end it is requisite, *first*, that its *parts* (as far as their existence and their form are concerned) are possible only through their relation to the *whole*" (*CPJ* V 373.4–6; second and third italics added).

The most important point about (R1) is concerned with the fact that Kant is not explicitly stating that the relationships between organic component-parts of an organized being and that being, considered as a whole, should be characterized in causal terms. More specifically, the relation that we establish between a component-part and the projected unity of an individual biological organism cannot be characterized in the same causal manner as the class of relations between, e.g., two component-parts themselves can. The latter class of biological causal relations, in a stricter sense, is posited by Kant's second requirement:

---

11 In my view, this is the clearest way to summarize Kant's elaborate opening of §65 in the "Analytic", in which an adequate notion of biological causation is viewed as the right continuation of the definition of the special character of natural end products previously introduced, i.e., individual biological organisms that we encounter in the world. Additionally, the above distinction clearly suggests that only these two kinds of causation exist (*CPJ* V 373.3). In the next section of this paper, I will deploy Kant's distinction between *nexus effectivus/real* causal connection and *nexus finalis/ideal* causal connection to analyze a case-study from molecular plant physiology. Until then, the remainder of Kant's account, as canvassed in §§64–6 of the third *Critique*, will be brought out in appropriate detail.

(R2) "[I]t is required, *second*, that its parts be combined into a whole by being *reciprocally* the cause and effect of their form" (*CPJ* V 373.17–9; second italics added).

Although I will further examine this requirement, in particular with regard to the above claim that component-parts themselves, namely, by their reciprocal causal relations, are in fact building up the corresponding whole pertaining to the individual organism, at this point the *ground-cause* distinction is more pressing within Kant's overall account of biological causation. Namely, to emphasize the differences between (R1) and (R2), the *ground-cause* distinction is recalled. Thus, the first requirement presupposes that the whole, as this notion is used in physiology, is nothing other than a "ground for the cognition [...] for someone who judges it" (*CPJ* V 373.22–4). If we go back to Kant's previously introduced classification of the two main kinds of causation, only the causal relations posited by the second requirement can be potentially subsumed under the group of '*real* causal connections'. The relations posited by (R1) cannot.

Finally, according to the textual evidence displayed by §65 of the "Analytic of the Teleological Power of Judgment", a third requirement relevant to a fuller understanding of Kant's account of biological causation can be singled out:

(R3) An organic component-part does not just exist "*for the sake of the others* and *on account of* the whole [...]; rather it must be thought of as an organ that *produces* the other parts" (*CPJ* V 373.35–374.4).

This particular requirement is most important for substantiating the reading endorsed here according to which Kant can be accommodated within the production approach to causation. It is something *in* the cause that *produces* or *brings about* the effect. However, the requirement of causal production, which is more directly stated by (R3), remains binding only at the organizational level of biological component-parts. The "idea" of a whole, as employed both in physiology and in more common usage, is, as it is understood, most elaborately, in the "Analytic of the Teleological Power of Judgment", *causally vacuous*. Moreover, we can claim this from the point of view of 'real causal connections' that hold between biological component-parts.

At this point, Kant's account ends. There is, nevertheless, in §66, the introduction of the notion of a biological maxim, which applies to the way we should appropriately judge the 'formative' feature that appears to be involved in the causal powers displayed by biological objects at all organizational levels. The above methodological twist in Kant's philosophy of the biological sciences is set out in more detail in the remaining sections of the "Critique of the Teleological Power of Judgment". This paper, however, remains concerned with the view on biological causation, at this point in particular, with the notion of a formative power as this notion is employed in understanding basic physiological processes. In order to get a clearer grip on this final aspect of Kant's account of causation in the domain of biology, I will relate it to a case-study that shows relevant similarities to the examples given in the elucidation part.

3 Kant's Account Molecularized

The general process of photosynthesis in green plants, as one of the crucial areas of research in modern molecular plant physiology, relates directly to the example of growth in Kant's elucidation with the tree. I will not refer to the overall process of photosynthesis, from the point of view of molecular biology, but rather to one of its so-called causal *pathways*. Namely, I will focus on the biosynthetic pathway leading to chlorophyll, more specifically, on the causal role played by the enzyme of *magnesium chelatase* (Mg-chelatase) in this pathway.[12]

This enzyme exhibits its causal efficacy within the overall process of photosynthesis, both in bacteria and higher plants, at the extremely important branchpoint that comes before the bifurcation towards either the group of proteins such as haems or the other group of proteins, such as chlorophylls in the area of interest of molecular plant physiology (for the position of the enzyme in question, see figure 3.1).

As shown in figure 3.1, Mg-chelatase catalyzes the insertion of magnesium (2+) ions into the protoporphyrin IX molecule. The resulting molecular precursor of the chlorophyll end-product is illustrated by the figure 3.2 below, which also focuses on the causal role of the enzyme in photosynthesis.

Now, the affinity with Kant's tree example and especially with the corresponding elements in the elucidation part of his account of causation in the "Analytic of the Teleological Power of Judgment" consists mainly of the following features:

(1) The ionic form of magnesium (Mg (2+)) corresponds to the items that are referred to in Kant's notion of "raw material" (*CPJ* V 371.24).[13] In brief, the chem-

---

12 For a comprehensive and detailed summary of the research done specifically on this important enzyme, see especially Walker/Willows (1997); see also, more recently, Axelsson (2006). Here, I will only discuss those features of Mg-chelatase which are most pertinent for the purposes of this paper.

13 Moreover, as will be explained in a moment, Kant says in a metaphorical manner that the formative power of biological entities "communicates" (*CPJ* V 374.21–3) itself to raw material

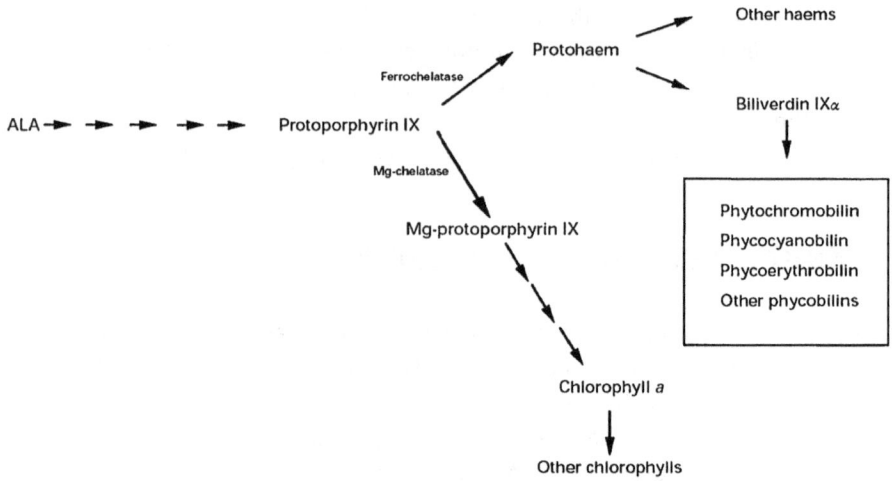

Figure 3.1 Position of the Mg-chelatase enzyme at the branch point of haems and chlorophylls biosynthesis (reproduced from Walker/Willows 1997, 321; figure 1).

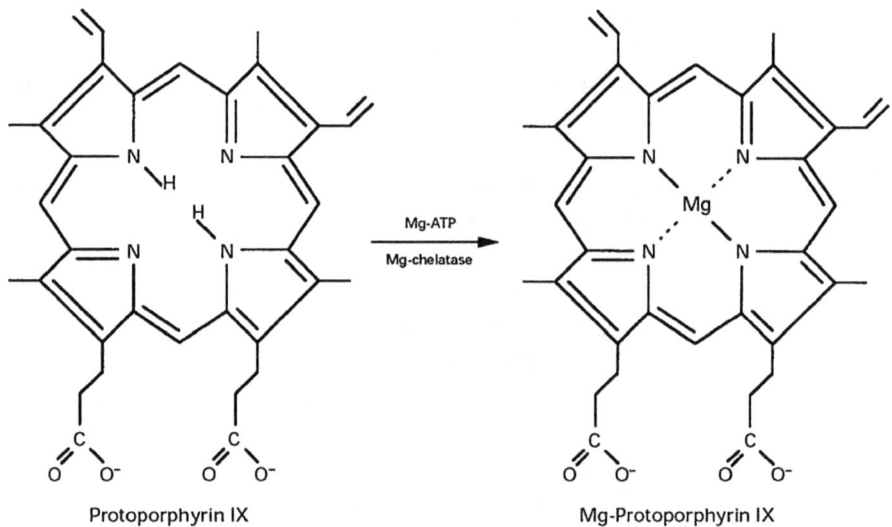

Protoporphyrin IX   Mg-Protoporphyrin IX

Figure 3.2 Reaction catalyzed by the Mg-chelatase enzyme. The figure shows the main substrate for the enzyme's activity, i.e., Protoporphyrin IX, and other targets of its activity, such as, Mg (2+) ion and ATP, which serves as an activation molecule for the enzymatic activity through which Mg-chelatase inserts the ion in the right molecular site within the Protoporphyrin IX substrate. Thus, the effect of its causal efficacy is marked as Mg-Protoporphyrin IX, an important precursor of the future chlorophyll molecule, which, consequently, initiates photosynthesis (reproduced from Walker/Willows 1997, 322; for more molecular details, see in particular Walker/Willows 1997, 321–3).

ical substance in question represents a substrate on which the enzyme of Mg-chelatase intervenes through its causal efficacy in the same way as, according to Kant's §64 in the third *Critique*, a raw material gets separated and reconfigured by the capacity displayed by natural end products.

(2) The enzyme under consideration, which has a macromolecular organization of three distinct sub-units, exhibits the same "capacity for separation and formation" (*CPJ* V 371.25–7) as already remarked in Kant's account. The corresponding Mg-chelatase's capacity to catalyze in a certain way the insertion of magnesium ion (2+) into a Protoporphyrin IX molecule overlaps significantly with Kant's understanding of the capacities shown by biological organisms in the process of growth.

(3) It is the above capacity—somehow associated with organized beings at all levels of biological organization—that by activating itself in certain circumstances establishes the expected causal relation. In this particular case, the causal production of Mg-Protoporphyrin IX, as this production is now understood, illustrates the distinctive features of causal dispositionalism in the context of biological sciences.

On the basis of these affinities, I will now examine Kant's overall account of biological causation, especially, with regard to the notion of the formative power, which is associated with the view of causal dispositionalism (H1). The reading that we obtain in that way can be summarized as follows:

(i) *The "Connection of Real Causes"*. The causal efficacy of the enzyme, through which, as already pointed out, Mg-chelatase produces a new biomolecule, is a genuine causal relation established by the corresponding disposition of the enzyme in question to behave in the standard way in a certain range of environmental conditions. In accordance with Kant's basic classification of 'causal connections', the behavior of Mg-chelatase is a '*real* causal connection' in which the enzyme inserts ions (2+) of magnesium into the right conformational positions within the Protoporphyrin IX molecule. Now, in spite of this *real* causal connection, consider again the causal efficacy of the enzyme depicted by figure 3.2, the connection at issue, as seen from the point of view of Kant's account, is not correctly defined without a distinctively formative aspect that causal powers in biology display. This particular aspect relates to the other basic kind of causal connection in the "Critique of the Teleological Power of Judgment".

---

taken from environment. For an overview of recent philosophical debate on so-called 'informational talk' in biology, see Godfrey-Smith (2007).

(ii) The causal efficacy of Mg-chelatase, by the same causal pattern at the higher level of biological organization referred to in Kant's elucidation, is not just an outcome of "*motive* power", but of a distinctive "*formative* power" (*CPJ* V 374.23–4). Taking into account the data collected before, we can interpret Kant's distinction of powers by saying that "*formative* power" (ibid.) refers to a kind of *programmed* activity through which the biological object in question brings about the corresponding effect within certain process.[14] More precisely, it is somehow expected in biology that Mg-chelatase makes its particular causal contribution to a containing physiological system in green plants, which is programmed to the extent that the enzyme is a component-part embedded within a complex system in which it *ought* to insert Mg (2+) ions into one specific molecular site. Moreover, this causal efficacy *ought* to be done in the right way, at the right time and with respect to other requirements bound to the containing individual organism. Now, going back to the formative feature of causal powers exhibited by objects studied in biology, Kant extremely carefully shifts to a methodologically oriented group of ideal causal connections. Properly speaking, according to the claim advanced by §66 in the "Critique of the Teleological Power of Judgment", these particular connections, specifically invoked throughout biology, are not causal in any of the two possible senses that causal relations can be grouped into according to the derivation part in Kant's account.

4 No Metaphysical Systems for Biology

In the concluding section, I will leave out Kant's further analysis of the notion of maxim by which he characterizes the previously mentioned non-causal claims in biology. Rather, I will complete the analysis of Kant's overall account of biological causation by relating the results obtained so far to the second interpretative hypothesis (H2).

The epistemic privilege that Kant assigns to the causal connection between *real* biological objects with respect to merely subjectively necessary, but non-causal connections between the projected unity of the whole—and its corre-

---

14 A similar line of thought can be found in Cummins (1975), which opposes the etiological selectionist strand of the debate on biological functions. For an analogous opposition to the latter strand of the debate, see Ginsborg (2001). I will not explore in this paper the applicability of Ginsborg's normative reading to the formative aspect pertaining somehow to causal relations in biology. For a more general analysis of the normative reading of some of the central tenets in Kant's philosophy of biology, see Šustar (2013).

sponding normative constraints—and component-parts is consistent with the following two claims in his philosophy of the biological sciences:

(1) *Eliminativism about the Traditional Metaphysical Systems in Biology.* As announced in §§64–6 of the third *Critique* and, then, worked out in more detail in §§72–3, Kant's unequal treatment of the two main kinds of causation when we deal with the biological domain is partially justified by the theoretical inadequacy of traditional metaphysical systems, at the time most notably, hylozoism and animism, that usually accompany biology. In the "Critique of the Teleological Power of Judgment" and in his other works concerned with his philosophy of the biological sciences, Kant nevertheless explicitly acknowledges the so-called "causality of nature" that brings about biological objects, such as in the case of the natural *"production of the eye"* (*First Introduction* XX 236.20 – 1; my italics). Despite acknowledging the productivity pertaining characteristically to biological organized beings, both concerning their individual growth and self-maintenance and their phylogenetical history, the overall historical view to which Kant is committed in the third *Critique* plays only an extremely marginal role in accounting for the distinctively formative feature of causal powers exhibited by biological objects, such as those instantiated in the tree example or, more clearly, in the test-case of Mg-chelatase.

(2) *Deflationism about Biological Metaphysical Systems.* The determining ground with regard to the formative feature read into biological causal talk basically stems from a specific inner constitution of our cognitive apparatus. In brief, this constitution yields both the formative feature in question and related normative constraints on the causal behavior of biological component-parts as their causal behavior has been described above.[15]

At this point, a challenging comparison comes to mind between Kant's peculiar account of causal connections that we should use in biology and Mayr's (1961) theory of *proximate* versus *ultimate* causation. As is well known, the latter kind of biological causation is firmly embedded within a Darwinian metaphysical research program, which completes our understanding of biological phenom-

---

15 As far as Kant's exact classification of the constitution in question is concerned, see, e.g., *CPJ* V 405–6, where he speaks about a "special character of *our* (human) understanding" or, similarly, about the "particular constitution of our understanding" (*CPJ* V 405.21–3). The specific reaction of our understanding to causal regularities in biology, which, then, results in the introduced ideal connections of the corresponding merely regulative kind, finds its initial textual evidence in §10 of the *CPJ*, where, for instance, an object "is called purposive merely because its possibility can only be *explained and conceived by us* insofar as we assume as its ground a causality in accordance with ends" (*CPJ* V 220.18 – 20; my italics).

ena through proximate causation.[16] Now, on the account of (H2), how should we interpret Kant's metaphysical deflationism endorsed in point (2)? As a justified philosophical response to the theoretical poverty of the traditional "metaphysical systems" (*CPJ* V 392.21–3) or, rather, as a claim that is not responsive to these and any other data in that regard? With this, however, Kant's biological causal dispositionalism, as interpreted more strictly by (H1), interestingly overlaps with the issues that go beyond his account of causation given in the "Analytic of the Teleological Power of Judgment".

References

Axelsson, Eva 2006, Magnesium Chelatase—a Key Enzyme in Chlorophyll Biosynthesis, Lund.

Carrier, Martin 2001, Kant's Theory of Matter and His Views on Chemistry, in: *Kant and the Sciences*, ed. by Eric Watkins, Oxford: Oxford University Press, 205–30.

Cartwright, Nancy 1999, *The Dappled World: A Study of the Boundaries of Science*, Cambridge: Cambridge University Press.

Cummins, Robert 1975, Functional Analysis, *Journal of Philosophy* 72, 741–64.

Friedman, Michael 1992, *Kant and the Exact Sciences*, Cambridge, MA: Harvard University Press.

Ginsborg, Hannah 2001, Kant on Understanding Organisms as Natural Purposes, in: *Kant and the Sciences*, ed. by Eric Watkins, Oxford: Oxford University Press, 231–58.

Ginsborg, Hannah 2004, Two Kinds of Mechanical Inexplicability in Kant and Aristotle, *Journal of the History of Philosophy* 42, 33–65.

Godfrey-Smith, Peter 2007, Information in Biology, in: *The Cambridge Companion to the Philosophy of Biology*, ed. by David L. Hull and Michael Ruse, Cambridge: Cambridge University Press, 103–19.

Hull, David L. 1999, The Use and Abuse of Sir Karl Popper, *Biology and Philosophy* 14, 481–504.

Mayr, Ernst 1961, Cause and Effect in Biology, *Science* 134, 1501–6.

Mumford, Stephen 2009, Causal Powers and Capacities, in: *The Oxford Handbook of Causation*, ed. by Helen Beebee, Christopher Hitchcock, and Peter Menzies, Oxford: Oxford University Press, 265–78.

Mumford, Stephen/Anjum, Rani Lill 2011, *Getting Causes from Powers*, Oxford: Oxford University Press.

Steigerwald, Joan (ed.) 2006, Kantian Teleology and the Biological Sciences, *Studies in History and Philosophy of Biological and Biomedical Sciences* 37, 621–792.

---

16 For this characterization of the role that Darwinian theories of evolution actually play or should continue to play with regard to middle-range scientific theories in biology and biologically-related scientific areas, see Hull (1999), in which Hull follows more closely Popper's changing attitude towards Darwinian theory, i.e., from a negative assessment as a pseudoscience to finally characterizing it as a viable metaphysical research program for biology.

Šustar, Predrag 2008, The Organism Concept: Kant's Methodological Turn, in: *Purposiveness. Teleology between Nature and Mind*, ed. by Luca Illetterati and Francesca Michelini, Frankfurt/M.: Ontos, 33 – 57.

Šustar, Predrag 2013, Normativity and Biological Lawlikeness: Three Variants, in: *Akten des XI. Internationalen Kant-Kongresses, Pisa 2010*, ed. by Claudio LaRocca, Stefano Bacin, Alfredo Ferrarin, and Margit Ruffing, Berlin/Boston: Walter de Gruyter.

Teufel, Thomas 2011, Kant's *Non*-Teleological Conception of Purposiveness, *Kant-Studien* 102 (2), 232 – 52.

Walker, Caroline J./Willows, Robert D. 1997, Mechanism and Regulation of Mg-chelatase, *Biochemical Journal* 327, 321 – 33.

Watkins, Eric 2005, *Kant and the Metaphysics of Causality*, Cambridge/New York: Cambridge University Press.

Wilson, Robert A. 2005, *Genes and the Agents of Life. The Individual in the Fragile Sciences: Biology*, Cambridge: Cambridge University Press.

Zuckert, Rachel 2007, *Kant on Beauty and Biology. An Interpretation of the* Critique of Judgment, Cambridge: Cambridge University Press.

Eric Watkins
Nature in General as a System of Ends

1 Introduction

Despite the many complexities, mysteries, and puzzles that one inevitably en-
counters reading the "Critique of the Teleological Power of Judgment", the
basic structure of Kant's main line of argument would seem to be quite straightforward. What prompts Kant to write about teleological judgment in the first
place is organisms, since organisms involve purposes and purposiveness is a
unifying theme of the *CPJ* as a whole. Thus Kant begins his discussion of organisms by first providing analyses of what (material, intrinsic, natural) objective
purposiveness is and of what an organism is, before asserting that organisms
are natural purposes. In this way, his analyses reveal that organisms are different
from inorganic matter, which can be explained according to the mechanical principles already laid out in, e. g., the *Metaphysical Foundations*. But since organisms are in some sense mechanically inexplicable, despite also involving matter,
the existence of organisms presents a problem that can be expressed in the form
of an antinomy concerning mechanical and teleological modes of explanation.
Kant then solves the antinomy by invoking a supersensible ground that (for
some reason) allows one to (somehow) privilege teleological over mechanistic
explanation, though without thereby rejecting the possible applicability of the
latter to organisms. Kant concludes his argument by explaining how this solution fits with his discussion of the physico-theological and moral arguments of
the first and second *Critiques*, thus putting his solution to the antinomy into
the larger context of his critical philosophy. In accordance with this three-fold
structure of i) conceptual analysis, ii) antinomial conflict and resolution, and
iii) broader context-setting, Kant divides his treatment of organisms in the "Critique of the Teleological Power of Judgment" into i) an "Analytic of the Teleological Power of Judgment", ii) a "Dialectic of the Teleological Power of Judgment",
and iii) a "Methodology of the Teleological Power of Judgment". In this way, one
has, it seems, an account of Kant's overall argument that corresponds to the
load-bearing elements of this part of the third *Critique*.

Without denying any of these claims about the structure of the "Critique of
the Teleological Power of Judgment", however, I want to assert that such an account omits a crucial element of Kant's main line of argument. For in a series of
passages that have not received sufficient attention, Kant claims that once we
start to reflect on organisms, we are necessarily led beyond this initial topic to

think of nature in general.[1] That is, even if teleological judgment is prompted by our experience of organisms, it does not then restrict itself to judgments about organisms, but rather necessarily goes further to make claims about all of nature, which will include at least some things that are not organisms. But this immediately raises two questions. 1. What exactly are Kant's claims about nature in general (as contrasted with his claims about organisms in particular)? 2. What are his arguments for these claims? I first argue that according to Kant, reflection on organisms necessarily leads to two further claims about nature in general. The first claim is that not just organisms, but in fact every specific thing *in* nature must also be judged teleologically, while the second is that nature as a whole is a system of purposes that itself has a purpose. I then argue that Kant's distinctive conception of reason as a faculty that searches for the unconditioned condition of all conditioned objects provides the key for understanding Kant's arguments for both of these assertions. Although the *CPJ* does depend in essential ways on the faculty of the power of judgment, the faculty of reason also plays an ineliminable role in its overall argument, since it motivates and justifies Kant's moving beyond organisms to what is his ultimate concern, namely the final and unconditioned purpose of nature, which is the existence of human beings, not as natural organisms, but rather as free and rational noumenal agents.[2]

2 Two Claims

In the third paragraph of §67, whose title is "On the principle of the teleological judging of nature in general as a system of ends", Kant argues:

> It is therefore only matter insofar as it is organized that necessarily carries with it the concept of itself as a natural end, since its specific form is at the same time a product of nature. However, this concept now necessarily leads to the idea of the whole of nature as a system in accordance with the rule of ends, to which idea all of the mechanism of nature in accordance with principles of reason must now be subordinate (at least in order to test natural appearance by this idea). The principle of reason is appropriate for it only subjectively, i.e., as the maxims that everything in the world is good for something, that nothing in it is in vain; and by means of the example that nature gives in its organic products, one is

---

1 Guyer (2005, 314–42, esp. 327) similarly argues that the step from organisms to nature as a whole is both intrinsically important and crucial to the structure of Kant's overall argument.
2 Though scholars sometimes treat the third *Critique* as if it had to be about the faculty of judgment (exclusively), it can, I argue, involve reason as well, since there is, to my mind, no genuine conflict between the faculty of judgment and reason, even if they have somewhat different ends and functions.

justified, indeed called upon to expect nothing in nature and its laws but what is purposive in the whole (*CPJ* V 378.35 – 379.9).

This passage contains two crucial moves. The first is Kant's assertion that organized matter is a natural purpose. This assertion, which is obviously central to Kant's entire argument, is clearly intended as a conclusion that is supposed to follow from his analyses of purposes and organisms in the previous sections (§§61 – 3 and §§64 – 6, respectively). The second, which Kant signals (with "now" [*CPJ* V 379.1]) as the novel claim that is his focus in this §, is that the notion of an organism "necessarily leads" (*CPJ* V 379.1) to the idea of the entirety of nature. But how exactly should this second statement be understood?

Kant seems to have two different claims in mind here. First, he clearly asserts that one is not only justified, but even called upon ("*berufen*" [*CPJ* V 379.8]) to expect that everything *in* nature has a natural purpose or end. Accordingly, not only organisms, the very concept of which entails the concept of a natural purpose, but also inorganic matter must be viewed in terms of the purposes it serves. The force of this claim is thus that the *scope* of teleological judgment is not limited to organisms, but rather extends *universally* to each and every thing throughout nature. This is the clear import of his remarks that "everything in the world is good for something" and "nothing in nature is in vain" (*CPJ* V 379.5 – 7). I will call this claim the Claim of Universal Scope (CUS).[3]

Second, Kant seems to think not only that everything *in* nature has a purpose, but also that nature itself, or nature as a whole, must be a system of purposes. That is, there is a systematic connection between the different purposive things in nature. This point comes out not only in the title to §67, which makes mention of "nature in general as a system of ends" (*CPJ* V 377.25 – 7), but also in his phrase "the idea of a whole of nature *as a system*" (*CPJ* V 370.1 – 3) as opposed to simply 'the whole of nature *as such*'. Accordingly, though CUS clearly applies to 'the whole of nature', it does not obviously require thinking of all of nature 'as one system', that is, such that the various things in nature are related to each other in any systematic way. This second claim is also on display at the conclusion of the final paragraph of §67, when Kant clarifies that "the unity of

---

3 Kant also states CUS in the §66 of the third *Critique* (*CPJ* V 376.27– 8). In the first passage in particular, Kant asserts not merely the limited claim that "nothing in such a creature [an organism] is *in vain*" (*CPJ* V 376.28) (such that Kant is claiming only that everything in an organism has a purpose), but also "the general doctrine of nature that *nothing* happens *by chance*" (*CPJ* V 376.29 – 30), which I take to mean that nothing in the world at all is in vain, since otherwise there would be no contrast between the first and second claims (when I take it that Kant is trying to go beyond the first claim when he makes the second).

the supersensible principle must then be considered as valid in the same way not merely for certain species of natural beings, but for the whole of nature as a system" (*CPJ* V 381.5 – 7).[4] However, Kant wants to go even further than simply claiming that the things in nature come together to form a system. For he also wants to claim both that nature itself is a system *of purposes* (i. e., the systematic connection of things in nature involves their being purposes), and that nature as a whole must itself have a purpose, just as all of the individual organisms within nature do.[5] Let me label this constellation of positions (that nature as a whole is a system of purposes and that nature as a whole has a purpose) the Claims about Nature as a Whole (CNW).

In the second paragraph of §67 (which immediately precedes the first quotation above and lays the foundation for it), Kant articulates the context for, and the meaning of CNW more fully as follows:

> To judge a thing to be purposive on account of its internal form is entirely different from holding the existence of such a thing to be an end of nature. For the latter assertion we need not only the concept of a possible end, but also cognition of the final end (*scopus*) of nature, which requires the relation of nature to something supersensible, which far exceeds all of our teleological cognition of nature; for the end of the existence of nature itself must be sought beyond nature (*CPJ* V 378.12 – 9).

In this passage, Kant begins by drawing a distinction between judging the natural purpose of a thing according to its *form* and determining the *existence* of a thing as a natural purpose. We judge that a thing has the form of a natural purpose when we judge that it is an organism (by displaying the special kind of causality that is definitive of organisms). But to judge the form of a thing as purposive is distinct from passing judgment on the purpose of the existence of that thing. I can judge that this object in front of me is a tree (because its parts and its whole stand in a certain reciprocal dependency relation that matter as such does not display) without judging why the tree exists (what ends it might satisfy), or indeed judging that it exists for any purpose at all.

In light of this distinction, Kant then clarifies that once we start thinking about the purpose of the existence of things in nature, we end up being committed to: i) a purpose for the existence of nature as a whole, ii) the purpose of the existence of nature as a whole being a final purpose or end (*Endzweck*)[6] and iii)

---

4 The centrality of Kant's point is revealed by the fact that he introduces the passage with: "In this section we have meant to say nothing except [...]" (*CPJ* V 380.26).

5 Textual evidence for this claim can be found in §85 at, e. g., *CPJ* V 437.3 – 17.

6 For the sake of brevity, I am abstracting from the distinction between "ultimate end" (*letzter Zweck*) and "final end" (*Endzweck*), helpfully clarified by Guyer (2005, 318).

this final end being something supersensible that lies outside of nature. The first claim amounts to one aspect of CNW as it was originally introduced—nature forms a system of purposes and nature as a whole has a purpose of its own. The second and third claims, by contrast, significantly extend his commitments about the purpose of nature as a whole and thus reveal the full meaning of CNW.

Moreover, Kant's expression of commitment to all three components of CNW is no fluke; in several passages from later on in the third *Critique* he expresses the same constellation of claims. Thus in §86 he asserts:

> Now if we encounter purposive arrangements in the world, and [...] subordinate the ends that are only conditioned to an unconditioned, supreme end, i.e., a final end, then one readily sees, first, that in that case what is at issue is not an end of nature (within it), insofar as it exists, but the end of its existence, with all its arrangements, hence the ultimate *end of creation*, and in this, further, what is actually at issue is the supreme condition under which alone a final end (i.e., of the determining ground of a highest understanding for the production of the beings of the world) can obtain (*CPJ* V 443.29–37).

And in §88, he reiterates:

> But we certainly do find ends in the world, and physical teleology presents them in such measure that [...] we will ultimately have reason to assume as the principle for research into nature that there is nothing in nature at all without an end; yet we try in vain to find the final end of nature in nature itself (*CPJ* V 454.1–5).

As a result, Kant understands CNW to involve a robust commitment not only to nature as a system of purposes and to a purpose of nature as a whole, but also to a final purpose that lies outside of nature in the supersensible.

3 Kant's Arguments

If Kant thus maintains that reflection on individual organisms "necessarily leads" (*CPJ* V 379.1) to asserting CUS and CNW, how is one led beyond organisms to these claims and what arguments does he develop in support? Why not be a biologist who is committed to understanding (the functioning of) organisms and leave it at that? Kant's most explicit statement in regard to CUS is found in §66:

> For this concept [of a natural end] leads reason into an order of things entirely different from that of a mere mechanism of nature, which will no longer satisfy us here. An idea has to ground the possibility of the product of nature. However, since this is an absolute unity of the representation, while the matter is a multitude of things, which by itself can provide no determinate unity of composition, if that unity of the idea is even to serve as

the determining ground *a priori* of a natural law of the causality of such a form of the composite, then the end of nature must extend to *everything* that lies in its product. For once we
have related such an effect in the *whole* to a supersensible determining ground beyond the
blind mechanism of nature, we must also judge it entirely in accordance with this principle;
and there is no ground for assuming that the form of such a thing is only partially dependent on the latter, for in such a case, in which two heterogeneous principles are jumbled
together, no secure rule for judging would remain at all (*CPJ* V 377.1–16).

Kant is making two main points here. First, the discovery of even a single natural
purpose (in the form of an organism) leads the faculty of reason, Kant says, and
not the faculty of judgment, as one might expect, to "an order of things that is
completely different" (*CPJ* V 377.1) from the mechanistic order laid out in the *Metaphysical Foundations*.[7] For in a natural purpose, everything is both cause and effect of itself (*CPJ* V 370.36–7) such that the whole depends on its parts at the
same time that the parts depend on the whole, a distinctive causal ordering
that is evident, Kant thinks, in the maintenance and growth of an organism as
well as in the preservation of the species through reproduction. It is because
of this special kind of reciprocal dependence that Kant speaks in this passage
of a unity that matter, which essentially involves a plurality (but not a principle
that necessarily unifies the plurality), does not have. Given the analyses of the
previous §§, this point is not fundamentally new, even if it does stress the importance of the distinctive kind of order and unity found in organisms.

Second, and more importantly, Kant suggests that if we posit a supersensible
ground outside of nature as responsible for the distinctive order and purpose
found in an organism, then we are committed to viewing everything in that organism as ordered in that same kind of way. However, if the appeal to a supersensible ground requires the application of a single criterion in order to explain a
given object (including its constituent parts and structure), then one can reconstruct an argument for CUS based on CNW by noting that CNW requires that the
object we hope to explain is simply the world in its entirety. Specifically, once
one has claimed that nature as a whole forms a system of purposes and posited
a supersensible ground to account for the final purpose of nature as a whole (as
CNW maintains), it follows that one must consider all specific things *in* nature
with respect to what purpose they have, because we consider them to be essential members of the system of purposes of nature that is brought about by this
supersensible ground. But investigating whether each and every thing has a pur-

---

7 Again, though there is no conflict between reason and the faculty of judgment, in this case the
distinctive features of reason make it more appropriate for Kant to refer explicitly to it.

pose is simply tantamount to CUS. So it is plausible to think that CUS follows straightforwardly from CNW.[8]

However, it is worth noting that Kant does at least suggest a line of thought in support of CUS that runs independently of CNW, at least to a certain extent. At the end of the quotation, he notes that if one attempted to explain the form of a thing as depending in part on its relation to the supersensible principle invoked in CNW and in part on something else, then we would have two heterogeneous principles that worked independently of each other, which would entail that we would have "no secure rule" (*CPJ* V 377.16) for judging what makes such a form possible. That is, such a scenario would require that one judge some features of an object in nature according to mechanistic principles and others features according to teleological principles, which would be confusing since one would have no criterion to determine which principle should be used for any given feature. At the same time, it is not clear that it is supposed to be an argument that carries much independent weight. For it is not obvious (at least not without more explicit argument than Kant provides here) why it is necessary that we can have only one criterion for judging the different features of things. There are all sorts of cases in which we use multiple criteria in a given explanatory context, even if there is a hierarchy among them. Granted, it is useful to have a single criterion for all cases, and also convenient, but its necessity is not particularly perspicuous. Fortunately, CUS follows from CNW, so this result is not damaging to Kant's position.

In the very next paragraph, Kant clarifies that the conclusion of his argument is still consistent with the possibility that some of the things in nature might not themselves have purposes, but rather are explicable according to purely mechanical laws. For example, CUS does not entail that everything in nature is itself an organism, which is important given that water, for example, clearly is not. Nor does it even necessarily imply that everything in nature must serve as a purpose for something else. Water serves as a purpose for the existence of other organisms in nature insofar as it is necessary for the sustenance of plants

---

8 I should note that Kant's text does not explicitly formulate this argument. His explicit argument is only that the appeal to a supersensible ground requires the application of a single criterion to an organism. I am proposing that this argument be extended to the world as a whole, however, because it seems to be the most promising way to justify CUS. I do not rule out the possibility that there are other ways of justifying CUS. (Perhaps one might argue that CUS follows instead simply from nature having an end, and then nature forming a system of purposes follows from CUS plus some other assumptions.) What the textual justification for such arguments might be cannot easily be anticipated. The interpretation presented above is at least based on the text, even if it requires extending the argument's scope so as to encompass the world in its entirety (which, given CNW, Kant is committed to).

and animals, but it is not metaphysically impossible that things serving no purpose might exist, such as, e.g., small bits of matter in a distant galaxy that have no effect on us or on any other rational living being.[9] Neither the fact that a whole has a purpose (or is viewed as having a purpose) nor the fact that there is a system of purposes within nature as a whole immediately entails that absolutely everything within nature must actually have a purpose (or must be viewed as having a purpose). In short, the system of natural purposes need not be completely coextensive with all of the objects in nature. Instead, the demand is that one *consider* whether all things in nature have either intrinsic or extrinsic purposes. CUS is, in short, no more than a regulative principle, and its truth depends, as we have seen, on the truth of CNW, just as Kant claims.

So what is Kant's argument for CNW? The passage that comes closest to containing an explicit argument for CNW is in the latter half of the second paragraph of §67 (the first half of which was quoted above). After claiming that the purpose of the existence of nature must be sought beyond nature, Kant provides the following by way of justification:

> The internal form of a mere blade of grass can demonstrate its merely possible origin in accordance with the rule of ends in a way that is sufficient for our human faculty of judging. But if one leaves this aside and looks only to the use that other natural beings make of it, then one abandons the contemplation of its internal organization and looks only at its external purposive relations, where the grass is necessary to the livestock, just as the latter is necessary to the human being as the means for his existence; yet one does not see why it is necessary that human beings exist (a question which, if one thinks about the New Hollanders or the Fuegians, might not be so easy to answer); thus one does not arrive at any categorical end, but all of this purposive relation rests on a condition that is always to be found further on, and which, as unconditioned, (the existence of a thing as a final end) lies entirely outside of the physical-teleological way of considering the world. But then such a thing is also not a natural end; for it (or its entire species) is not to be regarded as a natural product (*CPJ* V 378.12–34).

Kant's argument seems to be that once one distinguishes between the inner form of natural purposes in organisms and the purpose of the existence of things and then looks beyond the former, one will see that it is not possible to explain the purpose of the existence of things by way of external purposive relations within nature.[10] Even if the existence of grass is necessary for the existence of cattle and

---

9 I understand regulative principles such that what they presuppose is not that the relevant feature that is governed by the content of the principle is metaphysically possible, but rather that it might be metaphysically possible (or is metaphysically possible for all we know).

10 By distinguishing between these two cases Kant is in effect operating with two different instances of purpose, one pertaining to the properties of a thing, the other concerning the

the existence of cattle is in turn necessary for the existence of human beings, it is not necessary that human beings exist. Indeed, there are, Kant maintains, no necessarily existing things within nature. So if one searches for the purpose of the existence of things, external purposive relations within nature cannot suffice. Instead, the only thing that could possibly suffice to explain the purpose of the existence of things is, Kant wants to argue, an unconditioned final end that lies outside of nature. But Kant's argument raises two questions: 1. What is it that moves one to seek a sufficient explanation of the purpose of the *existence* of things? 2. Why must a sufficient explanation take recourse specifically to an unconditioned final end that lies *outside* of nature?[11]

I want to suggest that the answer to both of these questions derives from Kant's distinctive conception of reason. For current purposes, Kant's conception of reason can be summarized in three claims.[12] First, reason is a faculty that searches for the conditions for whatever is conditioned that is given. For example, in syllogistic logic (as Kant understands it), reason searches for the conditions for a conditioned judgment, since if successful, it can then formulate a syllogism, with the premises serving as the conditions for the conclusion, which is 'conditioned' by its premises. However, reason is not restricted to judgment and the realm of logic, but rather applies to any object of experience; for any conditioned object that is given to us, reason necessarily searches for the conditions that would explain it, such as when reason searches for the cause of some change of state that we experience. Reason is thus the faculty that is interested in identifying the conditions for anything that is conditioned.

Second, reason searches not simply for conditions, but also for the *totality* of conditions and thus the *unconditioned*. There is obviously an analytic connection between anything that is conditioned and its conditions, because to characterize something as conditioned entails conditions on that thing. However, reason's interest in the unconditioned, which alone offers it a satisfactory resting place (since if it obtained the object of its inquiry, it would have nothing else to pursue), goes beyond this analytic connection and can be satisfied only by seeking the totality of conditions. For the totality of conditions cannot be a totality if it

---

existence of the thing. While one might think that Kant is thus illegitimately sliding from the one notion to the other, one might think that both are instances of conditioning relations and thus both are equally of interest.

11 That is, in effect, the second question asks: why is CNW true?

12 Both Guyer (2005) and Breitenbach (2009) suggest that reason is important in drawing this inference due to its desire for unification. I argue that one can take this point further by showing that it is not just any unity that reason wants to produce, but one that would culminate in an idea of something unconditioned that serves as the condition for all conditioned things.

does not contain *all* conditions, but if it contains *all* conditions, then it must be unconditioned, because if this totality were itself conditioned, it would not have included the condition that conditions it and would thus not in fact be the totality of conditions. As a result, if reason could find all of the conditions for something conditioned, it would necessarily have found the unconditioned as well. Moreover, because reason seeks the totality of conditions for what is conditioned by starting with something conditioned and moving to its conditions, which, because they are themselves conditioned, leads reason to yet further conditions, etc. until it reaches the unconditioned, the unconditioned also provides a principle of organization, Kant thinks, for all the conditions that fall under it, such that it leads to a systematic interconnection of conditioned elements under a single unconditioned principle. That is, the unconditioned serves as the principle for the system of condition-conditioned relations that reason discovers in its search for the totality of conditions.

Third, Kant holds that the unconditioned object that alone could provide reason with a satisfying resting place can never be given to us through the senses and thus can never be an object that we could cognize in nature.[13] Since the objects of traditional metaphysics that interest us most, such as God, freedom, and the soul, are characterized as unconditioned, it follows that we cannot have cognition of them.[14] At the same time, since reason does not, on that account, lose its interest in the unconditioned, our ideas of this kind of object function as regulative principles that guide our understanding's judgments such that we strive to come ever closer to approximating these ideals. So, even though reason does not find satisfaction in cognition of the unconditioned, it still functions as a regulative principle that unifies its subject matter in a distinctive and systematic way. In this way, it finds as much satisfaction as is possible for a faculty limited by the fact that objects must be given to it sensibly.

This account of Kant's conception of reason puts us in a position to understand his argument for CNW. First, as we saw above, Kant identifies reason as the faculty that leads us to move beyond the purely mechanistic order of inert matter to the distinctive order of organisms. Specifically, the experience of organisms reveals a special kind of conditioning relationship, where both the whole and its parts reciprocally condition each other according to some unified principle. It is precisely because a distinctive kind of conditioning relationship is involved

---

13 There are passages to this effect scattered all over Kant's corpus. There are not, however, clearly stated arguments that would justify this claim.

14 This claim coincides perfectly with transcendental idealism's claim that we cannot have cognition of things in themselves, such as God, freedom, and the soul. However, Kant's reasons in this case, whatever they are, turn out to be quite different.

in organisms that reason is the faculty that moves us beyond mechanisms to this new order. Moreover, it does so in several ways. On the one hand, reason, with its desire for conditions, is interested in the *internal form* of organisms, since this form involves a distinctive unity that involves complex conditioning relations. Yet reason also discovers that organisms, like all other objects in nature, are conditioned by external circumstances. For example, plants require sunlight, water, and nutrients from the soil in order to grow, maintain, and reproduce themselves.[15] In this way reason is able to discover vibrant ecosystems as well as understand how they might be endangered. In short, the internal form of organisms involves external conditions that reason must seek out.[16] On the other hand, and more importantly for present purposes, reason is also compelled to search for the purpose and thus the condition of the *existence* of organisms. In this case, it is an *external* or relative purposiveness that is pivotal. When reason seeks the purpose for the existence of one thing (organism A) and finds that it cannot lie in that thing, it seeks its purpose in another thing in nature (organism B). Thus, plants exist for animals, which exist in turn for humans, etc. The purpose of the existence of things, since it contains a conditioning relation, is of fundamental interest to reason as well.[17] So, just as Kant suggests in several places, it is reason that leads us to consider the special explanatory status of organisms, both in their internal form and with respect to purposive relations that condition their existence.

Second, CNW specifically asserts further that nature as a whole must itself have a purpose, and that it must, moreover, be a final and unconditioned purpose. Why this demand and why in this specific form? If reason's ultimate interest is with the unconditioned, it is clear that though reason begins by seeking the purpose, or condition, for the existence of one finite thing in another, it continues to seek ever further conditions such that it ultimately ends up inquiring into the purpose for the existence of nature as a whole.[18] And it is clear that the purpose of the existence of nature as a whole must be a final purpose, one not conditioned by anything else, i.e., it must be unconditioned. Put more informally,

---

15 See Goy (2013) for a detailed discussion of this issue.

16 Indeed, because organisms involve both internal and external conditions, reason's search for conditions provides a direct justification of CUS.

17 In a later passage, Kant adds that in this way, one can discover "many laws of nature which, given the limitation of our insights into the inner mechanisms of nature, would otherwise remain hidden from us" (*CPJ* V 398.20 – 2).

18 There are significant similarities between this account of why nature as a whole involves the notion of an unconditioned purpose and Kant's explanation of the ideas of reason in "On the transcendental ideas" in the first *Critique* (*CPR* A 321/B 377– 9).

the purpose of nature as a whole cannot be something that exists for the sake of something else, but rather something that exists for its own sake. Kant's repeated references to a "final end" (*CPJ* V 378.15), a "condition that is always to be found further on, and which, as unconditioned, [...] lies entirely outside of the physical-teleological way of considering the world" (*CPJ* V 378.30–2), are clear expressions of the structure of reason's interest in finding the unconditioned.[19] Thus, Kant's understanding of reason reveals why nature as a whole must have a purpose that is both a final and an unconditioned purpose.

Moreover, taking these two points together, we can see why nature as a whole must also be a system of purposes. As the first point shows, reason seeks the conditioning relations between the purposes of things that exist in nature, but that alone would not require a unifying principle that would organize these purposes into a single system. (Perhaps some ecosystems are completely distinct from others or perhaps things are more like an aggregate rather than a system.) As the second point shows, however, reason, in seeking the totality of conditions for some conditioned, seeks something unconditioned that subordinates everything that it conditions. As a result, the final, unconditioned purpose subordinates (or conditions) the purposes of the things that exist in nature such that they form a system in Kant's specific sense of the term.[20] That is, it is precisely because of the unconditioned purpose of nature that nature as a whole must also form a system of purposes and not a mere aggregate.

Third, not only is the final end of nature unconditioned, for Kant, but it also lies beyond the limits of our cognition, in the supersensible. Why? (In particular, why is the supersensible introduced?) Given that we cannot experience anything unconditioned and given that the final purpose of nature is unconditioned, it follows immediately that we cannot have experience of the final purpose of nature. Indeed, Kant explicitly asserts that nature as a whole is not given to us *as organized* (*CPJ* V 398.24–5). At the same time, because reason does not lose interest in the unconditioned simply because it cannot cognize it, it posits the unconditioned as something that lies beyond what we can experience, that is, beyond

---

19 Another relevant passage can be found at *CPJ* V 436.5–7: "*Physicotheology* is the attempt of reason to infer from the *ends* of nature (which can be cognized only empirically) to the supreme cause of nature and its properties".

20 Though Kant's position here may (or may not) be plausible, he would need an additional argument to show that everything in nature must be related as a single system of purposes. For, at least prima facie, one could imagine several causal chains that were all subordinate to a single unconditioned condition, but that were nonetheless distinct from each other. For example, perhaps there could be a plurality of ecosystems that are conditioned by some further purpose, yet completely separate from each other.

the sensible world, and thus in the supersensible. As a result, it functions not as an object in nature that we could know as such, but rather as a regulative ideal that we use to organize what we do experience into a systematic whole. In this way, an appreciation of Kant's conception of reason allows one to understand what his justification is for both CNW and CUS.

4 Conclusion

One might naturally start reading the "Critique of the Teleological Power of Judgment" with the expectation that Kant will (simply) try to explain the distinctive status of living organisms. As philosophers of biology will attest, providing such an explanation is no mean feat and if Kant has accomplished such a significant task, we should be glad. However, what we have come to see is that reason, on Kant's distinctive understanding of that faculty, leads us to expectations that have an even grander scope and even more fundamental ambitions. For reason, as the faculty that searches for any and all conditions until it finds the unconditioned, has legitimate interests not only in the inner form of organisms, but also in the external conditions on these organisms and in the purpose for the existence of objects in nature. However, given its essential interest in the unconditioned, reason does not stop there. It also seeks systematic connections within nature as well as a final unconditional purpose for nature as a whole that must itself lie outside of nature. It is at this point that Kant's grandest ambition becomes apparent.[21] For this question is simply the question of why the world exists at all. And in line with the fundamental results of his moral philosophy, his answer is that only a human being, or any being that has the supersensible ability to act freely and thus morally, can be both unconditioned and yet still necessary in itself, that is, man considered not as an organism within nature, but rather as a noumenon (*CPJ* V 435.20).[22] For a free, rational, and spontaneous being is the only kind of entity of which one cannot ask why it exists, given that its existence already contains the highest end itself and is the only kind of entity that could be "a final end to which the whole of nature must be subordinated" (*CPJ* V 436.21–3).[23] Identifying human beings as the final end of creation has im-

---

21 For an excellent discussion of Kant's intentions toward the end of the third *Critique* (and in the third *Critique* as a whole), see Ameriks (2009, 165–7).
22 It is at this point that reason, which had been theoretical in investigating the conditioning relations that obtain between organisms, becomes practical as well.
23 For a more extensive account of Kant's views on this point, see Guyer's contribution to this volume.

plications, in turn, for the kind of systematic relations that obtain within the members of the system of ends in nature, such as for the kinds of laws that obtain, because they must, as Kant argues in the second *Critique*, make possible the highest good that rational agents presuppose is possible in acting morally.[24] In this way, we see how Kant unifies major elements of his entire critical project, while addressing one of the most basic questions that we can raise about our existence.[25]

References

Ameriks, Karl 2008b, The End of the *Critiques:* Kant's Moral 'Creationism' in: *Rethinking Kant*, ed. by Pablo Muchnik, Newcastle: Cambridge Scholars Publishing, 165–90.

Breitenbach, Angela 2009a, *Die Analogie von Vernunft und Natur. Eine Umweltphilosophie nach Kant*, Berlin/New York: Walter de Gruyter.

Goy, Ina 2013, On Judging Nature as a System of Ends. Exegetical Problems of §67 of the *Critique of the Power of Judgment*, in: *Akten des XI. Internationalen Kant-Kongresses, Pisa 2010*, ed. by Claudio LaRocca, Stefano Bacin, Alfredo Ferrarin, and Margit Ruffing, Berlin/Boston: Walter de Gruyter, vol. 5, 65–76.

Guyer, Paul 2005, *Kant's System of Nature and Freedom*, Oxford: Oxford University Press.

Watkins, Eric 2009, The Antinomy of Teleological Judgment, in: *Kant Yearbook 1*, ed. by Dietmar Heidemann, Berlin/New York: Walter de Gruyter, 197–221.

Watkins, Eric 2010, The Antinomy of Practical Reason: Reason, the Unconditioned, and the Highest Good, in: *Kant's 'Critique of Practical Reason': A Critical Guide*, ed. by Andrews Reath and Jens Timmerman, Cambridge/New York: Cambridge University Press, 145–67.

---

24 For discussion of this issue in two different contexts, see Watkins (2009 and 2010).

25 I thank Karl Ameriks, Hannah Ginsborg, Ina Goy, Paul Guyer, Peter McLaughlin, James Messina, Günter Zöller and all of the participants at the international symposium on Kant's theory of biology held in Tübingen in December 2010, for helpful comments on or discussion of an earlier version of this paper.

Angela Breitenbach
Biological Purposiveness and Analogical Reflection

1 Introduction

In the "Critique of the Teleological Power of Judgment", Kant claims that in the realm of living nature we encounter phenomena that appear to display a peculiar purposiveness. The wings of a bird, for example, seem to be conducive to the bird's capacity to fly and thereby to the survival of the bird as a whole. The whole organism, moreover, appears to be the result of an end-directed developmental process. Some natural objects, namely the living organisms, thus appear to us as if they were characterized by a purposive organization of the whole and its parts, and by a particular end-directedness. In order to account for this, Kant introduces the concept of biological purposiveness, that is, an objective, material, and internal purposiveness.[1]

On Kant's account, this concept of biological purposiveness grounds judgments that have a rather peculiar status. On the one hand, these judgments are elicited by particular experiences of biological phenomena. It is the experience of organisms that leads us to employ the concept of a means-ends relation as characterizing the living world.[2] On the other hand, judgments about biological purposiveness are purely regulative and, hence, make no determinate claim

---

1 In §61 of the *CPJ*, Kant introduces the concept of objective purposiveness as the third of three types of the purposiveness of nature. The first is the subjective purposiveness of nature as a whole for its "comprehensibility" by the human intellect (*CPJ* V 359.4 – 5). See, for example, Kant's discussion at *CPR* A 653 – 4/B 681 – 2, *CPJ* V 179.19 – 186.21, and *First Introduction* XX 211.6 – 216.26. The second is the subjective purposiveness of the objects of aesthetic experience, which "contain a form so specifically suited for [the human power of judgment] that by means of their variety and unity they serve as it were to strengthen and entertain the mental powers" (*CPJ* V 359.9 – 12); see also *CPJ* V 219.26 – 236.11. In §§62 and 63 of the *CPJ*, Kant further distinguishes the third type, that is, the objective purposiveness of living beings as "material" from the "formal" (*CPJ* V 362.4 – 5) objective purposiveness of geometrical figures, and as "internal" from the "relative" (*CPJ* V 366.25 – 6) or "external" (*CPJ* V 368.32) material objective purposiveness of things that are merely "useful" or "advantageous" (*CPJ* V 366.8) for something else. My discussion of Kant's notion of biological purposiveness will focus on this concept of objective, material, and internal purposiveness.

2 See *CPJ* V 366.27 – 8. In this respect, judgments of objective purposiveness importantly differ from those of the subjective purposiveness of nature (see *CPJ* V 193.24 – 194.2).

about the character of the biological phenomena themselves. For, as Kant also claims, by means of these judgments we reflect about nature as if it were purposive without determinately explaining it as purposive. This raises the question of how we should make sense of the peculiar status of judgments about biological purposiveness, given that those judgments are in some form dependent on particular experiences without, however, making any determinate claims about the objects of those experiences.

My aim in this paper is to propose an answer to this question which focuses on the analogical character of this type of teleological judgment that Kant introduces at the beginning of the "Critique of the Teleological Power of Judgment". As Kant argues, teleological judgments are employed in order to reflect about living nature "in *analogy* with the causality according to ends" (*CPJ* V 360.23), that is, in analogy with the end-directed causality of an intelligent agent. Teleological judgments may thus be regarded as analogical considerations about particular experiences. And yet, teleological judgments do not make any claims about the purposive nature of the objects of those experiences themselves, but only about the way in which we reflect about those objects by means of an analogy.

While the analogical status of claims about biological purposiveness has been recognized in the literature, its particular import may remain unclear.[3] For if, as Kant argues, we cannot *know* that living nature *is* purposive, what help is it to *reflect* about nature *as if it were* purposive? Answering this question is, I think, complicated by the fact that Kant's discussion of the role of teleological reflection expresses two apparently diverging tendencies. Kant argues not only for the heuristic use of teleological judgment as a helpful guide for the study of nature, but he also claims that teleological considerations are indispensable for our very understanding of something as a living being. According to the first suggestion, teleological judgments, while not themselves directly explanatory, provide a useful means for discovering explanations in terms of the laws of nature. On this account, the aim of thinking about living beings by analogy with the "causality according to ends" lies in discovering non-teleological, causal explanations. According to the second claim, by contrast, teleological judgments are presented not merely as a guide for the discovery of non-teleological explanations but as themselves necessary for our understanding of living beings. The aim of teleological judgments, on this account, does not consist in the discovery of causal explanations, but in making possible a conception of the living

---

3 See, e.g., Ginsborg (2001, 237–8), Guyer (2001, 264–7), McFarland (1970, 111), and McLaughlin (1989, 39).

world that would not have been possible without the use of teleological considerations.

The "Critique of the Teleological Power of Judgment" thus presents two accounts of the function of teleological reflection that, I believe, have not been sufficiently distinguished in the literature. I argue that we need to take account of both conceptions in order to make sense of the peculiar status of teleological judgments. Rather than reading Kant as wavering between two distinct accounts of biological purposiveness, however, I suggest that we should understand him as putting forward one coherent conception that has implications for both our scientific practice and our very thinking of the living world. This claim can be substantiated, I believe, by taking a closer look at the different roles that Kant attributes to analogical thinking. I argue that the analogy with the "causality according to ends" provides a heuristic guide to the study of living nature while constituting a necessary condition for representing something as an organized being. It is this insight, I suggest, that sheds light on the question addressed in this paper.

In order to argue for this claim, I begin in section 2 by giving a brief sketch of the analogy that I take to ground teleological judgments. I propose that it is not conceived by Kant as the well-known analogy between organisms and artifacts, but consists in an analogy between the special character of living beings and the capacity of reason. In section 3, I spell out the two functions that Kant attributes to this analogy. Moreover, by clarifying the form and distinctive roles that Kant ascribes to analogical reflection, I show in section 4 how the teleological analogy can function not only as a heuristic tool for the study of nature but also as a symbolic representation that grounds our very conception of living beings. It is this two-fold account of teleological judgment as analogical reflection, as I argue in section 5, that sheds light on the peculiar status of claims about biological purposiveness. Thus, although further questions will need to be addressed, I conclude that Kant's introduction of the concept of an objective, material and internal purposiveness of nature gives a first account of the special character of teleological judgment.

2 Kant's Analogical Conception of the Organism

How, then, should we conceive of Kant's analogical conception of the organism? As Kant spells out in the opening paragraphs to the "Critique of the Teleological Power of Judgment", our experience of living beings is importantly different from our experience of non-living parts of nature. In particular, Kant claims that we experience organisms as distinguished by a purposive organization of their parts within the whole. The parts of a tree, for instance, seem purposively arranged to maintain the survival of the tree. The leaves, branches and roots each perform a specific function, contributing to its survival. Moreover, the form and functioning of the individual parts of an organism depend not only on the whole but also on each other. In its generation, growth, and regeneration of damaged organs, we can observe how the parts of a tree stand in mutual interaction with one another, reciprocally influencing and maintaining each other. In this way, organisms appear to organize themselves. In seeming to purposively strive for their own existence and survival, organisms display, as Kant argues, not only a purposive organization of parts, but also a capacity for goal-directed self-organization.

Kant is quick to add, however, that we have no reason to expect to encounter this type of purposiveness as an objective feature of the natural world (see *CPJ* V 359.14–17). For the concept of a purpose, on Kant's account, is essentially tied to intelligent agency, that is, to a conscious subject that intentionally sets something as a goal.[4] The "general idea of nature", however, is that of "the sum of the objects of the senses" (*CPJ* V 359.16–7). By contrast with our own activity, which we can consider from the practical, first person perspective, we have no reason to expect that nature, regarded as the sum of all causally determined objects of possible experience, will display intentional purposiveness. In this sense, nature is such that "we do not assume [it] as an intelligent being" (*CPJ* V 359.22). According to the general idea, particular experiences will thus lead us only to a more and more detailed account of the causal mechanism of nature, but will not provide evidence for a purposively acting intelligent cause as the origin of that mechanism. This is why Kant claims that the purposiveness that we seem to observe in living beings must have been "projected" (*CPJ* V 360.1) onto nature by means of an analogy: we merely consider organisms *as if* they were purposive by "*analogy* with the causality according to ends" (*CPJ* V 360.23).[5]

---

4 See *CPJ* V 369.33–370.15. As Kant puts it in his writing *Teleological Principles* (VIII 182.11–2), purposes have "a direct relation to reason". This assumption has been criticized, for instance, by Illetterati (2008) and Toepfer (2008), who argue that the purposiveness of living beings should not be interpreted on the model of intentional agency but as a type of circular causality. Ginsborg (in this volume) argues for a conception of teleology that is independent of intentionality, but attributes it to Kant himself.
5 In the *CPJ* the need to understand and explain the natural world in mechanical terms, on the one hand, and our inability to account for the apparent purposiveness of living nature mechanically, on the other, raises the notorious difficulty of reconciling the two principles of mechanism and teleology, discussed by Kant in the "antinomy of judgment" (see *CPJ* V 385.1–388.19). Taking seriously the analogical character of biological purposiveness is, I believe, a first step in making sense of the compatibility of teleological judgments with explanations of nature

In the literature, this analogy is commonly construed as the analogy between nature and design, and between the creator of nature and an intelligent designer.[6] According to this reading, we regard living beings as if they were the products of design. For it is by analogy with an artifact that is the product of a purposefully acting "will" (*CPJ* V 370.13), as Kant suggests, that we can think of the parts of an organism as being *there for* the whole and, hence, as contributing in a determinate way to the organic body as a whole. And yet, although Kant acknowledges that the artifact analogy can be taken to elucidate the apparently systematic organization of parts within the whole, I believe that this account is only a first component of Kant's analogical conception of the organism. The artifact analogy, as Kant himself recognizes, is ultimately insufficient for shedding light on the special character of nature that we seem to experience in the organic realm. As Kant concludes, the "inner natural perfection" of living beings, that is, their internal organization and purposive directedness,

> is not thinkable and explicable in accordance with any analogy to any physical, i.e., natural capacity that is known to us; indeed, since we ourselves belong to nature in the widest sense, it is not thinkable and explicable even through an exact analogy with human art (*CPJ* V 375.10 – 6).

The problem is that the analogy between nature and the product of intelligent design could only account for the first part of the two-fold teleological character of organisms. That is to say, it would only account for the characterization of organisms as displaying purposive organization. But it would not account for the apparent self-organization of living beings, that is, for the way in which organisms seem to bring about themselves and, in so doing, to strive for their own existence and survival. Thus, while artifacts are the products of a purposive agent that is external to these products, organisms seem to produce themselves. They appear to be the products of their own striving.

After rejecting the artifact model as an analogy of the particular character of living nature, however, Kant nevertheless goes on to claim that the concept of an organism is thinkable only through "a remote analogy with our own causality in accordance with ends" (*CPJ* V 375.19 – 20). This may be read as a weaker assertion of the same analogy, and hence as the claim that, even if the analogy with human art is ultimately insufficient, it is the best we can get. On a different

---

in terms of the laws of mechanics. As I argue below, it is our analogical conception of biological purposiveness that can both make the representation of something as a living being possible and guide our study and explanation of nature in terms of the laws of mechanics (see section 5).
6 This is how, for instance, Ginsborg reads the analogy.

and, I believe, more plausible interpretation, however, we can read Kant as drawing an analogy not with the products of human activity but with the very capacity for that activity, namely, the capacity of practical reason itself. On this reading, it is the goal-directed and self-organizing features of human reason, that is, our capacity to set ourselves ends and to try to realize them by a coherent and unified employment of our rational faculties, that provides the analogon for living nature.[7]

If this reading is correct, then on the one hand Kant rejects the claim that we can draw an analogy between living beings and "human art" which is the result of a "physical, i.e., natural" human activity in the phenomenal world (*CPJ* V 375.16, 13–4). On the other hand, he explicitly affirms an analogy between the apparent purposiveness of living beings and our own "causality in accordance with ends", that is, the non-physical "practical faculty of reason in us" (*CPJ* V 375.20, 24). Thus, according to this proposal, it is not the external construction of artifacts according to a pre-conceived plan, but the purposive organization and end-directedness of our rational capacities themselves that can provide the ground for an analogy with the particular character of living nature. When we think of living things as purposively organized and self-organizing beings, Kant seems to suggest, we thus regard them by means of an analogy with the purposiveness with which we are familiar from ourselves. By analogy with our own rational activity, we consider organisms as if they were the purposively organized products of their own goal-directed striving and, thereby, as "natural ends" (*CPJ* V 369.32).

3 Two Functions of Teleological Judgment

According to Kant's conception of teleological judgment, we may consider nature by means of the analogy with our own capacity for purposive activity, as long as we do not make any determinate claims about the purposive character of nature itself. But if we cannot know that nature is purposively organized and goal-directed, what is the function of this analogical conception? How can the analogy with human reason help us in making sense of the living world?

---

7 I assume here Kant's conception of human reason as not only characterized by the ability for free and end-directed activity, but also presenting a complex capacity whose different functions are purposively related to realizing and maintaining the capacity of reason as a whole. See *CPR* B xxii–iii, and *CPR* B xxxvii–viii. My argument is based on my account of the organism analogy in Breitenbach (2009a, 84–108).

On Kant's account, teleological reflection is suitable only for certain empirical objects. Thus, Kant argues that the

> concept of the combinations and forms of nature in accordance with ends is still at least *one more principle* for bringing its appearances under rules where the laws of causality according to the mere mechanism of nature do not suffice (*CPJ* V 360.26–9).

Teleological judgment comes into play, Kant here suggests, when the laws of mechanical causality are insufficient for subsuming particular experiences under "rules" (ibid.). This statement can, I think, be read in two ways. First, Kant can be understood as referring to the insufficiency of natural mechanism for *explaining* given appearances.[8] Where the attempt to provide an explanation in terms of the laws of mechanics fails, teleological considerations provide a heuristic guide for the study of natural phenomena. As Kant puts it in the "Dialectic of the Teleological Power of Judgment",

> [it is a] necessary maxim of reason not to bypass the principle of ends in the products of nature, because even though this principle does not make the way in which these products have originated more comprehensible, it is still a heuristic principle for researching the particular laws of nature (*CPJ* V 411.1–5).

When objects of experience do not appear to be amenable to mechanical explanation, Kant suggests, thinking about those objects as if they were purposively organized and directed at their own ends may guide scientific research. According to this conception, teleological judgment performs the role of a heuristic research tool, where the ultimate end of employing such a tool is the discovery of natural laws.

This first account presents a rather weak reading of the function of teleological judgment as a helpful means for the study of nature. In other passages, however, Kant makes a second and stronger claim. He argues that mechanistic causality is insufficient not only for formulating explanations in terms of the laws of nature, but also for making sense of our experiences of living nature at all. Kant suggests that we cannot even think of living beings without judging them teleologically: they are products of nature "that can only be conceived by us in ac-

---

8 I shall leave to one side here the exact nature of Kant's account of mechanical explanation in general, and the mechanical inexplicability of living beings in particular. See the different readings presented by McLaughlin (1989, 137–61), Ginsborg (2004), and Quarfood (2004, 196–205). I discuss the mechanical inexplicability of living nature in Breitenbach (2008, 355–62).

cordance with the concept of final causes" (*CPJ* V 380.28).[9] Thus, it is by reference to the idea of biological purposiveness, Kant argues, that we can first make sense of the possibility of organisms that cannot otherwise be comprehended. By regarding a tree as alive, for instance, we think of it as a unity of systematically organized parts that are purposively arranged within the whole and purposively striving towards the survival of the whole. In this sense, teleological judgment has a function over and above its heuristic use in science. Judgments of biological purposiveness are not only useful for the study of nature, but also necessary for considering parts of nature as living beings at all. As Kant puts it in the "Dialectic of the Teleological Power of Judgment", "even the thought of them as organized things is impossible without associating the thought of a generation with an intention" (*CPJ* V 398.29 – 31).

Under the title of the 'objective purposiveness' of living beings, Kant thus refers to our teleological conception of organisms as, on the one hand, presenting a heuristic tool for the study of nature and, on the other hand, grounding our very conception of the particular nature of organic beings. Considered in isolation, however, neither of these two conceptions seems to give a satisfactory answer to the question of the peculiar status of teleological judgment. According to the heuristic conception, judgments about biological purposiveness are purely regulative, but it remains unclear why such a regulative consideration is elicited by the experience of particular living beings and, indeed, required for the consideration of those beings. According to the second conception, teleological judgments are necessary in order to conceive of certain experiences as representing living beings at all. And yet, if teleological judgments ground our very thinking of organisms, one may wonder in what sense they may be regarded as purely regulative rather than constitutive of the living world.

The question then is whether, among these two apparently diverging tendencies in Kant's account of biological purposiveness, we can find a coherent conception that makes sense of teleological judgment as linked to the experience of particular natural objects while providing a purely regulative reflection on those experiences.[10] As a first step towards answering this question, I suggest

---

9 See §62 where Kant contrasts the "merely formal" objective purposiveness of geometrical figures with the "material" objective purposiveness of living beings (*CPJ* V 362.4 – 5). While, in the case of geometrical forms, Kant argues that the principle of formal purposiveness "does not make the concept of the object itself possible" (*CPJ* V 362.13 – 4), he seems to imply that, by contrast, the principle of material purposiveness does make the concept of an object, namely that of a living being, possible.

10 Quarfood (2006, 736) gives a clear expression of this problem when he writes that "the difficulty lies in balancing the claimed indispensability of teleology with its regulative status"

that Kant is not wavering between two alternative accounts, but presents one conception of teleological judgment that has two different but related functions. Kant seems to refer to these two functions when, regarding the teleological principle that in organisms that "nothing [...] is in vain" (*CPJ* V 376.13), he states that

> in the case of the abandonment of the [...] principle there would remain no guideline for the observation of a kind of natural thing once we have conceived of it teleologically under the concept of a natural end (*CPJ* V 376.33–6).

The principle to search for the purposive organization and directedness of living beings, Kant seems to suggest, provides a guideline for the investigation of those things that we have already judged teleologically as natural ends. In other words, in thinking about living beings we not only reflect about them by means of the concept of biological purposiveness, but also, by means of this reflection, are led to observe and investigate the particular structures and processes that characterize living nature according to teleological principles.

The proposed differentiation of these two inter-connected functions of teleological judgment clarifies, I believe, the first, heuristic, role of judgments about biological purposiveness. It shows that teleological judgments are elicited by particular experiences of living beings precisely because we already conceive of those living beings as natural ends. This reading leaves open, however, how we should construe the second function of teleological judgment. If, in other words, our very thinking about the living world is grounded in a teleological judgment, then one might wonder whether such judgment is not in fact constitutive of living nature. Other commentators have proposed to answer this question by arguing that teleological judgment, on Kant's account, enables "a level of special experience" (Quarfood 2006, 736) of living beings. Moreover, insofar as teleology is regarded as an enabling condition for such experience, it has been considered to be "constitutive for the identification of biological objects" (Goy 2008, 230, my translation).[11] Questions can be raised about these readings, however. Most importantly, in what exactly does this special experience of the living world consist? And how should we conceive of its relation with experience proper, which grounds our cognition of the objects of biology? What, in other words, is the connection between the phenomena that can be studied in biology and the level of special experience that we may associate with these phenomena? In order to answer these questions, I believe that we need to take a closer

---

without, however, interpreting Kant as making either "constitutive" or "trivial" statements about teleology.

11 See also Toepfer (2004, 382–4).

look at the form of analogical reflection on Kant's account. In particular, we need to examine more closely the character and role Kant attributes to the analogical reflection that grounds our understanding of living nature.

4 Teleological Judgment as Symbolic Representation

In his lectures on *Logic*, Kant characterizes analogies, on a par with induction, as one of "the two kinds of inference of the power of judgment" (*Logic* IX 132.21–2). Analogies, he claims, are "functions not of the *determining* but the *reflecting* power of judgment" that infer "from *many* determinations and properties, in which things of one kind agree, *to the remaining ones, insofar as they belong to the same principle*" (*Logic* IX 132.6–7, 26–8). By comparing two things that share certain properties we can thus infer by analogy that certain other properties, known to hold for only one of the two objects, also hold for the other. In this way, we can arrive at general concepts that subsume different phenomena. Analogical judgments are thus "useful and indispensable for the sake of the extending of our cognition by experience" (*Logic* IX 133.24–5). Insofar as analogies can "give only empirical certainty" (*Logic* IX 133.25–6), however, Kant argues that they provide merely "crutches of human reasoning" (*Lect. Log. Busolt* XXIV/1.2 680.9–10). They present a methodological device that can help us in the search for empirical truth. This, I believe, sheds light on the heuristic function that Kant attributes to teleological reflection in the *CPJ*. We may, for example, reflect about the apparently purposive development of an animal body by analogy with the goal-directedness of our own purposive activity. And we may investigate the parts of an organism by analogy with the purposively arranged parts of an intentionally designed product. These analogical considerations may help us in focusing our research into the causal processes that determine the development of organisms, and the causal relations that hold between their individual parts.

Insofar as the thought of living beings as natural ends does not itself present an empirical knowledge claim, however, it cannot be identified with the kind of analogical inference that Kant characterizes in his lectures on *Logic*. It is thus crucial that, in the *CPJ*, Kant presents a different characterization of the role of analogies as providing not a heuristic tool for empirical investigation, but an indirect, symbolic representation of concepts that cannot be represented directly.[12] This symbolic representation is made possible, Kant explains, by judgment performing

> a double task, first applying the concept to the object of a sensible intuition, and then, second, applying the mere rule of reflection on that intuition to an entirely different object, of which the first is only the symbol (*CPJ* V 352.13–6).

By applying a concept to an object experienced in intuition, Kant argues, we can thus transfer the way we think about the first object to a second object not itself experienced in intuition.

Kant presents this account of symbolic representation as grounded on analogical reflection in §59, the penultimate paragraph of the "Critique of the Aesthetic Power of Judgment". There, he argues that "the beautiful is the symbol of the morally good" (*CPJ* V 353.13). The concept of moral goodness and the related concept of freedom are rational concepts that cannot be represented directly by application to experience. Instead, Kant claims, they can be represented only indirectly, that is, analogically by means of the beautiful, as a symbol. In particular, it is our mode of reflection on the beautiful, a reflection that involves the free play of the faculties and, thereby, grounds a disinterested appreciation of the object, that can be transferred over to our reflection on the morally good.

Similarly, Kant argues two paragraphs later at the beginning of the "Critique of the Teleological Power of Judgment" that the concept of the objective purposiveness of nature is not applicable to experience itself. The concept of a purpose cannot be found in nature, but can only be used to represent the character of living nature by means of an analogy. I suggest that this analogical elucidation of living beings, too, must be understood according to Kant's conception of symbolic representation. Analogies, according to this conception, present not "an incomplete similarity between two things, but rather a complete similarity between two relations of wholly dissimilar things" (*Prolegomena* IV 357.27–9; see also *CPR* A 179/B 222). Just as in our rational activities we set ourselves ends and strive for the realization of those ends, so we view living beings as purposively directed towards their own ends. And just as our rational activities are purposively related to realizing and maintaining our rational agency, in the same way the working of the parts of living beings are purposively related to ensuring the existence and survival of the organism as a whole. Thus, it is the analogy with human reason and its end-directed and self-organizing activity that provides a symbolic repre-

---

12 Kant's different discussions of the nature and roles of analogies can, I believe, be read as presenting a coherent conception of analogy that performs varying functions in different epistemic contexts. A defense of this claim will have to await another occasion.

sentation of the objective purposiveness in biological objects. Just as moral goodness can be represented symbolically through the analogy with aesthetic experiences of the beautiful, in the same way the objective purposiveness of nature
can be represented symbolically by experiences of the end-directed causality
with which we are familiar from our own rational activities.[13]

What is crucial about this reading is that it presents the analogical thought-
process of transferring reflections from one object to another as the very ground
of our understanding of organisms as natural ends. The important insight here is
that the analogical reflection is itself creative in making the representation of
something as a living being possible. It does so by picking out certain structural
parallels between organic nature and our own rational capacity, and by transferring associations of purposiveness with which we are familiar from our rational
activity into our consideration of the natural structures. Certain parts of nature
that strike us as structurally similar to human reason are thus considered as displaying internal purposiveness according to the analogy with reason. This analogical consideration does not, therefore, simply draw out existing similarities
between the apparent purposiveness of living beings and our own purposive activity, but first constitutes a representation of some parts of nature as purposively
organized and end-directed. By projecting thoughts that we associate with our
own rational purposiveness onto our consideration of certain natural objects,
we thus reflect on those objects in a way that would not have been possible without the analogy. We first make sense of these objects as parts of living nature. It
is this "double task" (*CPJ* V 352.13) of our faculty of judgment that thus enables
us to represent natural objects as purposively organized and self-organizing living beings.

---

13 One may worry that neither of these symbolic representations consists in a simple analogy of
an empirical concept (such as that of a "handmill") with a rational idea (such as that of a state
that "is ruled by a single absolute will" [*CPJ* V 352.18 – 9]). Neither the beautiful nor the capacity
of reason and its causality of ends can straightforwardly be cognized in nature. But even if
representation of these concepts is itself dependent on further indirect, and hence symbolic,
representation, I believe that it may nevertheless form part of an—albeit more complex—analogical reflection. Indeed, the analogy between human reason and the organism may be sym-
metrical to some extent, in elucidating both sides of the analogical reflection. I argue for this in
Breitenbach (2009a, 84 – 108 and 154 – 72).

5 Reflective Judgment and Empirical Cognition

This account of the analogical character of teleological judgment as providing a symbolic representation of organisms can now shed light on the peculiar status of claims about biological purposiveness. First, it can illuminate the way in which the teleological judgment of something as a natural end constitutes a reflective representation of certain natural objects as alive, even though it does not constitute the objects of that representation themselves. And it can clarify, second, the relationship of this reflective representation of living nature with particular experiences that provide the basis of our knowledge of the empirical world.

As we have seen, Kant argues that we can represent parts of nature as purposively organized, end-directed living beings only by means of reflecting on the empirically given by analogy with the causality of ends. This representation of something as a living being is not a determinate representation of a natural object but an analogical reflection on certain aspects of nature. The important insight of this is that it is this analogical reflection that makes the symbolic representation of biological purposiveness possible and, thereby, enables us to have what may be called a reflective awareness of some parts of nature as alive. Thus, when other commentators speak of teleological judgment as enabling a "level of special experience" (Quarfood 2006, 736) this should be understood, I suggest, as the claim that our representation of living nature is a particular reflective awareness that is made possible by teleological judgment.

A comparison with the relationship between aesthetic judgments and experiences of the beautiful can, I believe, illuminate this characterization of the relationship between teleological judgments and our representation of biological purposiveness. In aesthetic judgment it is our non-conceptual reflective response to certain formal features of the object that generates our feeling of aesthetic pleasure and, thereby, the experience of beauty.[14] It is this reflection that grounds the experience of aesthetic pleasure, without representing any particular property, such as the property of beauty, in the object itself. Similarly, in the case of teleological judgment, it is our reflection on certain aspects of nature that constitutes a representation of something as a purposively organized and goal-directed living being, without attributing to nature any particular property, such as the property of purposiveness, itself. In contrast with aesthetic judg-

---

14 Kant argues that in reflecting about these forms our understanding and imagination are in harmonious "free play": they interact freely, reflecting about certain forms unconstrained by the employment of any particular concepts; and they interact in "harmony", reflecting about these forms as if they could be subsumed under a concept just as in determining judgments (*CPJ* V 217.22, 218.10).

ments that are unconstrained by the employment of any particular concept, teleological judgments do, of course, make use of the concept of purposiveness. And yet, this concept is not employed to subsume given experiences, but only to reflect about nature analogically. Rather than determinately ascribing to nature the property of purposiveness, it is the non-determining reflection that constitutes a representation and a reflective awareness of the objects as natural ends.

In parallel with aesthetic experience, there is thus no independent standard against which we could measure the truth or objective validity of our teleological judgments about the living world. We cannot, in other words, look behind our analogical considerations in order to check whether living nature really is, for instance, striving for, or purposively directed towards, its own existence and survival. The reflective consideration of empirically given objects, in both aesthetic and teleological judgments, constitutes a representation of nature *as* beautiful and, respectively, *as* purposive, without determinately claiming nature to *be* beautiful or, respectively, to *be* purposive.[15]

And yet, if considerations of the living have the same status as considerations of the beautiful as non-constitutive, purely reflective representations, one may wonder how these representations relate to particular experiences of the natural world and to knowledge of the objects of those experiences. One may question, moreover, where this leaves biology as the study of the living. Do we need to conclude, in other words, that on Kant's account biology is as incapable of making determinate judgments as aesthetics?

Despite the similarities between aesthetic and teleological judgment, this conclusion is not entailed by Kant's analogical account of biological purposiveness. For although our teleological judgments make only a reflective awareness of nature as purposively organized and end-directed possible, it is this reflection that can, in turn, function to pick out certain objective structures and processes in nature to be studied by the biologist. These objective structures and processes are those that must be given in experience in order to be reflected on by means of the analogy with human purposiveness. As Kant argues

> the teleologically employed power of judgment provides the determinate conditions under which something (e. g., an organized body) is to be judged in accordance with the idea of an end of nature (*CPJ* V 194.12–5).

---

15 A more detailed discussion of the relationship between aesthetic and teleological judgments would go beyond the scope of this paper. See Ginsborg (1997) and Zuckert (2007, in particular 23–86) for in-depth accounts.

In teleological judgments, one may read Kant as claiming, we pick out particular phenomena that are considered by analogy with our own rational purposiveness. Even though the reflection made possible by the teleological judgment goes beyond the determinate representation of the objects of experience, it is a reflection that is suitable for the consideration of certain phenomena. And it is these phenomena that can also be considered in abstraction from our teleological reflection and can be studied by the biologist. In judging teleologically, we may thus regard the parts of a tree, to use Kant's own example, as contributing purposively to the working of the organism as a whole. And we may consider the tree's generation, and its capacity for growth and regeneration of damaged organs, as aspects of the tree's goal-directed striving for its own survival (see *CPJ* V 371.7–372.11). It is through this teleological consideration that we regard the natural object as a living organism. Abstracting from this teleological reflection, moreover, we can also investigate the causal laws that determine the natural structures thus picked out. We may, for instance, examine the causal connection between the photosynthesis of the leaves and the energy consumption of the organism as a whole. Or we may study the causal processes that determine the division of meristematic cells that give rise to the different parts of the tree and keep it growing. It is this relationship between our reflective awareness of living beings as natural ends, on the one hand, and our empirical cognition of the natural world, on the other, that leaves room for the possibility of studying organisms in the biological sciences.

The idea that teleological judgment is "constitutive for the identification of biological objects" (Goy 2008, 230) should thus, I believe, be understood as the claim that, although teleological judgment makes possible a representation of living nature that is not itself constitutive of the objects of biology, it can function to pick out those very objects in experience. By reflecting about certain natural structures, such as non-linear, holistic causal structures, as objectively purposive, teleological judgments distinguish some parts of nature from their environment as those parts that are to be studied by the biologist. Moreover, once we have identified something as a living being we can use teleological considerations as regulative principles for the study of the causal processes that determine living nature. Employed as heuristic tools, teleological considerations may then guide us, for example, in investigating the evolutionary history of the traits of an organism, or the causal roles that the parts of an organism play within the organic system as a whole.[16] We can thus investigate, in biology,

---

16 I elaborate on the heuristic use of teleological reflection in biology in Breitenbach (2009b).

the causal histories and structures of those objects that, through analogical reflection, we represent to ourselves as living beings.

6 Conclusion

This paper set out to make sense of the peculiar status of judgments of biological purposiveness on Kant's account. Its aim was to shed light on the question of how we should conceive of teleological judgments as, on the one hand, being purely regulative while, on the other hand, being at least in some sense linked to the character of particular experiences. I have argued that we should answer this question by considering the particular analogical character of judgments about biological purposiveness. More specifically, I have proposed we should make sense of the analogical character of teleological judgment as having two functions. Teleological judgments present analogical inferences that serve as heuristic tools for the study of nature, and they function as symbolic representations that constitute a reflective representation of parts of nature as natural ends. As a heuristic device, teleological judgment is purely regulative while being useful for the study of nature insofar as we already experience living beings as natural ends. As symbolic representation, teleological judgment constitutes a reflective awareness that is non-constitutive of the objects of experience while consisting in analogical reflection on empirically given structures.

This account raises many questions that are the subject of Kant's "Critique of the Teleological Power of Judgment". More will need to be said, for example, about the compatibility of considering parts of nature as objectively purposive and the explanations of natural objects in terms of efficient causality. Ultimately, as I have indicated in this paper, I believe that Kant's account of our conception of living beings as purposively unified and striving for their own existence and survival is something that ought to be understood in the more comprehensive context of our own human nature. It must be comprehended against the background of our nature as beings that are not only cognizers but practical agents who conceive of themselves as both free and part of the causal structure of the natural world. Although a more detailed exploration of this claim goes beyond the scope of this paper, I hope to have shown that Kant's introduction of biological purposiveness as a type of analogical reflection gives an important first account of the peculiar status of teleological judgment.[17]

17 I would like to thank Hannah Ginsborg, Paul Guyer, Luca Illetterati, Sasha Mudd, and all participants of the international symposion on Kant's theory of biology in Tübingen for in-

References

Breitenbach, Angela 2008, Two Views on Nature: A Solution to Kant's Antinomy of Mechanism and Teleology, *British Journal for the History of Philosophy* 16, 351–69.

Breitenbach, Angela 2009a, *Die Analogie von Natur und Vernunft. Eine Umweltphilosophie nach Kant*, Berlin/New York: Walter de Gruyter.

Breitenbach, Angela 2009b, Teleology in Biology: A Kantian Approach, in: *Kant Yearbook 1*, ed. by Dietmar Heidemann, Berlin/New York: Walter de Gruyter, 31–56.

Ginsborg, Hannah 1997, Kant on Aesthetic and Biological Purposiveness, in: *Reclaiming the History of Ethics: Essays for John Rawls*, ed. by Andrews Reath, Barbara Herman, and Christine Korsgaard, Cambridge: Cambridge University Press, 329–60.

Ginsborg, Hannah 2001, Kant on Understanding Organisms as Natural Purposes, in: *Kant and the Sciences*, ed. by Eric Watkins, Oxford: Oxford University Press, 231–58.

Ginsborg, Hannah 2004, Two Kinds of Mechanical Inexplicability in Kant and Aristotle, *Journal of the History of Philosophy* 42, 33–65.

Goy, Ina 2008, Die Teleologie in der organischen Natur (§§ 64–68), in: *Immanuel Kant. Kritik der Urteilskraft*, ed. by Otfried Höffe, Berlin: Akademie Verlag, 223–39.

Guyer, Paul 2001b, Organisms and the Unity of Science, in: *Kant and the Sciences*, ed. by Eric Watkins, Oxford: Oxford University Press, 259–81.

Illetterati, Luca 2008, Being-for: Purposes and Functions in Artefacts and Living Beings, in: *Purposiveness: Teleology Between Nature and Mind*, ed. by Luca Illetterati and Francesca Michelini, Frankfurt/M.: Ontos, 135–62.

McFarland, John 1970, *Kant's Concept of Teleology*, Edinburgh: Edinburgh University Press.

McLaughlin, Peter 1989a, *Kants Kritik der teleologischen Urteilskraft*, Bonn: Bouvier.

Quarfood, Marcel 2004, *Transcendental Idealism and the Organism: Essays on Kant*, Stockholm: Almquist and Wiksell.

Quarfood, Marcel 2006, Kant on Biological Teleology: Towards a Two-Level Interpretation, *Studies in History and Philosophy of Biological and Biomedical Sciences* 37, 735–47.

Toepfer, Georg 2004, *Zweckbegriff und Organismus: Über die teleologische Beurteilung biologischer Systeme*, Würzburg: Königshausen und Neumann.

Toepfer, Georg 2008, Teleology in Natural Organised Systems and Artefacts: Interdependence of Processes versus External Design, in: *Purposiveness: Teleology Between Nature and Mind*, ed. by Luca Illetterati and Francesca Michelini, Frankfurt/M.: Ontos, 163–81.

Zuckert, Rachel 2007, *Kant on Beauty and Biology: An Interpretation of the* Critique of Judgment, Cambridge: Cambridge University Press.

structive discussions of an earlier version of this paper. I am particularly grateful to Ina Goy and Eric Watkins for very helpful comments on the penultimate draft.

Peter McLaughlin
Mechanical Explanation in the "Critique of the Teleological Power of Judgment"

In the "Antinomy" of judgment Kant speaks of merely mechanical laws as being sufficient to explain the production of an organism, but also as being insufficient to explain such a product. Most recent interpretations take the maxim that all products of nature are to be judged as possible according to "merely mechanical laws" (*CPJ* V 387.4) to be merely regulative for our cognitive activities and not constitutive of experience or of the objects of experience (Watkins 2009, 200; see Quarfood 2004, 166–8 for dissent). On the other hand, the regulative maxim of mechanism does not merely suggest a heuristic method that might be useful for biological research: it is supposed to be *necessary*. Thus the interpretation of what Kant means by 'mechanical laws', the 'mechanism' of nature, 'mechanical' explanation, and other expressions of this sort is crucial to an understanding of Kant's project in the "Critique of the Teleological Power of Judgment" and of his position on the explanation of the organism in particular. Whatever he means by these expressions, the fact that we are supposedly incorrigibly mechanistic in our scientific explanations of nature is said to be due to a *peculiarity* of our understanding, a peculiarity that is binding for us but not constitutive of the objects of our experience. This is where the problem lies: how to be necessary without being constitutive. The question is whether Kant provides the resources to produce a convincing explanation of how this is possible.

In the *CPR* Kant had already appealed to the peculiarity of our cognitive faculties, which depend on both (spontaneous) understanding and (receptive) intuition, to explain why the world of experience is constituted the way it is. This peculiarity of our understanding, which makes it depend on spatiotemporal sensibility and makes it apply precisely the twelve categories that it applies, is binding for us—but also for the objects of our experience. But when Kant reintroduces this figure of argument after §75 of the *CPJ*, he asserts that mechanism is merely regulative, not constitutive, that is, that it is in some sense necessary, but not necessary in the same way as space, time, and causality, which are of course constitutive of phenomena. Thus we have two interlocked problems: what is mechanism? And how does the structure of our cognitive faculties (understanding and intuition) force us to be regulative mechanists without *thereby* forcing nature to be mechanical? A satisfactory interpretation of the "Critique of the Teleological Power of Judgment"—at least within the current consensus— must not only clarify what "merely mechanical laws" are, but also explain why the relation of intuition to understanding in our cognitive faculties makes these kinds of laws necessary for scientific explanation without their being constitutive of the objects of experience.

In the following three sections, I shall first analyze a number of candidates for the meaning of 'mechanical' law in the *CPJ* and argue that the determination of a whole by the properties and interactions of the parts is the strongest candidate. In the second section, I will take up a number of difficulties with this interpretation as it has been advanced in the past and dispense with some of them. In the third section, I will explore some possibilities of reconstructing a Kantian argument for a part-whole determinism that is subjectively necessary for us, but not constitutive of the objects of experience.

1 What Does Kant Mean by 'Mechanical' Laws or 'Mechanical' Explanation?

What Kant meant by 'mechanistic' or 'mechanical' explanation in his discussion of teleology and the organism is either what most people at the time meant or else it has a definite technical meaning defined by Kant somewhere in his writings (or in the ideal case, both of these). In the first case we are dealing with a question of intellectual history to be answered by investigating the context in eighteenth-century science. In the second case we are dealing with an internal question of Kant-interpretation. A number of candidates for a technical meaning have been proposed, which are best characterized by the respective contrast concepts involved: spontaneous, non-physical, dynamic, holistic. It turns out that none is completely satisfying.

1) One possibility for a technical meaning, based on the *CprR* and the "Preface" to the second edition of the *CPR*, is the opposition of mechanism and freedom (spontaneity). The second manuscript of the so-called *Progress* gives some credence to this interpretation inasmuch as Kant's discussion moves directly from natural purposes to freedom (*Progress* XX 294.10 – 4). Zumbach (1980, 99), for instance, sees a kind of "free cause" acting in the organism. But if freedom is the counterpart to mechanism, then mechanism would simply be the same as the causal determinism of the phenomenal world—and the thesis of the antinomy of judgment would just assert that all production in nature should be judged as causal (which it of course is).

2) We could also interpret 'mechanical laws' as referring simply to empirical physical laws: from the *Theory of Heavens* (1755) onwards, Kant uses 'mechanical' to refer to empirical laws or regularities and even explicitly doubts whether

these mechanical laws are sufficient to explain the production of organisms. However, although it is true that already in the *Theory of Heavens*, Kant doubted the possibility of giving a mechanistic explanation of the production of even a worm or a weed, he nonetheless seems to have been convinced that these processes were completely *natural*, and he had no qualms about propagating something like the spontaneous generation of the first inhabitants of countless planets as soon as the physical conditions were right (*Theory of Heavens* I 230.14–6, 351.5–7). He does not give exact dates for these events, as does Buffon (1775, 513) two decades later, but he does subscribe to some pretty wild naturalistic speculations in the appendix to the *Theory of Heavens* ("On the Inhabitants of the Celestial Bodies"). And we should not forget that, while Kant might also despair of ever giving a mechanistic explanation of the production of a table or a chair (in as much as these things are *purposes*, that is, they involve the causality of, or according to, a concept), nonetheless, the production of tables and chairs is certainly considered to be completely compatible with the causal closure of the natural world. The arguments of the "Third Antinomy" make it clear that Kant believes that even human moral action is compatible with the causal closure of the material world.[1] And Kant reaffirms in the *Metaphysics of Morals* that our reason, as long as it is embodied in a living creature, can have causal powers: a human being as a "*natural being* that has reason (*homo phaenomenon*)" can be a cause in the world of appearances—by means of his "reason as a theoretical faculty" (*MM* VI 418.14–6, 418.8–9). Thus, though Kant never spells it out in detail, there is a form of causality by embodied minds outside the realm of mechanical laws, but this technical human action is always compatible with physical conservation laws.[2]

3) The third possibility is that Kant means in the *CPJ* what he meant in the *Metaphysical Foundations*, where he made a distinction between the areas of mechanics and dynamics: here, mechanics can be contrasted with another sub-discipline, dynamics, or distinguished from the more comprehensive discipline, physics. Mechanics has to do with the transfer of motion or force; dynamics has to do with the sources of force and the constitution of matter. This is the po-

---

1 See *CPR* B 565. In *CprR* (V 96.28–30), Kant contrasts "psychological" with "mechanical" causality according to whether it is a representation or a bodily motion that produces an action.
2 Kant wrote to Kiesewetter: "No motion can be produced in the world either through a miracle or through a spiritual being without causing an equal motion in the opposite direction, thus in accordance with laws of effect and counter-effect in matter, for otherwise a motion of the universe in empty space could arisec" (*Notes and Fragments* XVIII 320.12–6).—Even a disembodied mind would have to conform to the conservation of momentum. If the momentum of the world system is not conserved, then the center of gravity of the system must shift.

sition Kant took in the mid 1780s, as documented also in the Danzig physics lectures (XXIX/1.1 106.8–10) and various papers that Adickes published in the fourteenth volume of the "Academy Edition":

> The mechanical kind of explanation explains by the mere communication of motion not by its original natural production, that is, not completely based on the nature of bodies (*Notes and Fragments* XIV 470.2–4).

I think it is highly unlikely that Kant is using the term 'mechanical' in the "Critique of the Teleological Power of Judgment" to contrast one area of physics with another. The term 'dynamics' is not used at all in the discussion of teleology and the one contrast between the mechanical and the dynamical in the "Critique of Aesthetic Judgment" (*CPJ* V 234.19–23) has nothing to do with the distinction in physics.

However, we could also interpret Kant's distinction in the *Metaphysical Foundations* as intending a contrast between *internal* and *external* causal relations: dynamics deals with the generation of forces in matter and the constitution of bodies; mechanics on the other hand deals with the force-governed interactions *between* bodies. In this sense we could say that mechanism involves how one piece of matter exerts influence on another piece of matter. And insofar as the interacting bodies in mechanics constitute a material system, mechanism essentially involves the causal interactions of *parts* of a system. Thus, interpreting mechanistic causation as external causation might be seen to introduce a sort of implicit mereology into the causality of nature. A mechanical cause could be seen as an external cause, and insofar as we are dealing with objects in space, they consist of parts external to one another. This makes aspects of the third possibility of understanding mechanism look very similar to the compositionality problem involved in the fourth option to be discussed below.[3]

In the *Metaphysical Foundations* (IV 543.16–7), Kant distinguishes between the proposition that every change has a cause and the proposition that every change has an *external* cause[4], and in the "Introduction V" to the *CPJ* (V 181.15–20), he repeats this distinction, characterizing the first proposition as *transcendental* and the second as *metaphysical*. This suggests at least the possibility of a distinction such that causality itself is a (transcendental) condition of the possibility of an object and *external* causality is metaphysically necessary for the causal relations of bodies in space. Thus *mechanism* could be seen as "the

---

3 See Breitenbach (2006) (implicit) and Teufel (2011) (explicit).
4 See Teufel (2011, 254). In this context, the introduction of inertia as a basic property of matter, Kant also indulges himself with a polemic against hylozoism similar to those found in the *CPJ*.

category of causality under conditions of corporeality" (Teufel 2011, 254). This mechanistic principle is, however, constitutive of external experience. We would still have to explain why the maxim of mechanism is merely regulative, since in the study of organisms we are dealing only with bodies—for which mechanism, so understood, is constitutive. Furthermore, if 'mechanism' denotes the external causality of the corporeal, it would seem to have to include the actions of embodied reason on bodies, that is, the causality of concepts, which is usually taken to distinguish artifacts from mechanical productions. In any case we need some serious clarification of the causal status of art and technique, which Kant seems to exclude from the realm of the mechanical.

4) Kant could also mean by mechanical explanation the kind of reductionistic explanation of wholes by their parts that he mentions in passing in the *Metaphysical Foundations* (IV 532.34–6): "The mode of explaining the specific variety of matters by the constitution and composition of their smallest parts, as machines, is the *mechanical natural philosophy*". This view is also articulated at length in §77 of the *CPJ*.

I take it that the default setting for the meaning of 'mechanism' in its various forms in the *CPJ* should agree with that particular definition that comes closest to what the terms meant in eighteenth-century science—what Kant could expect his readers to think.[5] And there is, I think, some consensus in the history of science that the term has two basic meanings, which overlap in significant ways. 'Mechanics' in the sense of the 'study of machines' explains things by showing how they can be produced out of their parts. You can take a machine (material system) apart and reassemble it so that it performs the same tasks as the original machine (system). This is one sense of what was called the *analytic-synthetic method*. The other meaning of 'mechanics' has to do with the fundamental science of very small particles, which have very general or even universal ('mechanical') properties (or 'qualities' as they were called at the time). The particular objects of nature can be subsumed under the universal properties of these particles and manufactured out of them. *Analysis* in the sense of going from the particular to the general and *analysis* in the sense of separating a system into its parts are done at the same time. In §77 Kant says a number of things that make it plausible that this is what he, too, meant by mechanical explanation. Most clearly:

> Now if we consider a material whole, as far as its form is concerned, as a product of the
> parts and of their forces and their capacity to combine by themselves (including as parts

---

5 For a detailed discussion of the options in eighteenth-century science, Wolffian philosophy, and Kant, see van den Berg (2011, 87–153).

other materials that they add to themselves), we represent a mechanical kind of generation (*CPJ* V 408.24 – 7).[6]

Many of Kant's remarks suggest that he means by mechanism a specific kind of causal relation, namely the determination of a whole by its parts. On this reading 'mechanism' or 'mechanistic explanation' is basically a synonym for reductionism. And on this view the "Antinomy" of judgment characterizes a fundamental contradiction that defines the science of biology: the biologist accepts only reductionist (mechanistic) explanations as satisfactory explanations of the origin and workings of the organism. But she also accepts only basically holistic descriptions of the organism as phenomenally satisfactory. The contradiction with which the biologist is supposed to live is that the characterization of the system to be explained must be holistic, whereas the explanation of the system must be reductionist. That is, only a description that precludes explanation is acceptable. There seems to be a basic tension between what Kant called "*Naturbeschreibung*" and "*Naturgeschichte*"[7]. Kant's solution to the antinomy is a sort of *modus movendi* for the working biologist, which ensures that the phenomenon is always adequately described and that explanations, even if they are relativized by 'as-if' qualifications, always have the correct reductionistic form—and of course should we one day achieve a mechanistic explanation of the organism, all these elements of science would still be valid because they are mechanistic.

If we read Kant this way, many of the things he says begin to make some sense. But the interpretation also gives rise a number of problems that I want to take up now.

2 Problems with Part-Whole Determination

First of all, although many good historical arguments could be made to the effect that this position is widespread and more or less self-evident in seventeenth- and eighteenth-century science, and although we would be hard pressed to find someone important who did not subscribe to this mechanistic presupposition,

---

6 This passage could technically be read not as a definition of mechanism, but as merely stating that part-whole determination is one of possibly many forms of mechanism; see Teufel (2011, 255).

7 Kant contrasts "natural history" (*Naturgeschichte*), a causal historical discipline, with "natural description" (*Naturbeschreibung*), a primarily classificatory discipline, in a number of writings (most importantly *Races,* II 435.34), associating natural history with Buffon and natural description with Linnaeus.

nonetheless this is simply a contingent historical fact of the matter and does not provide a systematic philosophical argument—and especially not a compelling argument immanent to Kant's system. This is one formulation of a widespread misgiving about the interpretation suggested in McLaughlin (1990).[8] Kant's published writings give no hint of an *argument* as to why reductionistic explanation should be binding, or absolutely necessary for us and should thus force us to reject holism out of hand.

Secondly, although in §77 Kant characterizes the *discursive* nature of our understanding by the fact that it explains wholes by their parts, not parts by their wholes, he seems nowhere else in the book to adhere to this definition of discursivity—not even in those passages of §77 immediately preceding the part-whole commitment, where he characterizes our discursive understanding in the usual way: simply as being bound by concepts. We would have to come up with an argument as to why the fact that our understanding needs judgment to apply concepts to the manifold apprehended by intuition, forces us to explain wholes by their parts (without necessarily forcing phenomenal wholes to be determined by their parts).

Thirdly, it can be argued (see Teufel 2011, 253) that the conceptualization of compositionality, or the part-whole relation, as a kind of *causality* is metaphysically and logically objectionable. Mereology is not a causal discipline but a logical one. To the extent that Kant does in fact interpret the part-whole relation as causal, he has conflated a purely formal constitution relation with a kind of causation. In any case (according to this argument), compositionality is a constraint on causation insofar as it applies to bodies in space; it is not itself a causal relation. On this view we would, however, have to distinguish between mechanism as (metaphysically—not transcendentally) constitutive of empirical bodies and mechanism as a merely regulative maxim for research.

These three objections accept that part-whole relations are involved in mechanism but question whether they involve a special *kind* of causality, whether they really define our understanding as "discursive", or whether their "necessity" has more than contingent historical import.

In a fourth objection, Ginsborg (2004, 44) has rejected the very proposition that part-whole relations are essential to the notion of mechanism, pointing out that it is not just organisms that "cannot be judged as possible according to merely mechanical laws" (*CPJ* V 387.4–5). Machines, too, are mechanistically un-

---

8 From the first referee, who rejected the book for Oxford University Press, down to the work of most of the authors in this volume, no one has ever liked the interpretation that Kant is just making the method of modern science epistemically binding. Even McLaughlin (1990, 171–6), who takes this interpretation to be the best available, does not really like it.

explainable. She argues further that Kant's most explicit characterization of mechanistic explanation as part-whole determinism (§77 cited above, V 408.24–7) does not actually apply to machines—the properties and interactions of a machine's parts alone are insufficient to explain the production of the machine. Ginsborg (2004, 44) takes the "contingent unity" of the organism (and the machine) to characterize an aspect of the explanatory failure of mechanism that organisms have in common with machines. Thus, the fact that organisms are not explainable by merely mechanical laws does not even distinguish them from machines. Organisms do differ from machines, but this (second) kind of mechanical inexplicability is not (she believes) that involved in the antinomy: "this means that—*unless implausibly the need for teleology in the two cases stems from quite different sources*—it cannot be the non-machinelike character of organisms which makes them mechanically inexplicable" (Ginsborg 2004, 37; my italics). If I understand this right, I take the "implausible" position: the teleology involved in a machine postulates an idea or representation (formal cause) to explain the system because the properties and interactions of the parts are not sufficient to produce the whole, but this idea or ideal cause is nonetheless a 'real ground' (*Realgrund*) of the machine; the teleology of the organism, on the other hand, involves the idea of the whole, not as a *cause*, but as a "ground for the cognition [*Erkenntnisgrund*] of the systematic unity of the form and the combination of all of the manifold" (*CPJ* V 373.23) because we would otherwise have to view the functioning of an organism to involve holistic causal relations.

Ginsborg has thus pointed out that there are two quite different ways in which mechanical laws can fail to determine a system completely: in artifacts and in organisms. Let me explicate in my own words: *artifacts* involve some kind of mental causation—that is, physical causation in some way guided by a concept. (There is a formal cause or blueprint. But this is all compatible with the arguments of the "Third Antinomy".) *Organisms* involve a form of causation that bears no analogy to any other form of causality that we know of (reciprocal production of the parts/holism).

Clearly Ginsborg is right that the significant difference between an organism and a machine lies neither in the mere fact of the contingent unity of the system, nor in the need for a formal or final cause to explain its production, but rather in the peculiar kind of *efficient* causality (*CPJ* V 375.5–7) seen in the organism— which I have been referring to as holism—and which Kant wants to avoid having to acknowledge as a real causal dependency of the parts on the whole. The answer, I think, is: machines are mechanically inexplicable because they involve a form of concept-mediated causality; and this kind of causality is (purportedly) compatible with scientific naturalism as long as there is an embodied mind that has the concepts and acts on them. Organisms, on the other hand, seem to involve a causality *sui generis* that we cannot recognize as *real*; and our only alternative is therefore to treat them as if they were machines and as if at least one aspect of their causality were ideal—knowing full well that the idea of the whole is not the cause or real ground of the organism, but merely a *ground of cognition* or "marker" (*Logic* IX 58.9 – 60.37) that indicates that we are dealing with a natural purpose.

Let me introduce some qualifications to Ginsborg's objections, which will also address (and at least partly answer) the other objections. Then I will try to supply a sort of Kantian argument for the necessity of part-whole determinism. Both are offered in the spirit of theses for discussion. I do not think the issue is easily settled.

Kant notes that the "inner natural perfection" of a natural purpose that leads us to call them 'organisms' bears no "analogy to any physical, i.e., natural capacity that is known to us; indeed, since we ourselves belong to nature in the widest sense, it is not thinkable and explicable even through an exact analogy with human art" (*CPJ* V 375.13 – 6). Nonetheless, the fact that human agents act in the material world and instantiate causes that depend on or take account of a concept does not present a metaphysical problem for Kant (at least not in the *CPJ*): human actions are all part of the natural world. And in the *MM* (VI 418.14 – 8), with the distinction between *homo noumenon* and *homo phaenomenon*, Kant reaffirms the notion that human reason can in some sense be a natural cause. Just how the idea of an artifact manages to be part of the efficient cause of that artifact's production is left unexplained. But it is clear that Kant sees no violation of the principle of the "Second Analogy" or of the conservation of force in the world system in the fact that humans—or bees and beavers for that matter —make things that are otherwise underdetermined by the laws of nature. Thus, although machines are mechanically unexplainable, they present no special metaphysical or epistemological problem for Kant, since they are purposes (not products of nature), and the understanding that has these purposes can even tell us what they are. The mechanical inexplicability of artifacts (their contingent unity) cannot be the kind of inexplicability that leads to the antinomy. As mentioned earlier, Kant maintained in the *Theory of Heavens* that the origin of organisms on various planets seemed to be unexplainable by mechanical laws, but he seems to have assumed it somehow happened in the course of nature—that is, without mobilizing reason as part of the explanation. Thus the contingent unity of products of nature must be different from that of artifacts.

3 What is the Peculiarity of our Understanding that Makes us Mechanists?

The best argument for the assertion that Kant means part-whole determinism when he says 'mechanism' would be to show that the peculiar structure of our intellect, which uses intuitions and concepts, necessitates a certain part-whole determinism and makes it binding for us—we cannot do otherwise—but is not constitutive of the objects of experience. That is, we should strive for something more than a merely historical argument to the effect that this was a widespread phenomenon at the end of the eighteenth century and could be taken as binding scientific method. There seem to be three options for a reconstruction: (1) We might in an interpretation of the appendix to the "Transcendental Dialectic" try to construct a parallel between the transcendental presupposition of unity or specification of empirical laws and some kind of presupposition involved in the experience of organisms. I see no promising avenues there and will not pursue this option. (2) On the other hand we could argue that the specifically *spatial* character of the part-whole relation in bodies makes mechanism not a transcendental condition of experience, but a necessary principle that in some (to be specified) sense, while not constitutive of experience, is nonetheless binding— and still has some reasonable connection to reflective judgment. I shall explore this possibility in some detail, since it looks for a solution in the relation of understanding to intuition. (3) A third possibility is suggested by some scattered (unpublished) remarks by Kant about the essentially compositional nature of the understanding itself—with no appeal to spatial intuition.

Taking up the *second* possibility: in §76 of the *CPJ*, Kant explains that the peculiar structure of our cognitive faculties has serious consequences for the world of appearances. If we human agents were not so constituted that we separate understanding and intuition, we would not (in theoretical reason) have the distinction between the possible and the actual. Nor would we (in practical reason) have the distinction between what ought to be and what is. Finally, if our understanding were not such that it must go from the general to the particular, we would not have to make a distinction between the *mechanism* and the *technique* of nature when considering particular empirical laws and unifying them in a system of empirical laws of nature. Thus, Kant concludes at the opening of §77 that since the peculiar structure of our cognitive faculty in regard to theoretical understanding, practical reason, and determinate judgment has such far-reaching consequences, we should not be surprised if reflective judgment, too, were also subject to the consequences of these peculiarities.

Just like the distinction between possibility and necessity, just like the distinction between *ought* and *is*, and just like the transcendental presupposition that nature is orderly, the concept of natural purpose, too, according to Kant, is only possible due to a "peculiarity" of our understanding—apparently a peculiarity at least similar to that above. However, in the case of a natural purpose, the source of our difficulties is an actual object of experience: a real organism. This makes the problem quite different from those problems dealt with in §76:[9] the peculiarity in this new case is not constitutive of experience (theoretical reason), nor of the possibility of realizing *ought* in *is* (practical reason), nor is it a transcendental presupposition of empirical research. It is occasioned by the experience of particular entities; that is determining judgment and the understanding have already constructed an object of experience out of the sensible manifold, but this object somehow bothers us.

Our understanding, says Kant, is characterized in its relation to judgment by a contingency of what diverse particular objects may be given in nature and then have to be subsumed under concepts. In constructing perceptions out of the sensible manifold, the understanding is dependent on what is contingently given to empirical intuition. And since our sensibility or intuition is passive or receptive, it must take what happens to come to it. We can however, says Kant, conceive of a non-receptive—a spontaneous—kind of intuition. Now, because an intuition that is spontaneous would (by definition) be a kind of understanding, this comparison allows us (in a somewhat abstract way) to imagine a non-discursive, that is, non-conceptual spontaneity or understanding—at least by negation. Once we have an idea or a 'non-idea' of such an alternative understanding (a spontaneous intuition), we can ask how this 'understanding' (which apparently still has to go from the universal to the particular—and, I insist, is *finite*[10]) would relate to judg-

---

9 In §76 Kant gives three examples illustrating the peculiar relation of intuition and understanding in theoretical reason, practical reason, and judgment. The problems associated with those peculiarities in the first two examples are resolved by the distinction between appearances and things in themselves. The third example is quite rudimentary and does not obviously appeal to this distinction or deal in any way with the subjective purposiveness of aesthetic judgment. It can be read as dealing with regulative principles on the unity of nature as a whole, or as a misplaced preliminary to the discussion of the teleology of the organism in §77.
10 The purpose of introducing a negatively characterized alternative understanding is to teach us something about the peculiarity of our understanding that makes us mechanists or reductionists. If the alternative "intuitive" understanding is not just holistic but also infinite, its lack of difficulty with the explanation of organisms can be due either to its non-discursive character or to its infinitude; this would tell us very little. The figure of argument itself only makes sense if just one property ascription is being negated.

ment—a relation which for us is a constant source of contingency. For this other understanding, however, the particular is fully determined by the universal. But how is the universal, given in intuition to this new and intuitive understanding, actually given? Unfortunately, this primarily negatively characterized understanding only illuminates our own understanding if we know exactly what aspect of our understanding is being negated. Thus we need first to look at the positive characterization of our understanding in the *CPR*.

The logical places to look are the "Deduction", where the notion of a "peculiarity of our understanding" is first brought up (*CPR* B 145), and the "Second Antinomy", in which the *conditions* of a body are self-evidently characterized as the *parts* of the body and the *regress* in the series of conditions is presented as the division of a whole into preexisting parts. Let's see how far we can get in reconstructing a Kantian line of thought that explicates the part-whole relation.

We go out and perceive a sandy beach, a house, and a blade of grass. When we transform the empirical intuitions of these things into perceptions, we do this based on the *necessary unity* of space, and we, so to speak, sketch the shape of the object in accord with this synthetic unity in space.In Kant's words (in the "Deduction"):

> Thus if, e.g. I make the empirical intuition of a house into perception through apprehension of its manifold, my ground is the *necessary unity* of space and of outer sensible intuition in general, and I as it were draw its shape in agreement with this synthetic unity of the manifold in space (*CPR* B 162).

For perception, the objects need only have (necessary) spatial unity; that is enough unity to become an appearance. Determinate judgment can then provide for a physically necessary unity by adducing empirical laws for the sandy beach and purposes in the mind of an artisan (plus empirical laws) for the house. In

---

Kant admittedly (e.g., *Lect. Rat. Theol. Pölitz* XXVIII/2.2 1017.5 and 1111.26; *Lect. Rat. Theol. Volckmann* XXVIII/2.2 1214.33 and 1219.11) often characterizes God's understanding as "intuitive" (as opposed to discursive), but he does not seem to assert the converse: that an intuitive understanding has to be divine. In this section of the *CPJ*, Kant distinguishes clearly between the quantitative and qualitative aspects of our understanding: the "limits" [*Schranken*] and the "constitution" or "character" [*Beschaffenheit* or *Eigentümlichkeit*] of our understanding (*CPJ* V 395.14 – 5; 398.7 – 8; 400.34). A few lines later (*CPJ* V 409.34 – 6), Kant is explicit that a different (but finite) understanding also cannot explain the organism so long as it is *qualitatively* similar to ours: "and absolutely no human reason (or even any finite reason that is similar to ours in quality, no matter how much it exceeds it in degree) can ever hope to understand the generation of even a little blade of grass from merely mechanical causes". On the other hand, a qualitatively different understanding (even if it is finite) need not have problems with the organism.

the first case, empirical laws that govern the behavior of the parts lead them to produce the whole beach (I give you a differential equation for the center of mass of each sand corn); in the second case (the house), these laws are insufficient and must be constrained by an idea or concept of the house as a whole in the mind of an artisan (an embodied mind). But in the case of the grass blade, judgment (in its determining form) fails at its task and begins to reflect on what it should do in this case, because the physical unity of the object remains contingent (only the spatial unity is necessary). As Kant noted in §61 of *CPJ* (V 360.15 – 7), discussing the anatomy of a bird, the *same* parts could easily have been arranged differently into a *different* whole. The production of the whole is not completely determined by the properties and interactions of the parts, and there is no natural understanding or embodied reason in sight to provide the missing determination.

In the resolution to the "Second Antinomy", Kant confirms the merely spatial unity of perceptions, telling us that while an object of experience is given as a whole to our understanding, it is given only as a *quantum continuum* (*CPR* B 554), all the parts of which are given in the *intuition* of the whole. But as discrete entities, the parts are given to the understanding only to the extent that the whole has been analyzed by the understanding into its parts. As a whole, an object of experience possesses at first only spatial unity: it can be divided into parts *in infinitum* because it is extended in space—without, however, being composed of infinitely many parts. The notion of the composition of a whole out of the parts is presupposed in many such passages but scarcely explicated or even discussed by Kant.[11]

For Kant's newly invented intuitive (non-conceptual) understanding, on the other hand, a composite would presumably be given as a *quantum discretum*, all of whose parts are determined by the given whole. In this understanding, the particular is completely determined by the general: there is no contingency in the relation of judgment to understanding. Thus it is possible that this other kind of understanding might experience no contingency in the relation of understanding to judgment when dealing with particular already-given objects. We have to explain given objects mechanistically or reductionistically (or to specify their physical unity by constructing them from their parts), but we cannot determine the particular by the universal and must allow for contingency there. But

---

11 In the "Third Analogy", where an explication of composition would have been appropriate, Kant is so busy with the question of interaction that he barely mentions composition and fails to say anything significant about it. In a letter to Beck (July 1, 1794), he remarks that on reading over his own explanation of composition, he does not understand himself: "I notice as I am writing this down, that I do not even entirely understand myself" (*Correspondence* XI 515.30 – 1).

there is contingency only because (due to our peculiarity) we have to perceive the whole at first as a merely spatial unity or *quantum continuum*. The intuitive understanding, on the other hand, starts not with a merely spatial unity but with a physical unity, and thus experiences neither contingency, nor a need for teleological judgments.

Imagine that such an intuitive understanding confronts a complex system: just as a space is not determined by its subspaces, but is rather limited and determined by the space that encloses it (space is not a *compositum* but a *totum*, *CPR* B 554), so too, this intuitive understanding would go from a given whole to its parts. But since this understanding does not experience any contingency of the particular with respect to the general, the parts would be determined by the intuited whole. Whereas, for our understanding, the same parts could make up different wholes and the same whole as a *quantum continuum* could be divided up into different parts, for this intuitive understanding, the same whole must determine the division into the same parts.

So what would Kant really get here, if this somewhat tentative reconstruction reflects what he meant? At least an analogy between the relation of mechanism to contingency and the relation of holism to necessity: as Kant noted in §61, discussing the anatomy of a bird, the *same* parts could easily have been arranged differently into a *different* whole (*CPJ* V 360.15–7). However, the *same* whole can only be separated into *different* sets of parts as a *quantum continuum*; as *quanta discreta* only different wholes can be separated into different (discrete) parts. It is at least possible to interpret Kant's seemingly arbitrary commitment to part-whole determinism as coherent with the dependency of our understanding on spatial intuition and with the understanding's need for judgment in order to get from the general to the particular. The relation of our understanding to judgment has a peculiar characteristic: judgment's particular is not determined by or derivable from the understanding's general concept. Nonetheless, the particular is supposed to conform to and be subsumable under the general. It would seem to be contingent (good luck) that it works out so well. Kant then asserts that in order to conceive that this conformity of a natural object to our faculty of judgment is even possible, we have to imagine a different understanding, for whom the conformity of natural laws to its power of judgment is necessary. For this understanding the parts are not given to intuition with the whole (as *quanta continua*), but rather (as *quanta discreta*) only given as a task in an indefinite progressus.

This briefly sketched approach to deriving the non-constitutive but nonetheless binding character of the part-whole relation from the relation of the understanding to spatial intuition is not overwhelmingly convincing. There is just too little material to work with, so that it is impossible to pin Kant down on any one position: something related to spatial intuition might by way of perception commit us to some form of mereology. But these texts only support the assertion that there is some evidence that Kant might have been thinking in this direction.

The *third* possible solution to our problem can be found in some of Kant's (unpublished) later work, in which he takes an entirely different direction on the part-whole relation. In *Progress*, written a few years after the *CPJ*, for instance, Kant seems to be contradicting the main thrust of the famous §76 of the *CPJ*, which focuses on the relation of intuition and understanding, by insisting that composition does not belong to the receptivity of intuition at all, but rather only to the spontaneity of the understanding: *it is a concept a priori.* Not one of the categories, it is unique among the a priori concepts, in as much as it is "the only basic concept [*Grundbegriff*], that in the understanding originally underlies all concepts of sensible objects" (*Progress* XX 271.19–21; see XX 275.34–276.8). This includes not just spatial composition, but also temporal. If mereology is originally involved in all empirical concepts, this would seem to make the compositionality relation a conceptual issue, not a causal one—contrary to Kant's tendency in the *CPJ*.[12] In *Progress*, Kant asks:

> Whence do sensible objects acquire the connection [*Zusammenhang*] and regularity of their coexistence so that it is possible for the understanding to bring them under general laws and discover their unity according to principles (*Progress* XX 276.28–30)?

Such passages indicate that Kant later explored other interpretations of compositionality that do not recur to the relation of understanding to intuition, but rather to the nature of the understanding itself. He seems to have been looking for a different way to give the part-whole relation a special status, so that reflective judgment could use the notion of composition to construct the unity of empirical objects:

> All representations which constitute an experience can be assigned to sensibility, with one solitary exception, namely that of the composite, as such. [...] [Composition] belongs, not to the receptive nature of sensibility, but to the spontaneity of the understanding, as an a priori concept (*Progress* XX 275.36–276.2).

Furthermore, such a conceptualization of compositionality would allow us to explain retrospectively why Kant thinks that our specifically *discursive* understanding, precisely because it is bound by concepts, is also essentially committed to

---

12 This would seem to support at least the purely systematic part of the argument proposed by Teufel (2011, 253–4).

part-whole determinism. But this would also give us more than we want, since, if composition is an a priori concept that underlies all empirical concepts, then all bodies (organisms included) would be *constitutively* mechanistically (compositionally) determined, and the appearance of a Newton of the blade of grass (*CPJ* V 400.18–9) would not be an absurdity, but rather almost inevitable. After all, under this assumption, organisms (like all objects of experience) are in fact compositionally determined mechanisms.

4 Can Mechanism be Necessary and Still be Merely Regulative?

Our point of departure was the question of how the interpretation of 'mechanism' in the *CPJ* as referring primarily to the part-whole determinism of causal relations can be both a necessary and a regulative principle, how it can be binding for our cognitive activities but not constitutive of the objects of experience. If 'mechanism' simply refers to reductionistic explanation as practiced in science since the seventeenth century, it seems to be too historically contingent to deserve to be characterized as *necessary*. On the other hand, any attempt to make part-whole determination more than historically contingent as the method of modern science tends to make it constitutive of experience. The results of the considerations reviewed in the past three sections (see also Watkins 2009) seem to reinforce the dilemma: Kant does not provide the resources needed to argue that the purported peculiarity of our understanding is genuinely necessary *and* merely regulative.

References

Breitenbach, Angela 2006, Mechanical Explanation of Nature and its Limits in Kant's *Critique of Judgement*, *Studies in History and Philosophy of Biological and Biomedical Sciences* 37, 694–711.
Buffon, Georges-Louis Leclerc comte de 1775, *Histoire naturelle, générale et particulière*, Supplément vol. 2, Paris: Imprimerie Royale.
Ginsborg, Hannah 2004, Two Kinds of Mechanical Inexplicability in Aristotle and Kant, *Journal of the History of Philosophy* 42, 33–65.
McLaughlin, Peter 1990, *Kant's Critique of Teleology in Biological Explanation. Antinomy and Teleology*, Lewiston: Edwin Mellen Press.
Quarfood, Marcel 2004, *Transcendental Idealism and the Organism: Essays on Kant*, Stockholm: Almquist & Wiksell.

Teufel, Thomas 2011a, Wholes that Cause their Parts: Organic Self-reproduction and the Reality of Biological Teleology, *Studies in History and Philosophy of Biological and Biomedical Sciences* 42, 252–60.

van den Berg, Hein 2011, *Kant on Proper Science. Biology in the Critical Philosophy and the Opus Postumum*, Graz: Styria.

Watkins, Eric 2009, The Antinomy of Teleological Judgment, in: *Kant Yearbook 1*, ed. by Dietmar Heidemann, Berlin/New York: Walter de Gruyter, 197–221.

Zumbach, Clark Edward 1980, The Lawlessness of Living Things: Kant's Conception of Organismic Activity, Rutgers University.

Marcel Quarfood
The Antinomy of Teleological Judgment: What It Is and How It Is Solved

The title of this paper perhaps promises more than I will be able to deliver, considering the intricacy of the "Dialectic" in Kant's "Critique of the Teleological Power of Judgment", but at least it indicates two main puzzles for commentators: which pair of principles makes up the antinomy, and what is its solution?[1] Among older commentaries, one often finds the view that the antinomy consists of a pair of opposed constitutive principles concerning the generation of material things, and that the solution lies in replacing these principles with the corresponding regulative maxims, which do not conflict.[2] This view is resisted in more recent commentaries, which point out that it neither explains Kant's characterization of the conflict as an antinomy of judgment, nor fits the structure of the "Dialectic".[3] Instead the antinomy is construed as being between the regulative maxims, and its solution is sought in the analysis of the structure of our understanding and the remarks on the supersensible basis of nature that Kant offers in the second half of the text.

I will propose an intermediary reading of the antinomy, one which affords the constitutive principles and their replacement by regulative maxims a larger role than recognized in the recent interpretations, but which also accepts that the recent view has a point when it claims that the maxims are central for the antinomy, as well as in stressing the need to account for the structure of the entire text.[4]

---

[1] See Watkins (2009), where the antinomy is analyzed with regard to these two questions.

[2] For lists of commentators taking this view, see McFarland (1970, 120–1) and McLaughlin (1990, 138).

[3] McLaughlin (1990) is the most influential proponent of this alternative view. But it is found already in Marc-Wogau (1938, 225).

[4] For a more detailed account of the antinomy, taking this approach, see Quarfood (2004, 160–208).

1 The Regulative Maxims and the Constitutive Principles

According to *CPJ* §69, "there can be a conflict, hence an antinomy" between maxims of the reflecting power of judgment (*CPJ* V 386.5 – 6). In §70, these maxims are specified.

> *Thesis:* All generation of material things and their forms must be judged as possible in accordance with merely mechanical laws.
>
> *Antithesis:* Some products of material nature cannot be judged as possible according to merely mechanical laws (judging them requires an entirely different law of causality, namely that of final causes) (*CPJ* V 387.3 – 9).

After presenting these regulative maxims of the power of judgment, Kant introduces a pair of principles for the determining power of judgment. These principles omit any reference to how we judge, stating constitutive claims about the generation of things:

> *Thesis:* All generation of material things is possible in accordance with merely mechanical laws.
>
> *Antithesis:* Some generation of such things is not possible in accordance with merely mechanical laws (*CPJ* V 387.13 – 6).

Kant adds that this pair of conflicting constitutive principles would not qualify as an antinomy of the power of judgment, but rather of reason (though the conflicting claims would be unprovable).

There is thus an antinomy of the reflecting power of judgment; it consists of conflicting regulative maxims, and they should not be confused with the somewhat similar constitutive claims about the generation of things. Given this, how can there be any doubt about which pair of propositions that makes up the antinomy? It is Kant's further explanations that cause problems for the interpreter. Kant seems to say that the maxims of the reflecting power of judgment "do not in fact contain any contradiction" (*CPJ* V 387.26),[5] making it unclear why we should

---

5 Admittedly, this may be an artifact of the English translations of the text. Literally, as pointed out by McLaughlin (1990, 149) and Watkins (2009, 201–2), Kant just says that the maxim of mechanism does not contain any contradiction: "Was dagegen die zuerst vorgetragene Maxime einer reflectirenden Urtheilskraft betrifft, so enthält sie in der That gar keinen Widerspruch" (*CPJ* V 387.25–6). This seems rather pointless to say, but Watkins (2009, 202) proposes a plausible reading: that mechanism contains no contradiction means that it does not contradict the constitutive thesis or antithesis, so that even if the constitutive antithesis were true, this would not

assume that there is an antinomy of judgment in the first place. And in §71 he tells us that the "appearance of an antinomy between the maxims [...] rests on confusing a fundamental principle of the reflecting with that of the determining power of judgment" *CPJ* V 389.20 – 7). So it now seems that the antinomy after all is between the constitutive principles, and that the solution consists in abandoning their claims about the generation of things in favor of regulative maxims for our judging of things. This was the standard interpretation in the older literature. Recent commentators have attacked this interpretation. They construe Kant's statements at the end of §70 and in §71 so that they do not conflict with his initial claims in §69 and the beginning of §70, concluding that the antinomy is between the maxims of reflecting judgment.

The view I will defend in this paper is that there is some truth in both approaches. The recent interpretation plausibly claims that a pair of unproven and contradictory constitutive principles does not amount to an antinomy of judgment, and that replacing it with a corresponding pair of regulative maxims provides no solution—for those maxims are contradictory too. The old interpretation nevertheless seems to agree very well with at least some of the texts, namely the numerous passages where Kant stresses the importance of not taking maxims of the reflecting power of judgment for constitutive principles determining things objectively; these could be taken to confirm that this is what the antinomy is about. Can we combine the virtues of both interpretations while avoiding their respective pitfalls? I propose roughly the following: in some sense, the antinomy may well be between the maxims of judgment (just as the recent interpretations claim). But this needs to be qualified, since the maxims are compatible when used reflectively, according to what Kant says in §70 (even though it remains to be explained how he can claim this, see section three below). What constitutes an antinomy, I suggest, is the maxims dialectically taken as implying constitutive (ontologically determining) principles.[6] However, this in a sense just amounts to confusing the genuine maxims with the spurious constitutive principles, bringing us close to the old interpretation. The old interpretation, however,

---

invalidate the regulative maxim of mechanism. Another possibility, perhaps a little strained, is that "die zuerst vorgetragene Maxime einer reflectirenden Urtheilskraft" refers to the conjunction of the two maxims, treating them as a single complex one, of which it is claimed that it is consistent. In any case, it is still plausible to take Kant's view to be that the regulative maxims do not contradict each other, since a few lines after this sentence he says that reflecting according to the mechanistic maxim "is not an obstacle to the second maxim" (*CPJ* V 387.35 – 4), and a few pages on he claims that the maxims "could well be united with each other" (*CPJ* V 391.13).

6 See Allison (1991, 31– 2), who stresses the dialectical tendency to understand the maxims as ontological principles. My interpretation of the antinomy has much in common with Allison's.

neglects to explain why the constitutive principles are held to be true. By focusing on the dialectical tendency to interpret the necessary maxims for reflecting on natural forms as entailing constitutive principles about the very possibility of these forms, this is clarified and the difference between the old and the recent types of interpretation is minimized.

One often noticed weakness of the old interpretation is that if solving the antinomy amounts to replacing constitutive principles with regulative maxims, the argument would not seem to need a continuation after §71. And yet §71 is titled "Preparation for the resolution", and a lot follows: a discussion of metaphysical systems that attempt to explain natural teleology (§§72–3), sections on the critical use of teleology (§§74–5), the very dense §§76–7 which elucidate the transcendental source for our need to regard some natural objects teleologically, and finally §78 that treats of the supersensible and the unification of mechanism and teleology. I will attempt to show that the length of Kant's argument makes sense on the present interpretation. By stating in §71 that the "appearance of an antinomy" (*CPJ* V 389.20) consists in taking a principle of reflecting judging for a determining principle, Kant has given precisely a "preparation for the resolution" (*CPJ* V 388.21). For the resolution can hardly be just a flat assertion of a result; it also requires a richer account of the grounds for there being a dialectical illusion and of the consequences of not seeing through it, as well as an explanation of how the maxims can co-exist. And these are the kinds of issues Kant tackles in the succeeding paragraphs.

2 The Dogmatic Systems

The discussion in *CPJ* §§72–3 of "the various systems concerning the purposiveness of nature" (*CPJ* V 389.29–30) has been considered to be somewhat out of place by proponents of the recent type of interpretation (e. g., Watkins 2008, 257–8). If the antinomy is between the regulative maxims, why would Kant immediately after the introductory sections offer a detailed account of the dogmatic systems (according to which it is objectively determinable whether teleology exists in nature or not), instead of attempting right away to resolve the conflict between the maxims? On the present interpretation, attacking the systems makes more sense, since they are based on adopting a constitutive interpretation of the principles, something that results from the dialectical illusion that is to be dispelled by critique.

CPJ §72 begins with the (perhaps exaggerated) claim that no one has doubted that organized beings "must be judged in accordance with the concept of final causes" (*CPJ* V 389.32–3). When such judging is taken as explanatory, however, the various systems "controvert one another dogmatically" (*CPJ* V 391.9 – 11), giving rise to "*contradictorily opposed* principles" which "cancel each other out and cannot subsist together" (*CPJ* V 391.14 – 5). The dogmatic stance is based on a dialectical temptation: the "presentiment" of reason that the concept of final causes could allow us to "step beyond nature" (*CPJ* V 390.12 – 4). Dogmatic explanatory principles about final causes, whether affirmative or negative, correspond to the contradictory constitutive principles of "the possibility of the objects themselves" described in §70 (*CPJ* V 387.11). The opposing principles give rise to the systems of "idealism of purposiveness" (*CPJ* V 391.24) and "realism of purposiveness" (*CPJ* V 392.6) respectively. Idealism asserts that all apparent purposiveness in nature is unintentional, i.e., the result of accidental processes, while realism takes some purposiveness to be intentional and thus real. Idealism and realism each have two variants, depending on whether God is admitted in the system or not. Idealism without God is "casuality" (*CPJ* V 391.29 – 30) (or accidentality), associated with the Greek atomists, whereas Spinoza's "fatality" (*CPJ* V 391.32 – 6) is the idealistic position that admits a hyperphysical ground of nature, an original being that produces nature by blind necessity and thus qualifies as a "lifeless God" (*CPJ* V 392.29). Realism, on its side, also takes a physical and a hyperphysical form. The first one accords life to matter (hylozoism), while the second one derives the purposiveness in nature from an intentionally productive original being (theism).[7] Theism is deemed the most satisfactory of the systems, as it allows for the concept of purpose (which, if used regulatively, is a useful guide for the observation of nature). Kant takes hylozoism to be inconsistent because it goes against the essential lifelessness of matter (its inertia), by allowing matter a capacity for life (i.e., for spontaneous movement) which would destroy natural philosophy (*Metaphysical Foundations* IV 544.25 – 6). The less radical idea of an animated matter ("belebten Materie" rather than "lebenden Materie") can at best

---

7 Kant's classification of dogmatic systems in some respects resembles the critical divisions of atheistic systems made by Walter Charleton and Ralph Cudworth in the seventeenth century. Charleton (1652, 41) discussed Epicurus' views on "the first Casual Emergency of the World", which suggests that "casuality" was a standard term for the ancient atomist's view on the accidental origination of order. Cudworth (1678, 105) distinguishes between "Atomick Atheism" and "Hylozoism" (a term he apparently introduced); the first of these views considers matter to be inert while the second takes it to be living. Admittedly, Cudworth makes further distinctions which do not quite correspond to Kant's; he criticizes attempts to explain the emergence of consciousness rather than purposiveness, and, most importantly, in Kant's view Cudworth's theism would also count as a dogmatic system. Nevertheless, there seem to be connections worth exploring between *CPJ* §§72 – 3 and this seventeenth-century tradition. Kant owned a copy of the Latin translation of Cudworth's *True Intellectual System of the Universe* (see Warda 1922, 47), a work presumably well known in the German philosophical world.

be used heuristically for conceiving of the whole of nature as animated. But if we attempt to derive the purposiveness of organisms from a postulated life of matter, we can conceive of such life only by referring to our experience of organisms, and then we move in a circle (*CPJ* V 394.33 – 395.2).

Theism, though preferable to the other dogmatic systems, cannot justify any objective assertions about organisms, since our cognitive capacity is limited with regard to the possibility of an ultimate explanation of their organization. Its appeal to divine intention serves only as "a ground for the reflecting, not for the determining power of judgment" (*CPJ* V 395.19 – 20), and that is not what it aimed to establish.

The claim that the organism cannot be explained mechanically from our cognitive vantage point (famously expressed in the passage on the Newton of the blade of grass in *CPJ* V 400.16 – 20) is connected to the account in the "Analytic of the Critique of Teleological Power of Judgment" of the special properties of organized beings. Their self-organizing capacity, exhibited in reproduction, growth, and regeneration, is something we encounter empirically and cannot explain by physics (taken in a wide sense as including chemistry).[8] The impasse in the theory of generation, where neither the mechanistic type of epigenesis (committed to *generatio aequivoca*) on the one hand nor the theory of evolution (or pre-existing involuted germs) on the other accomplishes its goal, is another indication of the failure of the dogmatic stance (see *CPJ* V §81). Kant's recommended version of epigenesis as "generic preformation" (*CPJ* V 423.5) is to be seen against the background of a regulative conception of the organism, according to which organization is a necessary and ineluctable condition for all description of organic processes. Kant's proposed research program is tailored to avoid dogmatism: neither an idealist nor a realist explanation for the origin of organization should be desired.

But the rejection of the systems still leaves open the possibility that there could be a capacity in nature for organizing matter. That this is inconceivable for us is not a proof that it is impossible: "for how could we know that" (*CPJ* V 400.27)? Such a capacity is not available for us even as a hypothesis, but it is logically possible. The inexplicability of organization is thus not an ontological contention, but rather an epistemic claim about the conditions for conceiving of organized beings, given the "peculiarity" (*CPJ* V 405.2) of our cognitive capacity. This is further explored in §§76 – 7. Before touching upon that in section 4 below, I will return to the question of the antinomy, to what extent it is a genuine one, and how the regulative maxims are supposed to co-exist.

---

8 As the aim of this paper is exegetical I will pass over critical discussion of this view.

3 From Contradictory Principles to Disparate Maxims

According to the interpretation here proposed, the antinomial conflict between the pair of regulative maxims consists in the tendency of taking them as constitutive. Though contradictorily opposed, none of the constitutive principles are provable, however (*CPJ* V 387.21–2), and therefore no antinomy would seem to arise between them. Were it possible to prove them dogmatically, they would amount to an antinomy, but as it is, they are just a pair of dogmatic statements.[9] Why then should there be an antinomy between the regulative maxims when misconceived of as constitutive?

Let us call the pair of regulative maxims, when interpreted as having some determining (or ontological) force, RC (for "regulative as constitutive"). Since judging is taken to be determining, the mechanistic thesis ("All generation of material things and their forms must be judged as possible in accordance with merely mechanical laws" [*CPJ* V 387.3–5]) will for RC amount to the claim that every material thing must yield to mechanistic explanation. This is very close to the constitutive thesis, but whereas that, as an isolated claim, would be an ungrounded ontological thesis, we have now reached a similar thesis by taking a rule for reflecting on nature as a rule of its determination.

The antithesis, on the other hand, namely that "[s]ome products of material nature cannot be judged as possible according to merely mechanical laws (judging them requires an entirely different law of causality, namely that of final causes)" (*CPJ* V 387.6–9), is for RC more than just a necessary way of viewing organisms, pertaining to reflecting judgment; this way of viewing them assumes objective teleological properties, and therefore opens the door to realism of purposiveness. Again, the slip to ontology comes from taking reflection to involve determination. Critics have persistently objected to Kant that if it is necessary to judge a thing teleologically, then we cannot escape ascribing such properties to it.[10] Whether or not this is right—for Kant it is just a transcendental illusion — the objection expresses the tendency to go from a regulative rule of reflecting judgment to a corresponding ontological commitment.

The RC thesis and antithesis are contradictorily opposed, since the antithesis claims that some things will resist mechanistic explanation, while the thesis is taken to presuppose that everything will yield to that explanatory strategy. The

---

9 But see note 15 below.

10 Among other incisive objections, a related point is made by the the anonymous author of an early attack on Kant's teleology which appeared in Eberhard's journal (anonymous 1794, 16).

thesis is provided by the understanding, the antithesis is "suggested by particular experiences that bring reason into play" (*CPJ* V 386.32–3). The role of reason is to provide the concept of purpose, which is defined as the object of a concept in so far as it is taken to be caused by that concept (*CPJ* V 220.1–3). Such causality by concepts pertains to the will. Objects, the generation of which we can explain only by assuming them to be caused by concepts, are deemed as purposive, regardless of whether there is a purpose or not, which is not always possible for us to have insight into (*CPJ* V 220.22–7). The possibility of purposiveness without there necessarily being a purpose is based on the heautonomous principle of the power of judgment, which prescribes purposiveness subjectively in order to handle empirical diversity, without determining nature objectively (*CPJ* V 185.35–186.7). For RC, however, the distinction between reflecting and determining is blurred, and therefore purposiveness will entail purpose. When reason provides the concept of purpose, its use by the power of judgment (as purposiveness) is mistaken for the realist claim of objective purpose, and the antithesis will contradict the thesis.

How would a purely reflective interpretation of the pair of regulative maxims avoid this contradiction? It is clear that Kant takes the maxims to be complementary. Such "subjective" (i. e., non-constitutive) maxims are "disparate principles" and they "could well be united with each other" (*CPJ* V 391.11–3). The term 'disparate' has a technical use in logic that may be worth taking a look at. Reflection 4300 contrasts disparates and contradictories: "*A et non A sunt contradictorie opposita; A et B disparata*" (*Notes and Fragments* XVII 500.10). This suggests that disparate principles differ without being contradictorily opposed. Also in the antinomy of reason of the first *Critique*, disparate pairs of statements are contrasted to contradictory pairs. Here the example is a disjunctive pair of statements: "every body either smells good or smells not good" (*CPR* A 503/B 531). The predicates seem to be opposed contradictorily, so that one must be true and the other false when said of a body. But this presupposes that a body must have a smell. If a body may lack smell, none of the alternatives has to be true. Kant calls this an opposition '*per disparata*'; it is possible that both statements are false. These disparate opposites can thus be construed as contraries, and this serves to illustrate the concept of the world in the antinomy. Both the theses and the antitheses of the mathematical antinomies can be false (since they rest on an illegitimate subject term), and so what for transcendental realism is an 'analytical' (contradictory) opposition can be seen to be a 'dialectical' (contrary) opposition.[11]

---

[11] See *CPR* A 497/B 525–A 507/B 535, and the helpful account in McLaughlin (1990, 64–80). See also Horn (2010), who provides an Aristotelian square of opposition for singular terms. This fits

The handful of places where Kant employs the term 'disparate' are not always easy to interpret, but they have in common that they refer to concepts or statements that are not contradictorily opposed, whether it be pairs of terms dividing a concept non-exhaustively ("A et B"), or, as in *CPR*, pairs of contraries.[12] Now, the disparate principles in the "Dialectic" of the third *Critique* are not construed as contraries but rather in analogy to subcontraries. Both maxims are supposed to be valid, but not as in the dynamical antinomies of *CPR* by taking the subject term in different relations (as appearance and as thing in itself). In *CPJ* both thesis and antithesis are rather to be taken as valid in the empirical domain of material nature.

Already in §70, Kant claims that rightly understood, the maxims are not contradictory.[13] He interprets the thesis as saying that "I *should* always *reflect* on [material things] *in accordance with the principle* of mere mechanism" (*CPJ* V 387.31–3), and he takes the antithesis to tell us that the principle of final causes should be used "on the proper occasion, namely in the case of some forms of nature" (*CPJ* V 387.35–6). According to the present interpretation, this would be contradictory for RC, but not for a purely reflective employment of judgment. RC takes the rule for reflecting on material things to guarantee the success of the mechanistic procedure. When some objects (the organized beings) seem to be impossible to investigate without the use of teleological concepts, the mechanistic rule appears to be contradicted: its universality conflicts with the particular cases where it does not suffice. On the purely reflective stance, the mechanistic thesis is understood a little differently. It is now taken to say that in each case, we are to investigate the generation of a thing mechanistically. When that is not

---

Kant's antinomy in *CPR* better than the more well-known square for universal and particular propositions.

12 These passages are *CPR* A 503/B 531; *CPJ* V 391.13; *Notes and Fragments* XVI 619.6; *Notes and Fragments* XVII 289.14, 500.10, 570.20, 631.1; *Notes and Fragments* XVIII 362.2; *Notes and Fragments* XIX 252.12; *OP* XXI 72.30; *Lect. Log. Dohna* XXIV/2.1 761.11–3. The word also occurs twice in the *Anthropology*, presumably in a non-technical sense (*Anthropology* VII 183.31, 215.25). Excepting *CPR*, Kant's use of 'disparate' seems to agree with the two traditional definitions adduced by Runes (1942, 82): "Boethius defined disparate terms as those which are diverse yet not contradictory" and "Leibniz considered two concepts disparate 'if neither of the terms contains the other' that is to say if they are not in the relation of genus and species". In *CPR*, 'disparate' just means non-contradictory and therefore includes contraries.

13 It is noteworthy that Kant has "all *generation* of material things" in the thesis and "some *products* of material nature" in the antithesis (*CPJ* V 387.3 and 6, my italics). The different subject terms could seem to signal the disparateness of the maxims, in contrast to the constitutive principles that have a common subject term and are clearly contradictory. But it is hard to see how this could keep the maxims disparate unless it is denied that a product must be generated.

successful, because of properties in organized beings which we cannot grasp mechanistically, we should adopt the rule of teleology. This rule, the antithesis, at first seems to hinder the thesis' prescription of mechanism, but Kant points out that it provides a possibility to pursue the mechanistic mode of investigation on a new level. For when some things are conceived of as natural purposes, mechanistic research can continue in this framework. Reflection then "always remains open for any mechanical explanatory grounds" (*CPJ* V 389.14 – 5), and "the first maxim is not thereby suspended, rather one is required to pursue it as far as one can" (*CPJ* V 388.3 – 5). In summing up the outcome of the "Dialectic" in the beginning of the "Methodology of the Teleological Power of Judgment", Kant states that the judging of organized beings should be based "on some original organization, which uses that mechanism itself in order to produce other organized forms or to develop its own into new configurations" (*CPJ* V 418.13 – 7). I take this to imply that with regard to organisms, mechanism can and indeed must be pursued, but within a framework of a teleologically conceived original organization. Kant holds that the two maxims are able to complement each other, provided that mechanism is subordinated to teleology (see *CPJ* V 415.12 – 22, 417.27 – 35).

If RC engenders an antinomy, what, more precisely, is the dialectical presupposition at the root of the thesis and the antithesis? The possibility of switching from a dialectical contradiction to a pair of disparate maxims would seem to demand that there is some such presupposition that can be cancelled by critique. Allison (1991, 32) proposes that this assumption is "that purposiveness entails intelligent causality or design"; on the one hand, the mechanist "adopts a reductive strategy in dealing with apparent purposiveness", on the other hand, the "teleologist takes the same apparent purposiveness as evidence of a distinct type of causality". While this nicely fits the opposed dogmatic systems, I think Allison's proposed presupposition is a symptom of a deeper root of the antinomy, namely the neglect of the heautonomy of reflecting judgment. Ignoring heautonomy, reflection will be taken as determining nature objectively, rather than prescribing a principle for itself, and the two principles employed for the systematization of the particular (i.e., mechanism and teleology) will be considered to have ontological import.[14]

We should also examine to what extent the antinomy taken in the RC sense satisfies other conditions for Kantian antinomies. Are there enough parallels to

---

14 A weakness of this account is that it lacks the formal neatness of the antinomy of *CPR*. It is tempting (but methodologically dubious) to explain this as due to a loosening of the requirements for antinomies in the post-*CPR* period.

deem the antinomy of teleological judgment to be a genuine one, considering that Kant does not explicitly offer proofs of thesis and antithesis?[15] In *CPR*, each instance of the antinomy of reason has a pair of indirect proofs for thesis and antithesis.

Following Watkins (2009, 206–7), a promising suggestion for a proof for the thesis of the antinomy would seem to be to connect the need for mechanism to the task of unification of particular laws under general laws. Mechanism plays a necessary role for the formulation of such particular laws, since these will be constrained by the general laws of the understanding, in particular the causal law (see Watkins 2009, 206). Now Watkins, as an advocate of the recent type of interpretation, takes the antinomy's thesis to be the regulative maxim. On the present interpretation, it is rather the regulative maxim conceived of as having some constitutive force that occurs in the antinomy (i.e., RC), but also on this view Watkins' proposal is a plausible interpretation of what it means that the maxim is given to reflection from "the mere understanding *a priori*" (*CPJ* V 386.31).

As for the antithesis, its "proof" is presumably the occurrence in nature of things we cannot fully explain mechanically. It might seem hopeless to try to base an antinomy on such an a posteriori fact. But after all this is an antinomy of the power of judgment, not of reason (as explained in §69). Whereas the latter would have to arise from some a priori consideration, the antinomy of judgment in a sense starts from experience. To be sure, it concerns principles, but the employment of such principles is occasioned by the particular experience that the power of judgment has to reflect on. Given Kant's views on the impossibility for our understanding to grasp organized beings by purely mechanical reflection and the concomitant need for conceiving of them as natural purposes, the antithesis maxim "is suggested by particular experiences that bring reason into play" (*CPJ* V 386.31–3). The proof of the antithesis is thus the fact that organisms happen to be found in experience and that reflection can find nothing better than the concept of purpose (belonging to reason) to handle them.

---

15 It might be objected to my interpretation that Kant explicitly says that the constitutive principles are unprovable (*CPJ* V 387.21–2). But here I assume that if there are proofs, they are for the RC principles rather than for the constitutive principles as such. Another option is to follow Effertz (1994, 229) in taking the unprovability of the principles to express the final verdict of critical philosophy, which would not preclude there being *dialectical* proofs from the standpoint of transcendental illusion. Effertz construes the first paragraph of §78 (*CPJ* V 410.16–35) as the dialectical proof of the thesis, and the second one (*CPJ* V 411.1–29) as that of the antithesis (see Effertz 1994, 229–33).

Antinomies are rooted in transcendental illusion, and require a critique of reason for its solution. As regards the requirement that the antinomy should involve us in a natural and unavoidable dialectic, there are clear statements to that effect in the text, for instance in §§69–70 (*CPJ* V 386.4–10, 386.34–387.2) as well as in the description in §72 of how the existence of organized beings is taken as "a hint, as it were, given to us by nature, that we could by means of that concept of final causes step beyond nature and even connect it to the highest point in the series of causes" (*CPJ* V 390.13–6). In §§76–7, a critique of our understanding is offered (as expected in an antinomy), bringing up fundamental issues about transcendental idealism in connection to the radical contrast between human discursivity and a non-discursive understanding. As Watkins (2009, 211) notes, transcendental idealism also seems to be involved in diagnosing the failures of the dogmatic systems in §§72–3, which accords with the present interpretation since the systems are the ultimate consequences of taking regulative maxims for constitutive claims.

Finally, there is a further objection against the genuineness of this antinomy if construed as involving a slide from regulativity to constitutivity. McLaughlin (1989, 360) points out that such confusion is a simple mistake that Kant cleared up quickly in *CPR* (e.g., in A 666/B 694) without treating it as an antinomy. But though this may be true in some cases, the distinction between constitutive and regulative nonetheless plays an important role in the antinomy of the first *Critique* (as it does throughout its "Dialectic"). In *CPR* A 509/B 537 Kant states that the "constitutive cosmological principle" of the world as an absolute totality "as given in itself in the object" is based on a transcendental subreption, and is to be replaced by a regulative rule that "does not anticipate what is given in itself in the object". Such subreption can plausibly be seen as a central feature of the antinomy, since it is through a regulative conception of the rule of synthesis that the transcendental realist's antinomial concept of the world can be avoided and contradictions replaced by contraries or subcontraries. If that is the case in *CPR*, an interpretation of the antinomy of teleological judgment in terms of subreption should not be dismissed out of hand.

4 The Role of Discursivity and the Final Worry about Unification

As noted above, Kant claims that the maxims are not really contradictory. According to the recent type of interpretation, which locates the antinomy between the maxims of the reflecting power of judgment, it is the need to remove contradiction and give some sense to the claim of compatibility that is the task of §§74–8. This is plausible also on the present view, even though, as noted, Kant in the preceding paragraphs has already sketched how the maxims can co-exist. However, there may remain a worry that judging a thing with the disparate maxims must ultimately involve ascribing incompatible properties to it. Among the various aims of the very complex discussion in §§76–7 are the removal of this objection, as well as accounting for the transcendental ground of the maxims.[16]

CPJ §76 attempts to show how even as fundamental a distinction as that between possibility and actuality hangs on the discursivity of our understanding, i.e., its dependence on a supply of intuitions given to it from an independent receptivity. The method for showing this is to contrast our discursive understanding with a non-discursive or intuitive one. In §77 this contrastive procedure is used for identifying further traits of our understanding's peculiarity, relevant to the reflecting power of judgment's unification of particular laws under the general laws of the understanding as well as to the concept of a natural purpose. Kant conceives of an intuitive understanding that

> does not go from the universal to the particular and thus to the individual (through concepts), and for which that contingency of the agreement of nature in its products in accordance with *particular* laws for the understanding [...] is not encountered (*CPJ* V 406.26–9).

This addresses a central theme in the third *Critique*, the unification of particular laws and the general presupposition of a purposiveness of nature needed for us (but not for a non-discursive understanding). Guyer (2001, 272) points out that this merely reiterates what Kant has already said in the "Introduction" to *CPJ*, that we have to conceive of nature's specification in particular laws regulatively as if these were arranged in a system by a higher understanding. The question arises as to how this relates to natural purposes, since it seems that the study of any material things, not just organisms, would involve the unification of a manifold of laws. But there is a further characteristic of the intuitive understanding that bears more directly on the organism:

> we can also conceive of an understanding which, since it is not discursive like ours but is intuitive, goes from the *synthetically universal* (the intuition of a whole as such) to the particular, i.e., from the whole to the parts, in which, therefore, and in whose representation of the whole, there is no *contingency* in the combination of the parts, in order to make possible

---

16 The following account of §§76–7 is quite selective and programmatic. See also Huneman in this volume and Quarfood (2004, 177–91). I pass over §§74–5, that discuss the inexplicability of natural purposes and the use of purposiveness as a subjective principle for studying nature.

a determinate form of the whole, which is needed by our understanding, which must progress from the parts, as universally conceived grounds, to the different possible forms, as consequences, that can be subsumed under it. In accordance with the constitution of our understanding, by contrast, a real whole of nature is to be regarded only as the effect of the concurrent moving forces of the parts (*CPJ* V 407.19 – 30).

Two contrasts are drawn here between our discursive understanding and the intuitive understanding: that the latter does not need concepts to determine the particular, and also that the former (but not the latter) has to use a meristic[17] procedure to cognize a given whole. More specifically, this means that it has to conceive of complex things part by part, and that the parts are to be taken as causally producing the whole (whereas the intuitive understanding would have immediate cognition of the whole as such, without need of causality or other concepts). Now, these claims are quite difficult to analyze and justify.[18] In the present context I just want to suggest that the discursive understanding's need to think a "real whole of nature" (*CPJ* V 407.29) in terms of causality from parts to whole is relevant for explaining why grasping an organism should differ from grasping an ordinary thing. Whereas all attempts to systematize special laws depend on the general principle of purposiveness, organized beings present a further challenge for our understanding, since we have to assume that a natural whole is to be composed of causally efficient parts. This assumption works well in mechanistic investigation of the generation of ordinary things, but in the case of organized beings, their holistic character (as described in the "Analytic" of *CPJ*) precludes the explanatory *modus operandi* of the discursive understanding. Mechanism seems sufficient for the task of systematizing the manifold of particularity as prescribed by the principle of the purposiveness of nature,[19] except when it comes to organisms.

---

17 Taking this term in the sense explained in von Wright (1993, 45): "A view, according to which the characteristics of a whole have to be explained on the basis of features of its parts, is often called *meristic*, from the Greek word μέρος which means part".

18 For discussion of this, see Breitenbach (2008, 355–61), and Quarfood (2004, 182–8).

19 This connects to the question why mechanism is regulative even though causality is a constitutive *Grundsatz*. In the context of *CPJ*'s problem about the agreement of the special laws of nature under the general transcendental ones, the principle of the purposiveness of nature, according to which all laws are expected to harmonize in a system, is already presupposed. Mechanism, as discussed in the "Dialectic" of the third *Critique*, is subservient to reflecting judgment's principle of the purposiveness of nature. Organisms pose an additional problem, since their internal purposiveness is *both* unforeseen from the point of view of the principle of the purposiveness of nature and intractable for its subservient mechanistic maxim. This account of the regulativity of mechanism is neutral as regards the further question of how exactly to construe its relation to causality.

If this goes some way to explaining why the organism poses a special dialectical problem, it also helps to see how Kant wants to locate the maxims in a transcendental framework. The need for mechanism is now presented as a consequence of the peculiarity of the discursive understanding, its merism. But the appeal to final causes in the antithesis is also taken to be based on this same peculiarity. For when things present themselves which the meristic procedure of discursivity cannot handle, an alternative way to reach unification has to be found. The concept of purpose gives such an alternative, for it allows us to think of a whole as caused by the representation of this whole (see *CPJ* V 408.2–13). Such causation by means of a prior representation of the end is compatible with the character of discursivity, and thus the analogy with intentional production (in the concept of a natural purpose) can be used regulatively, even though it is no more than a "remote analogy" (*CPJ* V 375.20) for an organization of nature that properly speaking is not "analogous with any causality that we know" (*CPJ* V 375.6–7).

By framing the idea of a non-discursive understanding for which neither mechanistic nor teleological modes of explanation are required, conceptual space is made for the thought of the generation of things being independent of the peculiarities of the discursive understanding. This last point is developed in §78. Nature can be researched by means of both of the maxims, and there need be no real conflict between them despite their "apparent conflict" (*CPJ* V 413.11). Reflection on an alternative, non-discursive understanding helps to give sense to the distinction between appearance and thing in itself (supersensible ground), and the thought of a supersensible ground makes it possible to consider maxims as merely guiding research into nature without taking them as ontologically determining. Any worry that the maxims' opposing orientations might entail a contradiction is disarmed by the thought of their unifiability in a supersensible ground "which is neither the one nor the other (neither mechanism nor connection to an end)" (*CPJ* V 414.28–9), even though we can have no determinate cognition of it.

Watkins points out that the claim in §78 that the maxims can co-exist due to their common ground in the supersensible would need "considerable explanation and justification", but that Kant, instead of providing that, merely tells us that we cannot form any determinate concept of the supersensible ground, and that the unifiability of the maxims can thus not be explained (Watkins 2009, 219, referring to *CPJ* V 412.36–413.4). While Kant's claims certainly could use some clarification, I will here just suggest that their main thrust is metaphysically thinner than it might seem. The appeal to the supersensible is a reminder that we are dealing with experience, and that therefore any remaining worry concerning our difficulty of seeing how the maxims could co-exist is not to be solved by providing an ontology for empirical things, but must be put aside as something beyond experience that we are unable to form a determinate conception of. This strategy cannot work if the maxims flatly contradict each other, but Kant has already attempted to show that they are disparate, if adequately construed. It is important to stress that §78 returns to the antinomial conflict from a constitutive perspective. When Kant says that the two principles "cannot be united in one and the same thing in nature", he immediately adds that this applies to them qua "dogmatic and constitutive principles of insight into nature for the determining power of judgment" (*CPJ* V 411.30 – 3). I think this remaining demand for a dogmatic solution is what drives the discussion here. Kant goes so far as to admit that ultimately, if the maxims are to "subsist alongside one another in the consideration of nature", they must "flow from" a higher principle, namely the supersensible basis of nature (*CPJ* V 412.25 – 6). I take this to mean not that the maxims literally flow from this supersensible principle (whatever that would mean), but rather that the characteristic set of features of the organisms (i.e., its self-organizing capacity) that the maxims are used for reflecting upon (though neither of them seems to capture it adequately) must have its ground in the supersensible (which in a sense is trivial, since appearances in general have their ground in the supersensible). So it is not so much the pair of maxims that is grounded in the supersensible as the organism itself. Kant's account of the impossibility for us to determinately grasp the possibility of the organism resembles his view on the impossibility of explaining how freedom and causal determination can co-exist. Transcendental idealism attempts to make conceptual space for such co-existence, but cannot give any insight as to how it ultimately works.[20]

References

anonymous 1794, Ueber die Kantische Teleologie, *Philosophisches Archiv* 2 (3), 1–16.
Allison, Henry E. 1991, Kant's Antinomy of Teleological Judgment, *Southern Journal of Philosophy* 30, Supplement, 25–42.
Breitenbach, Angela 2008, Two Views on Nature: A Solution to Kant's Antinomy of Mechanism and Teleology, *British Journal for the History of Philosophy* 16, 351–69.
Charleton, Walter 1652, *The Darknes of Atheism Dispelled by the Light of Nature*, London: Lee, http://eebo.chadwyck.com/home [last visited: November 18, 2013].

---

20 I am grateful to the participants at the interenational symposion on Kant's theory of biology in Tübingen, and especially to Eric Watkins and Ina Goy for valuable comments. I also thank audiences in Stockholm and Uppsala.

Cudworth, Ralph 1678, *The True Intellectual System of the Universe*, London: Royston, http://eebo.chadwyck.com/home [last visited: November 18, 2013].

Effertz, Dirk 1994, *Kants Metaphysik: Welt und Freiheit*, Freiburg: Alber.

Guyer, Paul 2001b, Organisms and the Unity of Science, in: *Kant and the Sciences*, ed. by Eric Watkins, Oxford: Oxford University Press, 259–81.

Horn, Laurence R. 2010, Contradiction, in: *The Stanford Encyclopedia of Philosophy*, ed. by Edward Zalta, http://plato.stanford.edu/archives/win2010/entries/contradiction/[last visited: November 18, 2013].

Marc-Wogau, Konrad 1938, *Vier Studien zu Kants Kritik der Urteilskraft*, Uppsala: A.-B. Lundequistska Bokhandeln.

McFarland, John D. 1970, *Kant's Concept of Teleology*, Edinburgh: University of Edinburgh Press.

McLaughlin, Peter 1989b, What Is an Antinomy of Judgment?, in: *Proceedings: Sixth International Kant Congress*, ed. by Gerhard Funke and Thomas Seebohm, Washington, D.C.: University Press of America, vol. II/2, 357–67.

McLaughlin, Peter 1990, *Kant's Critique of Teleology in Biological Explanation. Antinomy and Teleology*, Lampeter: Edwin Mellen Press.

Quarfood, Marcel 2004, *Transcendental Idealism and the Organism*, Stockholm: Almqvist & Wiksell.

Runes, Dagobert D. (ed.) 1942, *The Dictionary of Philosophy*, New York: Philosophical Library.

Warda, Arthur 1922, *Immanuel Kants Bücher*, Berlin: Breslauer.

Watkins, Eric 2008, Die Antinomie der teleologischen Urteilskraft und Kants Ablehnung alternativer Teleologien (§§69–71 und §§72–73), in: *Immanuel Kant. Kritik der Urteilskraft*, ed. by Otfried Höffe, Berlin: Akademie Verlag, 241–58.

Watkins, Eric 2009, The Antinomy of Teleological Judgment, in: *Kant Yearbook 1*, ed. by Dietmar Heidemann, Berlin/New York: Walter de Gruyter, 197–221.

von Wright, Georg Henrik 1993, *The Tree of Knowledge and Other Essays*, Leiden: Brill.

Philippe Huneman
Purposiveness, Necessity, and Contingency

This paper investigates the relationship between Kant's analytics of biology, and the metaphysical conceptions of contingency and necessity in §§76–7 of the *CPJ*. In general, the connection between purposiveness and contingency is attested within the very project of a critique of a power of judgment. Indeed, the many particular, empirical laws of nature seem contingent regarding the universal laws of nature[1] (like Newton's laws of mechanics) that are explicated transcendentally by Kant in the first *Critique,* and that account for what happens in nature, conceived of as a set of objects under laws. 'Reflective judgment', as explored in the third *Critique,* deals with such empirical laws, and seeks to discover them when some phenomena are identified: the power of judgment therefore deals with the contingency proper to these laws. Since *purposiveness* designates the principle proper to the power of judgment (*CPJ* V 182–4), a deep connection between purposiveness and contingency will be the object of the transcendental investigation of the power of judgment in the *CPJ*.

The present paper addresses more specifically the connection between contingency and purposiveness as a concept proper to life sciences, namely "natural purposes"[2]. My *Leitfaden* here is the general relation Kant maintains between necessity, purposiveness, and contingency in arguing that purposiveness is the "lawfulness of the contingent as such" (*First Introduction* XX 217.28). I argue that Kant's understanding of purposiveness as a specific kind of lawlike contingency makes sense of several features of biological judgments, and must in turn be conceived of in relationship with the finiteness as discursiveness of our understanding, because natural purposiveness provides the reflection of such finiteness in a specific concept. This finally provides groundings for the solution of the antinomy of teleological judgment, and casts a light on Kant's last views of the metaphysical issues about contingency.

I first summarize the Kantian theory of organisms, in the context of the emerging life sciences, as it is articulated by the two criteria he provides for natural purposes. Then I show how this conception accounts for the biological understanding, by unraveling the connection between contingency and mechanism

---

1 "We must think of there being in nature, with regard to its merely empirical laws, a possibility of infinitely manifold empirical laws, which as far as our insight goes are nevertheless contingent" (*CPJ* V 183.23–6).
2 I translate *Naturzweck* by 'natural purpose', even though the standard translation is 'natural end'—in order to keep the homogeneity with the term 'purposiveness'.

which is proper to biological explanations. Third, I turn to §77 and consider Kant's elucidation of this concept of purposiveness as grounded in the structure of our finite understanding. The last two sections draw consequences from this analysis, concerning respectively the status of the modalities and the metaphysical issue of contingency.

1 Organisms are Natural Purposes

This section lays out Kant's theory of organism against the background of his metaphysical question about purposiveness and contingency. In the second section of the *CPJ*, Kant puts forth the thesis that "things, as natural ends, are organized beings" (*CPJ* V 372.13); this thesis implies, as many commentators have taken it (Zumbach 1984, 4–8; Lenoir 1982, 10–30; McLaughlin 1990, 1–6), that he is dealing with the specificity of biological judgment and argues that it requires a new concept of purposiveness.

Kant's project stems here from the recognition that the life sciences, as they emerged in the late eighteenth century, require a specific kind of intelligibility compared to that which he elucidated in the first *Critique*. He argues that all forms of such intelligibility required an ascription of purposiveness to living beings and living nature[3], and the critical project implies explicating this concept of purposiveness.

Although Kant's own research dealt with natural history, generation and heredity, with the theory of adaptive germs and dispositions triggered by environments,[4] Kant's examples are also from comparative anatomy (*CPJ* V 376.26–36) and physiology (*CPJ* V 418.18–419.9), even if the most important set of citations comes from embryology, and includes mentions of Blumenbach. Embryology requires, through the theory of germs, the idea of *type* reached by embryological processes; comparative anatomy, as developed later by Cuvier, integrates *functions* and structures in a way that functions make us understand the range of structures realizing them; and the Kantian idea of varieties and species in the two essays on human races requires *preadaptation* (germs are preadapted to all environments, and a given environment should trigger the proper germ). Hence, under the forms of adaptation, function and conservation of form, purposiveness is always present for Kant in biological judgments (Huneman 2006). In-

---

3 Adickes (1923, II 478) states that this general apperception was the novelty of the *CPJ*.
4 See the two essays on human races, *Races* (1775) and *Human Race* (1785), and in this volume Fisher, Goy, and Zuckert; see also Sloan (2002).

terestingly, these explanatory practices involved, at those times, many new kinds of forces, such as the *vis essentialis* in Wolff's embryology, or the *nisus formativus* or *Bildungstrieb* in Blumenbach's, and the vital properties like irritability or sensibility according to Haller and the Montpellier vitalists such as Bordeu or Barthez. The *CPJ* can be understood as an attempt to comprehend all those forces, and in general the *Naturlehre* discourse which makes use of it, in a synthetic framework, allowing researchers to justify or dismiss all such forces.[5]

§64 establishes a phenomenology of those entities which seem to require a teleological judgment: they display a specific relation between themselves and their components, at the three levels of parts, individual, and species (parts being respectively: organic parts; time-instances of the individual; and individuals as parts of the lineage). The tree "generates itself" (*CPJ* V 371.7) with regard to its species (i.e., reproduction); it "*generates* itself as an individual" (*CPJ* V 371.13); and it generates itself in such a way that "the preservation of the one is reciprocally dependent on the preservation of the other" (*CPJ* V 371.30 – 2).

§65 therefore conceives of the concept of natural purpose as necessary to make sense of those phenomena:

> Organized beings are thus the only ones in nature which, even if considered in themselves and without a relation to other things, must nevertheless be thought of as possible only as its ends (*CPJ* V 375.26 – 8).

If something is to be seen as a purpose, it means that its possibility is not conceivable except as by design, meaning that a concept should have made it possible (§61). It is because of a concept that this entity is what it is and how it is— that is the very meaning of purposiveness. Therefore, if something is to be thought as a purpose, it has to present a specific connection between its parts and itself as a whole (§64), according to which the parts are what they are due to a concept of the whole. Kantian purposiveness then means the interrelation between determination through a concept, and specific part-whole relationship. In English, this is captured by the ambiguity of the word 'designed', which means both being designed and having a design. Such a relationship between parts and wholes was exactly realized by the phenomenon presented by organisms in §64. Thereby, if something is to be understood as a (natural) purpose, it has to be an organism. But the concept of a natural purpose is a concept *we* need in order to make sense of a set of phenomena that are already physical objects, i.e., objects of nature, so it does not constitute the object of knowledge. But if we

---

5 On the difference between *Naturlehre* and *Naturwissenschaft*, see Sloan (2006), and Fisher (2007).

want to understand organisms, we have to think them in those terms. The concept of a natural purpose provides a rule for biological judgment as such but it is a "regulative principle".

Fundamentally, this view of purposiveness emphasizes the relationship between wholes and parts as definitional for purposiveness, rather than the classical view that purpose relates means and ends. Philosophically, this means that the very meaning of purposiveness exceeds the idea of instrumentality and design, which was the ordinary sense of purposiveness and according to which one can see organisms as somehow engineered, traits of animals as organs having utilities—in the tradition of Galenic physiology. This enables Kant to develop a notion of purposiveness throughout the book that allows for "purposiveness without an end [*Zweckmässigkeit ohne Zweck*]" (*CPJ* V 226.27), a notion which would have no place if purposiveness were wholly determined by intention, instrumentality, and utility.

Teleology in this sense is opposed to the explanatory stance proper to the understanding, which is mechanism. Indeed, mechanism explains wholes from parts, as McLaughlin (1990, 152–4) has emphasized. Kantian teleology therefore works the other way round: from wholes to parts. The way living beings grow had been previously contrasted with the growth of brute matter, by Bourguet (1762, 71) in his *Lettres philosophiques:* it is an *intussusception* opposed to growth by aggregation, because here the added parts are made of the same stuff as the whole—a distinction acknowledged by Kant in §64 (*CPJ* V 371.17–9). Clearly, the mechanical way of thinking could not account for intussusception. Claiming that organisms are natural purposes implies that the mechanism of nature is not enough to account for them and their properties.

To be precise, natural purpose is a complex concept because it requires two criteria. The first one concerns the determination of parts by wholes, which I call the *design* criterion; the second is an *epigeneticity* criterion. The design criterion can indeed be satisfied by any purposive individual system, be it an organism or a machine. To be a *natural* purpose, a system should satisfy another requirement: it has to be such that the parts are causes and effects of themselves as well as the whole, according to the whole (or else they are designed by another entity, and we are in the domain of technology, instrumentality, etc.). It means that parts produce themselves according to the whole, something which is attested by histology and contemporary cell physiology:[6]

---

6 See Huneman (2007b) on the necessary relationship between epigenetic criterion, and regulative character of the principle of teleology.

For a body, therefore, which is to be judged as a natural end in itself and in accordance with its internal possibility, it is required that its parts reciprocally produce [*hervorbringen*] each other, as far as both their form and their combination is concerned, and thus produce a whole out of their own causality, the concept of which, conversely, is in turn the cause (in a being that would possess the causality according to concepts appropriate for such a product) of it in accordance with a principle [...].

In such a product of nature each part is conceived as if it exists only *through* all the others, thus as if existing *for the sake of the others* and *on account* of the whole, i. e., as an instrument (organ), which is, however, not sufficient (for it could also be an instrument of art, and thus represented as possible at all only as an end); rather it must be thought of as an organ that *produces* the other parts (consequently each produces the others reciprocally), which cannot be the case in any instrument of art, but only of nature, which provides all the matter for instruments (even those of art): only then and on that account can such a product, as an *organized* and *self-organizing* being, be called a *natural end* (*CPJ* V 373.26 – 374.8).

Considering the second criterion, one could say that even if machines are organized, organisms—*organisierte Wesen*—are self-organizing, in the sense that they intrinsically have this epigenetic character indicated by the criterion. This epigenetic character of organisms allows Kant to distinguish between formative and motive forces, the former applying to organisms and the latter to all physical bodies. Because the concept of natural purpose is a regulative one, those forces proper to organic entities are not on a par with motive forces, they are postulated to make sense of organisms.[7] To this extent, those forces name a rather unknown kind of causation in nature, because we have no analogue for it (*CPJ* V 375.10 – 6).[8] The analytics of biological judgment here rightly allows Kant to criticize the proliferation of weird forces in the *Naturlehre*, as it was the initial problem.

The epigeneticity criterion, however, can have several readings. If parts cause the form—i. e., the structure, internal organization and arrangement—of the other parts, for example having a muscle here will mean having the antagonistic muscle there, this is a sense of causation that amounts to mere counterfactual dependence. If 'causing' means producing, this is a rather different sense of causation. Actually, the metaphysician Hall (2004) argues that those two general senses of causation are irreducible, so that the concept of causation is ulti-

---

7 "An organized being is thus not a mere machine, for that has only a *motive* power [*bewegende Kraft*], while the organized being possesses in itself a *formative* power [*bildende Kraft*], and indeed one that it communicates to the matter, which does not have it (it organizes the latter): thus it has a self-propagating [*fortpflanzende*] formative power, which cannot be explained through the capacity for movement alone [*Bewegungsvermögen*] (that is, mechanism)" (*CPJ* V 374. 21– 5).

8 See Breitenbach (this volume), about analogical judgment in this case.

mately equivocal. This means that the second criterion—definitive of a natural purpose, and therefore essential in the teleological explanatory stance—may pertain to two different interpretations of causation. In this sense, physiology—where the issue is rather *dependence* between parts—will need the first reading of the criterion, whereas embryology will need the second one. Those two disciplines therefore instantiate the teleological standpoint in different ways and will therefore be related in different ways to the mechanistic stance. Yet in the weak, counterfactual reading of causation, the second criterion seems to conflate with the first: after all, being dependent upon other parts and the wholes as to its form and constitution reduces to being integrated into a general design. Thereby, it is only with embryology that the distinction between purposes and natural purposes really holds. This is not surprising, if we consider that *purposiveness* can be distinguished into several concepts, including function and embryological type: physiology uses the concept of function, and it is clearly not restricted to biological entities, because it is also a concept in engineering. So finally, the fact that the second criterion of natural purposiveness is—among biological disciplines—only required by embryology amounts to the fact that, among the purposive notions, function indeed concerns both organisms and artifacts.

After having explicated Kant's view that organisms are natural purposes, I turn to the implications of his approach to teleology, and to the way it involves the notion of contingency.

2 Teleology and Mechanisms: the Lawfulness of Contingent as Such

What is purposive cannot be sufficiently explained through the mechanistic laws of nature—which means that there is an essential feature of the object, as it interests us, which cannot be accounted for by these laws. More precisely, one cannot explain why such a feature is present rather than not, through the mere knowledge of the laws. This makes for an involvement of necessity and contingency within the concept of purposiveness.

For example, regarding the laws of physics, and as a result of an ontogenetic process, a normal chick and a "monster" or a non-viable chick are on a par: the laws of physics (related to the initial conditions) account for their production, and from this viewpoint there is nothing special concerning the normal chick. Hence the very notion of normality, of embryological type, the idea that the embryological process should lead to producing a chick, does not make sense from the viewpoint of the laws of matter as such. In this sense, the normal vs. abnormal[9], or the embryological types, are contingent regarding the universal laws of nature. More precisely, those categories indicate a certain intelligibility of such contingent entities, structures, and processes (regarding the laws of nature). That is exactly why Kant conceives of purposiveness as the "lawfulness of the contingent as such" (*First Introduction* XX 217.28). There is nothing lawlike, i.e., nothing necessary regarding the laws of physics, in a chicken (*qua* chicken)[10] producing a baby chick; an abnormal development would be just as necessary. However, this contingent embryological process occurs according to certain regularities, which are exactly what embryology tries to capture. It is a new level of necessity, intrinsic to some contingent processes and entities.

Thus, it is clear that issues about biological purposiveness involve metaphysical issues about contingency and necessity. Therefore *two* questions arise. First, how are those necessities—necessity in general and the necessity of contingency as such—related? Which means both: how can a teleological principle of judgments be *compatible* with ordinary principles of science, which are non-purposive? And how can they be *articulated* in scientific practice (from §79 on, and then in the whole "Methodology")? This issue is addressed in the remainder of this section. Second, how can these ideas of explanation and necessity impact upon the general concepts of necessity and contingency, especially considered in the §§76–7? The last section will tackle this issue.

I do not consider extensively the question of compatibility, i.e., the "Antinomy of Teleological Judgment", which is treated by Quarfood in this volume. It is sufficient to note that its solution lies eventually in the idea that both maxims are regulative, but not on the same domain[11]—and that they can be conceived of as being the same in the supersensible, since all of them concern only appearances.

> [T]he common principle of the mechanical derivation on the one side and the teleological on the other is the *supersensible*, on which we must base nature as phenomenon (*CPJ* V 412.33–6).

This commonness implies that we can presume that

> we may confidently research the laws of nature (as far as the possibility of their product is cognizable from one or the other principle of our understanding) in accordance with both

---

9 Ginsborg (2001) emphasized the role of normativity as proper to life sciences and ontology of living being, in Kant's texts.

10 Seeing the chicken's embryogenesis *as a chicken* development, hence aiming at producing a chick, requires assuming purposiveness.

11 See the solution here proposed by Quarfood and his reconstruction of the interpretative debates.

of these principles, without being troubled by the apparent conflict between the two principles for judging this product (*CPJ* V 413.9 – 12).

To be conceivable, such a solution requires a deeper understanding of necessity
and contingency, undertaken below.

Given that the two principles seem compatible, the "Methodology" (from
§79 on) intends to justify their *articulation*. §§80 and 81 for example consider
the cases of the archaeology of nature, and of the theories of generation,[12]
and show how they are complementary. But the complementarity Kant imagines
is already sketched in §66 when he explains that the mechanical explanations of
an organic process such as hair growth being completed, something would still
be missing for our understanding of such growth:

> It might always be possible that in, e. g., an animal body, many parts could be conceived as
> consequences of merely mechanical laws (such as skin, hair, and bones). Yet the cause that
> provides the appropriate material, modifies it, forms it, and deposits it in its appropriate
> place must always be judged teleologically, so that everything in it must be considered
> as organized, and everything is also, in a certain relation to the thing itself, an organ in
> turn (*CPJ* V 377.17 – 23).

In contemporary terms, mechanism and teleology can be conceived of as two
complementary explanatory stances. The first one uncovers processes at work
in all of nature, and therefore it is not proper to biology. However, when facing
a particular process taking place in an organism, it does not answer questions
such as: why is this process here? Typically, knowing how hair grows does not
tell us many things about its function (it tells us some things, because the process excludes some putative functions). Therefore, we need another explanatory
stance, which asks why hair is here, and this stance gives rise to many functional
explanations. This second explanatory stance clearly does not concern *non organized* entities. As Kant would have said, non living entities are explained by
laws of nature in general, so there is no contingency which would lead us to
ask: why is it the case that X rather than not?

Against doubts raised against Kant by Förster,[13] these remarks support the
view that indeed Kant's conception of purposiveness explicates it as the lawfulness of the contingent as such, accordingly to the formulation of the "First Introduction". Because both stances are explanatory, they uncover some lawfulness;

---

12 See Zammito (2007) and Fisher (2007) on generation and §81, and Huneman (2006) on §80
and the archaeology of nature.
13 Förster (2008, 267) pointed out, that "the description by Kant of the problem as a lawlikeness
of the contingent is hardly helpful and the real problem arose before" (my translation).

and because the teleological one is different from mechanism, it concerns a lawfulness which is necessarily contingent regarding such mechanism. Understood as a distinction of explanatory stances that instantiate grades of lawfulness, Kant's view of biological judgment has indeed captured a hallmark of biological sciences, as is attested to by later reflections about the specificity of this field. Mayr (1961), aiming at capturing what makes biology different from physics, argued that there are two kinds of biologies, with completing explanatory stances for the same phenomena: the biology of proximate causes, unraveling the development and functioning of organisms; and the biology of ultimate causes, namely evolutionary disciplines, which investigates (by considering populations of organisms) why those proximate processes and developments exist. A complete explanation needs both stances, but partial questions (like the function of some behaviors, or the developmental pathways) can be answered through a specific stance. In the Darwinian framework (unknown to Kant) this exemplifies a bipartition, proper to life sciences, of complementary explanatory stances.[14]

Finally, after having characterized how judging organisms involves the lawfulness of the contingent as such, the "Dialectic" and the "Methodology" sections of the *CPJ* display the cohesiveness of natural necessity (correlated with scientific explanation), articulating various degrees and levels of contingency within it. Yet the question remains of what the concepts of necessity and contingency themselves mean, and how their meaning is affected when one acknowledges purposiveness—necessity within the contingent—as a notion necessary for biological judgments. The last two sections address this metaphysical issue.

3 Why Purposiveness? Its Transcendental Genealogy

The last question raised by the analytics of biology is: if the concept of purposiveness allows us to experience some beings that we contingently encounter, thereby revealing a concept which is not, however, a condition of experience (i.e., in the 'transcendental subject'), why is such a concept available to us since it is neither from observation, nor given in the transcendental frame of experience? The latter question is handled in §§76–7, just before the conclusion of the "Dialectic".

---

14 Another prominent example is Bernard's (1878, 359–62) theory of the "prescription of evolution", which directs all physiological processes of metabolism, and which is ultimately physico-chemical, but is not graspable through experimental physiology.

Kant's answer relies on the distinction between two kinds of understandings: the discursive understanding, which is ours and cannot derive the particular from the general,[15] and the intuitive understanding, which could cognize the particular together with the general. It is a contingent fact that our power of thought has the structure it has (since an intuitive understanding is conceivable):

> What is at issue here is thus the relation of *our* understanding to the power of judgment, the fact, namely, that we have to seek a certain contingency in the constitution of our understanding in order to notice this as a special character of our understanding in distinction from other possible ones (*CPJ* V 406.7–10).

This statement entails several consequences. First, our understanding needs the faculty of reflective judgment as an ability to find the rule for the case, because it cannot derive the particular from knowledge of the universal, in other words all cases from the rules. Where our understanding has to go from the general to the particular through more and more particularized concepts, each of them being schematized in an intuition as an exemplar—(vertebrate/*Canis*/dog), the intuitive understanding would produce the intuition together with the universal. For such understanding, there would be no gap between universal laws and particular empirical laws of nature, a gap which precisely defines the task of the power of judgment (*CPJ* V 180.1–6), and prescribes the territory of the *CPJ*.

Second, why is the discursivity of the understanding relevant for what one could call the transcendental genealogy of the concept of purposiveness and its link to mechanism? The reason is that discursivity means that we have to proceed from the rules governing the parts (which are general rules) to the most particular behavior of the whole, as an individual system. Discursivity indeed could mean that the understanding has to use some rules, or that the parts are given to it rather than the whole. Kant focuses on this second characterization, even though he thinks that at some point they go together. The discursivity of our understanding entails some contingency in that it "must progress from the parts, as universally conceived grounds, to the different possible forms, as consequences, that can be subsumed under it" (*CPJ* V 407.26–8),[16] this progression not being governed by a universal law. It is the parts, indeed, that are subject to the general

---

15 "[F]or through the universal of *our* (human) understanding the particular is not determined, and it is contingent in how many different ways distinct things that nevertheless coincide in a common characteristic can be presented to our perception" (*CPJ* V 406.13–6). Strawson (1995, 20) interprets concepts as general and intuitions as particulars. In this sense the discursiveness of our understanding means exactly the separation between concepts and intuitions.
16 On this point, see McLaughlin, in this volume.

laws; our knowledge of laws concerns parts—think of Newtonian mechanics, which understands motion from the motion of material points—which allows a system to be understood by starting from the understanding of the lawlike behaviors of its parts[17].

> In accordance with the constitution of our understanding, by contrast, a real whole of nature is to be regarded only as the effect of the concurrent moving forces of the parts (*CPJ* V 407.28 – 30).

Therefore mechanism constitutes the explanatory stance which is natural to our discursive understanding. The problem for a transcendental analysis is the possibility of the *other* stance, which is triggered by our experience of organisms, those entities that, in order to be conceived of as organisms, have to be purposively understood.

Since those entities cannot be understood by our discursive understanding— our understanding is one kind of understanding, but does not exhaust what understandings in general can be—we have to think of them *as grasped by another understanding than ours*, that is, by an understanding which goes from wholes to parts. But this grasp is, in its turn, *conceived* of by our understanding. Therefore, what appears eventually is an entity whose parts are somehow understood as being the object of a holistic perception through an intuitive understanding; yet this perception is at the same time a concept, given that the intuitive understanding is not subject to the separation between intuitions and concepts. Therefore, the parts in the end appear as being prescribed in a concept.

> Thus if we would not represent the possibility of the whole as depending upon the parts, as is appropriate for our discursive understanding, but would rather, after the model of the intuitive (archetypical) understanding, represent the possibility of the parts (as far as both their constitution and their combination is concerned) as depending upon the whole, then, given the very same special characteristic of our understanding, this cannot come about by the whole being the ground of the possibility of the connection of the parts (which would be a contradiction in the discursive kind of cognition), but only by the representation of a whole containing the ground of the possibility of its form and of the connection of parts that belongs to that (*CPJ* V 407.30 – 408.2).

In other words, the intuitive understanding can represent the derivation of the parts, ruled by laws, from the whole, but we can only think of this derivation

---

17 Up to a point, these considerations on the mechanistic stance concern what we could call now non-linear dynamics and the science of dynamical systems and their emergent properties. It is a fact that many uses of Kant's theory of biology in contemporary contexts concern those kinds of approaches to organisms, e.g., Kauffmann (1993, 18, 543).

as captured by an intuitive understanding which is not ours. This means that for us, such derivation goes from the *representation* of the whole to the forms and links of parts. We cannot think of deriving parts from the whole, except by positing the concept of the whole as the source of the whole. But this concept is our grasp on the knowledge the intuitive understanding would have (since it has itself no concepts). The concept of the whole, the representation, which makes this experience of organized wholes something purposive (because to be a purpose is to be derived from a concept [*CPJ* V 408.2–6]) comes from this sort of duplication of the understanding: it is a thought by the *intuitive* understanding conceived of by *our* discursive understanding. The whole as captured by an intuitive understanding is, in our discursive understanding, represented as the concept of the whole prescribing the parts and their connections. This means that, for us, the unconceivable causation[18] of some products of nature is explicated by a causality going from the concept of the whole to the parts.

Interestingly, there are two characterizations of purposiveness which Kant uses throughout the book: first, a purpose is something whose possibility of the production is given in a concept (e.g., in §61) and second, a purpose is an entity displaying a specific relation between its parts and itself as a whole (e.g., in §65)—a duality that is characterized in the double meaning of 'design'. Here their duality becomes transcendentally grounded: it is the difference between intuitive and discursive understandings (as a difference between how parts and whole are related in their comprehension) and the reflection here described that makes it the case that a purpose involves the part-whole relationship and at the same time is a matter of having a concept at its root.

Intrinsic to the concept of purposiveness is reflection of the finiteness (i.e., the separation of intuitions and concepts) within the construction of the concept. In this sense, judgments of purposiveness are reflective because at the occasion of the encounter with some specific entities in the world they reflect the contingent structure of our power of knowledge, namely its finiteness. It is through purposiveness that our thought relates itself to the contingency of its finite structure to the point that a specific concept instantiates such relation. This relation explains finally why purposiveness is at the heart of reflexivity in our judgment.

Given that contingency pertains to the finite structure of thought, and that purposiveness is the lawfulness of contingency as such, what are the connec-

---

18 "Strictly speaking, the organization of nature is therefore not analogous with any causality that we know" (*CPJ* V 375.5–7).

tions between contingency and necessity uncovered by this last transcendental analysis of purposiveness? This is the object of the last section of this paper.

4 Contingency, Purposiveness, and Critique

4.1 Possibility, Existence, and Their Non-Difference

Such a question requires us first to consider Kantian views on modalities. In our power of knowledge, concepts (possible, universal, terms underdetermining their objects) are separated from intuitions, which give us the existence of objects. Therefore, for a discursive understanding, the contingency of the particular regarding the general also entails the separation between the possible and the extant.

Possibility and existence are the two main modalities. In Kant's view, modalities concern the relationship of the knowing subject to its object[19]; they do not characterize the constitution of the object, but the mode through which the subject relates to it. To this extent, all categories of modality in Kant are reinterpreted in terms of the structures of our power of knowledge: to be objectively possible is to be related to the possibility of experience; to exist means to be involved in an actual experience.

To this extent, it is not only that intuitions provide access to existing objects, but, reciprocally (and this is the transcendental thesis), the meaning of 'existence' or 'actuality' is given by the intuition; and correlatively, the meaning of 'possibility' is given by concepts as likely to have an objective validity (through the transcendental conditions of experience).

> It is absolutely necessary for the human understanding to distinguish between the possibility and the actuality of things. The reason for this lies in the subject and the nature of its cognitive faculties. For if two entirely heterogeneous elements were not required for the exercise of these faculties, understanding for concepts and sensible intuition for objects corresponding to them, then there would be no such distinction (between the possible and the actual). That is, if our understanding were intuitive, it would have no objects except what is actual. Concepts (which pertain merely to the possibility of an object) and sensible intuitions (which merely give us something, without thereby allowing us to cognize it as an object) would both disappear. Now, however, all of our distinction between the merely possible and the actual rests on the fact that the former signifies only the position of the

---

19 "The relationship (of an object) to perception is its existence; the relationship to mere thought, its possibility; the relationship to thought in as much it determines existence: necessity" (*Notes and Fragments* XVII 500.3 – 6; my translation).

representation of a thing with respect to our concept and, in general, our faculty for thinking, while the latter signifies the positing of the thing in itself (apart from this concept) (*CPJ* V 401.31–402.10).

But because our faculty of knowledge, by conceiving of the contingency of its structure, conceives of another understanding having another structure, we conceive of a power of knowledge for which possibility and existence are not different, namely, for which possibility and existence have no meaning.[20] In the *CPR*, conceiving of this understanding was required because of the principle of complete determination, which compels us to think of a ground for all determinations ("Transcendental Dialectic", concerning the "ideal of reason").[21] But here, it is from the background of the difference between possibility and existence that this instance arises: reason demands that we "assume some sort of thing (the original ground) as existing absolutely necessarily, in which possibility and actuality can no longer be distinguished at all" (*CPJ* V 402.22–4).

In §76 Kant thereby acknowledges the fact that an instance of non-difference between both possibility and existence emerges for our understanding, in exactly the same logical move as the move through which purposiveness emerges from the reflection of finiteness. But by definition such an instance is empty, because we cannot either know it through a concept, or intuit it; it is "an unattainable problematic concept for the human understanding" (*CPJ* V 402.32).

Likewise, as far as the case before us is concerned, it may be conceded that "we would find no distinction between a natural mechanism and a technique of nature, i.e., a connection to ends in it, if our understanding were not of the sort that must go from the universal to the particular" (*CPJ* V 404.18–21). If there is no difference between possibility and existence, there is no sense talking of contingency and lawfulness of contingency as such (i.e., purposiveness), etc.—then there is no difference between mechanism and teleology. Yet the solution of the "Dialectic" (in §78), as seen earlier, will be that the compatibility between mechanism and teleology rests on their possible equation in the supersensible substrate of reality. Here, we see that it is in the nature of our understanding to make conceivable a non-differentiation between mechanism and teleology. To this extent, the possibility of a solution of the "Dialectic" is demonstrated in the transcendental 'genealogy' of the meaning of 'purposiveness' (as an inquiry on the origins of the concept of purposiveness within the discursiveness of our understanding) within the structures of our faculty of knowledge, because

---

20 Since if 'possible' means 'existent', it means something at least completely different than what 'possible' usually means, which entails that the possible is larger than the existent.
21 See comments in Allison (^{2}2004, 397–405).

such a genealogy establishes the conceivability of an instance of non-difference. That's why such a transcendental genealogy in its relation to concepts of possibility and existence was required for solving the "Antinomy".

4.2 The Circle of Contingency and the Nature of Criticism

The reflection of finiteness and the emergence of the meaning of 'purposiveness', as well as the correlative elucidation of the difference between possibility and existence, have major consequences for the very concept of contingency, which was metaphysically involved in the conception of purposiveness in terms of lawfulness of the contingent as such. I conclude this study by explicating such consequences.

First, what is contingency? It is the possibility of not being, or of being different. Like all modalities, according to transcendental philosophy (*CPR*, "Anticipations of possible experience"), this concerns the relationship of our knowledge to its objects rather than its objects themselves. In the *CPR* Kant distinguishes the pure and the empirical concepts of contingency. The pure concept of contingency is the possibility of being different; however, to apply it to phenomena, namely things in time, raises the difficulty of being different at two different moments. The possibility of the opposite of X in the same time t is something not testable, not provable by the occurrence of the opposite at later time t'. (What contingency concerns is the possibility of not-X at t, not at t', which is the only testable fact.) Therefore what contingency means for phenomena is that one state came after another state, and therefore could not have been possible by itself (and therefore is not necessary). Empirical contingency is attested to by change in time (*CPR* A 460/B 488).

Metaphysics in the Leibnizian style asked questions about the contingency of what is (or, in other words, the contingency of laws of nature).[22] In the first *Critique* modalities changed their status, so asking this question no longer makes sense: contingency or necessity are not properties of phenomenal being *per se*, but properties of the knowledge of those beings; empirical contingency, Kant shows, simply means that the being belongs to a temporal (hence causal) series—a requisite which makes no sense regarding the whole of being. The fourth antinomy of the *CPR* demonstrated thereby the lack of meaning of the metaphysical question about the contingency of what is.

------

22 On this issue see Tonelli (1959).

In addition, there exists a radical contingency, to which the heart of the philosophical questioning turns in §77 of the *CPJ*, namely the contingency *within our understanding* between particular and universal, which in turn expresses the contingency of the structure of our thinking. Where the first *Critique* condemned interrogations about the contingency of everything, i.e., of all that is, the third *Critique* now recasts this question. Particular laws of nature, and, moreover, organisms, display a specific kind of contingency—and in order to be understood they require some necessity proper to them. Purposiveness is the concept which makes sense of those various kinds of contingencies, to the extent that they involve some lawfulness. The consequence is that Kant's analysis provides a transcendental elucidation for life sciences and natural history. But this concept of purposiveness in its turn is traced back to the contingency that affects our faculty of knowledge. As a general consequence, the metaphysical question of deciding on the contingency of all that is, is dismissed, and replaced by the transcendental question of making sense of the necessity of the contingent, as it is investigated by our best science. This last question then gives rise to the question of how the contingent structures of our thinking shape some meanings, through which we can make sense of the contingent as lawlike—first of all, the signification 'purposiveness'. Do philosophical questions of this kind now ultimately define *critique?*

5 Conclusion

In this paper I have considered the metaphysical aspects of the third *Critique*, which concludes a lengthy questioning about contingency and necessity, once they have been understood in a transcendental fashion. I related it to the *CPJ*, as an analytics of the emerging life sciences. My argument is that §§76–7 are located precisely at this juncture because they ask why we have the concept of purposiveness, namely a concept construed as likely to make sense of all phenomena in living world. With purposiveness being the necessity of the contingent as such, answering the question of its transcendental origin casts a light upon contingency and necessity in late critical philosophy.

Contingency as a question means precisely the transcendental issue of the possibility of our knowledge of whatever seems contingent, i.e., particular laws of nature and the determining features of organisms (normativity, functions, adaptation, types). Teleology appears as the explanatory stance that provides a scheme for the contingent as lawlike—a stance whose possibility is ultimately grounded in the reflection of the finiteness of our power of judgment (§77). The antinomy of teleological judgment made reference ultimately to the supersensible ground of appearances in order to be resolved, because here mechanism was no longer constitutive and could then be equated with teleology. The elucidation of the concept of purposiveness led in paragraphs §76 and 77 to the final justification of this appeal by showing that an instance of non-difference between mechanism and teleology intrinsically belongs to the structure of such concept.

References

Adickes, Erich 1923, *Kant als Naturforscher*, Berlin: Walter de Gruyter.
Allison, Henry 1981/²2004, *Kant's Transcendental Idealism*, New Haven: Yale University Press.
Bernard, Claude 1878, *Leçons sur les phénomènes de la vie communs aux végétaux et aux animaux*, Paris: Baillière.
Bourguet, Louis 1762, *Lettres philosophiques sur la formation des sels et de crystaux et sur la génération et le méchanisme organique des plantes et des animaux*, Paris: François L'Honoré.
Fisher, Mark 2007, Kant's Explanatory Natural History: Generation and Classification of Organisms in Kant's Natural Philosophy, in: *Understanding Purpose. Kant and the Philosophy of Biology*, ed. by Philippe Huneman, Rochester: Rochester University Press, 101–21.
Förster, Eckhart 2008, Von der Eigentümlichkeit unseres Verstandes in Ansehung der Urteilskraft (§§74–78), in: *Immanuel Kant. Kritik der Urteilskraft*, ed. by Otfried Höffe, Berlin: Akademie Verlag, 259–88.
Ginsborg, Hannah 2001, Kant on Understanding Organisms as Natural Purposes, in: *Kant and the Sciences*, ed. by Eric Watkins, Oxford: Oxford University Press, 231–58.
Ginsborg, Hannah 2004, Two Kinds of Mechanical Inexplicability in Kant and Aristotle, *Journal of the History of Philosophy* 42, 33–65.
Hall, Ned 2004, Two Concepts of Cause, in: *Causation and Counterfactuals*, ed. by John Collins, Ned Hall, and L.A. Paul, Cambridge (MA): MIT Press, 225–76.
Huneman, Philippe 2006, From Comparative Anatomy to the "Adventures of Reason", *Studies in History and Philosophy of Biological and Biomedical Sciences* 37 (4), 649–74.
Huneman, Philippe (ed.) 2007a, *Understanding Purpose: Kant and the Philosophy of Biology*, Rochester: University of Rochester Press.
Huneman, Philippe 2007b, Reflexive Judgment and Embryology: Kant's Shift Between the First and the Third Critique, in: *Understanding Purpose: Kant and the Philosophy of Biology*, ed. by Philippe Huneman, Rochester: University of Rochester Press, 75–100.
Huneman, Philippe 2008, *Métaphysique et biologie: Kant et la constitution du concept d'organisme*, Paris: Kimé.
Kauffmann, Stuart 1993, *Self-organisation and the Origins of Order*, New York: Oxford University Press.
Lenoir, Timothy 1982, *The Strategy of Life: Teleology and Mechanism in Nineteenth-Century German Biology*, Dordrecht: Reidel.
Mayr, Ernst 1961, Cause and Effect in Biology, *Science,* 134, 1501–6.

McLaughlin, Peter 1990, *Kant's Critique of Teleology in Biological Explanation. Antinomy and Teleology*, Lampeter: Edwin Mellen Press.

McLaughlin, Peter 2001, *What Functions Explain?* Cambridge: Cambridge University Press.

Sloan, Phillip 2002, Preforming the Categories: Eighteenth-Century Generation Theory and the Biological Roots of Kant's A Priori, *Journal of the History of Philosophy* 40 (2), 229–53.

Sloan, Phillip 2006, Kant on the History of Nature: The Ambiguous Heritage of the Critical Philosophy for Natural History, *Studies in History and Philosophy of Biology and Biomedical Sciences* 37 (4), 627–48.

Strawson, Peter F. 1995, *The Bounds of Sense. An Essay on Kant's* Critique of Pure Reason, London: Routledge.

Tonelli, Giorgio 1959, La nécessité des lois de la nature au 18ème siècle et chez Kant en 1762, *Revue d'histoire des sciences* 12, 225–41.

Zammito, John 2007, Kant's Persistent Ambivalence Toward Epigenesis, 1764–90, in: *Understanding Purpose. Kant and the Philosophy of Biology*, ed. by Philippe Huneman, Rochester: University of Rochester Press, 51–74.

Zumbach, Clark 1984, *The Transcendent Science. Kant's Conception of Biological Methodology*, Den Haag: Martinus Nijhoff.

Ina Goy

Kant's Theory of Biology and the Argument from Design

In this paper[1], I treat the question of whether and in what regard Kant's theory of biology contains a version of the argument from design, which is the question of whether Kant considers the purposive order of organized nature as a physico-theological proof for the existence of God, and in turn, the existence of God as the supersensible ground for the teleological order of organized nature. As an introduction to the topic, I name traditional examples of the argument from design (section 1). I then outline Kant's changing attitude towards the argument in his *Theory of Heavens*, his *Argument* essay and the three *Critiques*, high-lighting Kant's return to the argument from design in the *CPJ* after examining and rejecting it in his earlier writings. I elaborate in detail Kant's different uses of the argument in the "Critique of the Teleological Power of Judgment" (section 2). In section 3, I develop a consistent reading of Kant's references to the physicotheological proof in the "Critique of the Teleological Power of Judg-ment": in the "Analytic" he develops a teleological account of nature that makes no use of the argument from design but is consistent with it. In the "Dia-lectic" and the "Methodology", however, Kant discusses more ambitious system-atic questions: the unity of the theoretical laws of nature and the unity of the nat-ural and the supernatural moral laws. These are questions that require an explicit reference to the argument from design (section 3).

I The Argument from Design

The argument from design or physicotheological proof of the existence of God has a long tradition in the history of ideas. The scriptures of many major theistic religions contain passages that suggest evidence of divine design in the world. The *Old Testament* (Psalms 19.2) states that "[t]he heavens declare the glory of God; and the firmament shows the work of his hands". Similarly, Romans 1.19 – 20 of the *New Testament* claims:

---

1 Kant quotations for which the reference is not given immediately are covered by the following reference.

Since what may be known about God is plain to them, because God has made it plain to them. For since the creation of the world God's invisible qualities—his eternal power and divine nature—have been clearly seen, being understood from what has been made, so that people are without excuse (New International Version).

Further, *Koran* 31.20 remarks: "Do you not see that Allah has made what is in the heavens and what is in the earth subservient to you, and made complete to you His favors outwardly and inwardly"? Ancient Greek philosophers also formulated versions of the argument from design. Plato in his *Timaeus* (28a–31b) argues that the demiurge could not create matter *ex nihilo*, but is able to organize pre-existing matter into the rational order that we see around us in the world. Thomas Aquinas in his writing *Summa Theologiae* (1266–72, I 2.3) presents the argument from design as the fifth of five proofs of the existence of God. If natural beings lack intention but nevertheless behave with apparent order and visible goal-directedness, then their behavior cannot be the result of chance, he argues. An intentional being external to them, namely an intelligent divine architect, must be the cause of their goal-directed behavior. In David Hume's famous *Dialogues Concerning Natural Religion* (1779), the character Cleanthes is a proponent of the argument from design:

Look round the World: Contemplate the Whole and every Part of it: You will find it to be nothing but one great Machine, subdivided into an infinite Number of lesser Machines, which again admit of Subdivisions, to a degree beyond what human Senses and Faculties can trace and explain. All these various Machines, and even their most minute Parts, are adjusted to each other with an Accuracy, which ravishes into Admiration all Men [...]. The curious adapting of Means to Ends, throughout all Nature, resembles exactly, tho it much exceeds, the Productions [...] of human Design, Thought, Wisdom, and Intelligence. Since therefore the Effects resemble each other, we are lead to infer, by all the Rules of Analogy, that the Causes also resemble; and that the Author of Nature is somewhat similar to the Mind of Man; tho' possessed of much larger Faculties, proportion'd to the Grandeur of the Work, which he has executed. By this Argument *a posteriori* [...] do we prove at once the Existence of a Deity, and his Similarity to human Mind and Intelligence (Hume 1779 in 1976, 161–2).

Drawing an analogy between man-made machines and natural products, Cleanthes claims that the material universe resembles the intelligent productions of human beings (machines) in that it exhibits marks of design (adjustment, accuracy, purposiveness). Since the design in any human artifact is the effect of having been made by an intelligent artisan, and since like effects have like causes, the design in the material universe is the effect of its having been made by an intelligent creator.

William Paley, whom I will use as a final example, formulates what became known as the "watchmaker analogy" at the beginning of his writing *Natural Theology* (1802):

> In crossing a heath, suppose I pitched my foot against a *stone*, and were asked how the stone came to be there: I might possibly answer, that [...] it had lain there for ever [...]. But suppose I had found a *watch* upon the ground, and it should be inquired how the watch happened to be in that place; I should hardly think of the answer which I had before given [...]. [...] [T]here must have existed, at some time, and at some place or other, an artificer or artificers who formed [...] [the watch] for the purpose which we find it actually to answer: who comprehended its construction, and designed its use. [...] [E]very indication of contrivance, every manifestation of design, which existed in the watch, exists in the works of nature; with the difference, on the side of nature, of being greater and more, and that in a degree which exceeds all computation (Paley 1802 in ⁶1819, 1, 3, 16).

Paley's watchmaker analogy, like the discussion in Hume, emphasizes the analogy between the production of a watch by an artisan and the generation of nature by a creator, although the latter cause, due to the incomparable size, complexity, and intricacy of its effected object, must be conceived of as "greater" than the maker of a watch.

As the examples demonstrate, the starting point of the argument is a posteriori. It begins with the identification of specific empirical properties as design-indicative features of nature—for instance the beauty, fine-tuning, purposiveness, and order of things and rejects that chance could be the cause of these features of nature. It identifies them as evidence that nature was designed. The second step of the argument states that marks of design require the presupposition of a designer with intellectual capacities such as intelligence and intentionality, able to cause the design-indicative features of nature. A third step is the identification of the designer as God. The crucial premise of the argument from design is the analogy between a work of art (e.g., a watch) and its designer (e.g., a watchmaker), and organic natural products—which are considered to be complex works of art or a complicated machine—and their maker. Human designers produce houses, watches, and other artificial things. The natural world is *like* a house or a watch or a collection of houses, watches, and other artificial things; therefore it is produced by something like a human designer.[2]

---

2 For analyses of the structure of the argument from design, see Himma (2009), Mackie (1982), Ratzsch (2001), Sober (2004, 117–47), and Swinburne (1979/²2004, 153–91). For accounts specifically of Kant's views on physicotheology, see Wood (1970, 171–7; 1978, 130–45) and McFarland (1970, 1–2).

II Does Kant Give an Argument from Design in His Theory of Biology?

In his dissertation (1755), Kant describes a cosmology that contains a physico-theological argument for God's existence (see *Theory of Heavens* II 331.21–347.32). Several years later, in the *Argument* essay (1763), Kant discusses the ontological and the physicotheological proofs for God's existence. Whereas he argues on the basis of the "logical exactitude and completeness" of the ontological argument that it is the only possible proof for God's existence, he nevertheless praises the "accessibility to sound common sense, vividness of impression, beauty and persuasiveness in relation to man's moral motives" (*Argument* II 161.8–11)[3] of the physicotheological or argument from design, and leaves his readers in no doubt about his esteem for the physicotheological proof.

At the time of the first *Critique*, the physicotheological proof of the existence of God was one the three traditional arguments for the existence of God beside the ontological and the cosmological argument. New to Kant's position in 1781 is his strict rejection of the ontological argument and, on the basis of this rejection, the rejection of the physicotheological and cosmological argument, which for Kant now are grounded in the ontological argument. Further modifications are that the only possible proof for God is no longer the ontological but the moral argument (*CPR* B 838–9, *CPR* B 856–8), and that the realism of Kant's pre-critical discussion of the proofs for God's existence has changed into the criticism of a regulative approach proving God's existence as an idea only. But despite the changes, even in the first *Critique* Kant expresses estimation for the physicotheological argument with a clear reference to his position in the *Argument* essay (*CPR* A 625/B 653). He claims that the physicotheological argument

> deserves to be named with respect. It is the oldest, clearest and the most appropriate to common human reason. It enlivens the study of nature [...] [and] brings in ends and aims where they would not have been discovered by our observation itself, and extends our information about nature through [...] a particular unity whose principle is outside nature (*CPR* A 623/B 651).

In the second *Critique* (1788), the moral argument is presented as one of the three postulates, whereas the physicotheological and other traditional arguments for

---

3 In this passage Kant names the physicotheological argument "cosmological" (esp. *Argument* II 160.7), but based on the context of the passage it is clear that he refers to the physico-theological proof. For scholarly comments in line with my view, see Schmucker (1983, 45) or Theis (1994, 140).

the existence of God are not noted.[4] All the more astonishing is the reappearance of the physicotheological or design argument in the third *Critique* (1790), which now precedes the moral argument for the existence of God in the discussion of the unity of the critical system (*CPJ* V 436.3–442.10). Even though it is in principle entailed in the moral argument, the design argument is presented in its own right and is attributed a limited function for the unity of the theoretical laws of nature. Whereas the physicotheological and the moral proofs of God's existence had nothing to do with one another in the first *Critique*, in the third *Critique* the physicotheological proof is described as a preliminary stage to moral teleology.[5]

The specific object of this paper is a detailed analysis of Kant's presentation of the argument from design in the "Critique of the Teleological Power of Judgment", i.e., in his theory of biology.[6] At first glance Kant seems to express different, if not even contradictory views on divine design in the "Analytic", the "Dialectic", and in the "Methodology": in §§61–8 he constructs a theory of biology based on mechanical and teleological laws, which does not need but also not contradict an argument from design. The justification of a possible unification of mechanical and teleological laws in §§69–78, however, explicitly refers to an argument from design. And the discussion of the unification of natural (mechanical and teleological) laws and moral laws, which Kant treats in §§79–91, leads to a criticism of the reach of the argument from design, since it can guarantee only the unity of the natural, but not the unity of the natural and moral laws. I begin with a closer look at Kant's statements.

II.1 Biology without Theology: the "Analytic" (§§61–8)

In the "Analytic" Kant offers an account of biology that makes no use of the argument from design but that would not be inconsistent with it. This can be demonstrated at three points:

(1) In §§64 and 65 of the "Analytic", Kant presents an analogy and also a disanalogy between the production of art and the generation of natural things. The

---

4 Beck (1960, 275–9) has put into question whether the argument for the existence of God in the second postulate is a moral (practical) or a teleological (theoretical) argument analogous to the physicotheological argument. He argues for its theoretical status. Wood (1970, 171–6; 1978, 133–5) agrees with Beck's claim that the argument is teleological. It is an argument from design. But he thinks that Kant argues on practical rather than theoretical grounds.

5 Here my view is in line with Sala (1990, 431, 435).

6 For further discussions of the design argument in Kant's theory of biology, see my paper "The Antinomy of Teleological Judgment" (unpublished manuscript).

acceptance of the analogy would be required to give a direct and strong version of an argument from design. But Kant is more concerned to name differences between artificial things and natural things, and claims that one "says far too little about nature and its capacity in organized products if one calls this an *analogue of art*" (*CPJ* V 374.27–9). Even though the "capacity for separation and formation" in an organized being is not entirely distinct from an artisan's productions it remains "remote from all art" (*CPJ* V 371.24–6).

The comparison between the production of a watch and the generation of a tree demonstrates these disanalogies: whereas an organism such as a tree "generates itself as far as the *species* is concerned" and as an "*individual*" (*CPJ* V 371.9, 13), a wheel of a watch "is not the efficient cause for the production of the other". And "even less does one watch produce another" (*CPJ* V 374.10, 15–6). And whereas one part of an organic being reproduces itself "in such a way that the preservation of the one is reciprocally dependent on the preservation of the other" (*CPJ* V 371.30–2), a watch "cannot by itself replace parts", or "make good defects in its original construction", or "repair itself when it has fallen into disorder" (*CPJ* V 374.17–20).

In addition, Kant claims that the maker of a product of art creates it by the means of the causality of an idea, which lies outside of the artificial product. A work of art is a "product of a rational cause distinct from the matter" (*CPJ* V 373.10–1); it is possible only due to the "causality of the concepts of a rational being outside of it" (*CPJ* V 373.16–7, see *CPJ* V 374.29–30). A natural object, however, cannot be conceived of otherwise than by the causality of an idea that is causally effective in the thing. A being is to be judged as a natural end in itself in accordance with its "*internal* possibility" (*CPJ* V 373.26, my italics). This is possible only if the idea of the whole determines "the form and combination of all the parts" and if the parts are "combined into a whole by being reciprocally the cause and effect of their form" (*CPJ* V 373.17–21).

However—one could argue—the only thing that is required for the argument from design is an *analogy* between and not the identity of the generation of organic objects and the creation of artificial things. On this line, the fact that Kant stresses the differences between the production of art and the generation of natural organic objects in §§64–5 does not mean that he rejects the analogy between them. And one could add that Kant does not really reject the analogy between artificial things and natural organisms because he only claims that organisms "cannot be explained through the capacity for movement *alone* (that is, mechanism)" (*CPJ* V 374.25–6, my italics) and that one "says *far too little* about nature and its capacity in organized products if one calls this an *analogue of art*" (*CPJ* V 374.27–9, first italics are mine).

Inasmuch as a natural being can be considered as matter in motion, artificial things and organisms resemble each other and can both be described according to the laws of motion. Furthermore, both kinds of beings have formal features that are contingent with regard to the mechanical laws. And it is also true that the form of the contingent features of both types of beings show a lawlikeness or necessity because these contingent features are only contingent compared to the laws of nature, but they are necessary compared to the idea of the artist or the idea of the observer of nature who judges that the natural object is organized with regard to a purpose in the thing. Thus, although Kant stresses the distinctions between works of art and natural objects, it is nevertheless a distinction between *similar* fundamental features, which does not rule out entirely the possibility of an analogy between them.

(2) In §§66–7, Kant relates teleological judgments to the idea of a supersensible, but he leaves open the question of whether this supersensible has to be identified with a supernatural divine architect. If Kant had an argument from design in mind, he would have stated that a designer, who has to be identified with God, actually designs organisms. God, he would have said, like an artist, created nature by means of his idea as a sign of his intelligence and wisdom. Hence one could think that Kant's mentioning of a "supersensible determining ground" (*CPJ* V 377.11) in §66 or of a "unity of the supersensible principle" (*CPJ* V 381.5–6) as the cause of the natural object in §67 would be the placeholder for the divine being that the argument from design requires.

However, the term "supersensible" can have many meanings much weaker than the theological supersensible divinity: it could refer to a human idea of a single organism or a human idea of nature as a whole—both as a regulative principle of reflective judgment—or it could be considered as a causality such as supernatural human noumenal freedom (see *CPJ* V 436.19–20), or as the idea of God, or as God. And it is obvious that Kant does not allude to a divine being: rather, he holds the position that the internal purposiveness in organized beings is only an "idea of the one who judges" (*CPJ* V 376.20–1). It is a principle of reflecting human judgment, for our human understanding cannot explain organisms otherwise than on the basis of the presupposition that an "idea has to ground the possibility of the product of nature" (*CPJ* V 377.3–4). And, since an idea in a Kantian sense is an absolute unity of the representation, the unity of this idea has the same grounding as the idea of nature as a whole. But this grounding is not specified as God, and therefore the "Analytic" also in §§66 and 67 can be reconstructed without theological references.

Nevertheless, the reference to an ambiguous term of a supersensible does not rule out the possibility of a supernatural divine grounding of nature. It does not make use of a strong interpretation of the concept of a supersensible in the sense of a supernatural being, but leaves open whether there could be such a being (for other theoretical requirements).

(3) At the end of the "Analytic", in §68 (see also §79), Kant explicitly rejects the systematic grounding of biology as a science in a theological principle, and adds an entire section on the topic of the immanent and internal systematic principles of a science as a science, and also of the science of organic nature as a self-sufficient science. In this passage, Kant again clarifies his views on the independence of natural teleology from theology: if one "brings the concept of God into natural science", he warns us, "in order to make purposiveness in nature explicable, and subsequently uses this purposiveness in turn to prove that there is a God, then there is nothing of substance in either of the sciences" (*CPJ* V 381.25 – 9). Natural science should not be intermingled with "the consideration of God", and "a *theological* derivation" of natural science (*CPJ* V 381.34). For instance, in all sections of the "Analytic", Kant presupposes organized materials and a formative power, without raising the question where these power and materials come from. He is not interested to ask whether they are created or eternal, a question that might require theological explanations.[7]

These passages find support in the writing *Teleological Principles*, written in 1788, two years before the *CPJ* (1790). There Kant emphasizes that questions about the very first beginning of the generation of organisms including the question of the origin of organized materials and a formative power cannot be investigated and analyzed within the field of biology.[8] It must lie outside the realm of natural philosophy and belongs to the fields of "*metaphysics*" or theology instead:

> Since the concept of an organized being already includes that it is some matter in which everything is mutually related to each other as end and means, which can only be thought as a *system of final causes*, and since therefore their possibility only leaves the teleological but not the physical-mechanical mode of explanation, at least as far as *human* reason is concerned, there can be no investigation in physics about the origin of all organization itself. The answer to this question [...] obviously would lie *outside* of natural science *in metaphysics* (*Teleological Principles* VIII 179.8 – 18).

If we take Kant's remark in the *Teleological Principles* into account, one could say that Kant, in his philosophy of biology, consciously does not raise questions that transcend biological investigation: for instance, the question of the ultimate origin of organized nature, and the question of whether the creative elements of na-

---

7 For discussions of those points see my paper "Kant on Formative Power" (2012).
8 Kant calls it "*physics*", since biology was not yet established as a science at this time.

ture (formative power and organized materials) themselves are created. Again, this does not rule out the possibility of a theological grounding of organized nature. It simply marks the border between distinct fields of scientific investigation.

II.2 The Idea of an Intelligent Divine Architect as a Unifying Ground of Natural Laws: the "Dialectic" (§§69–78)

In contrast to the "Analytic", in the "Dialectic" (§§69–78) Kant *states* a version of the argument from design. Two points are in need of particular attention:

(1) In the "Dialectic", as earlier in the "First Introduction", Kant frequently uses the concept of a "technique [*Kunstfertigkeit*] of nature" (*CPJ* V 390.33–7, 391.16, 393.3, 395.23–4, 404.18, 410.15, 411.18). The repeated use of this phrase suggests that Kant consciously affirms the analogy between the production of art and the generation of nature—and thus the premise of the argument from design—without emphasizing its restrictions as he does in the "Analytic". He seems to be convinced that nature can be analyzed in terms of art-like features and he seems to suggest that natural production equals craftsmanship.

In his first version of the "Introduction" to the third *Critique*, Kant says that a certain concept arises from the power of judgment, namely that of a "technique of *nature*" or "nature as *art*" (*First Introduction* XX 204.13–4). It designates the generation of the systematic form and order of nature with regard to its particular empirical laws. In view of its "products as aggregates", Kant claims, nature proceeds "*mechanically, as mere nature*", but in view of "its products as systems" nature "proceeds *technically*, i.e., as at the same time an *art*" (*First Introduction* XX 217.29–32). According to Kant, a human observer cannot conceive of this systematic generation of natural objects otherwise than in comparison with intentional actions that are directed to the idea of an end or a purpose: the "*technique of nature*" is a "*causality* of nature with regard to the form of its products as ends". It is opposed to the "mechanics of nature", a causality "through the combination of the manifold without a concept lying at the ground of its manner of unification" (*First Introduction* XX 219.18–23). Also in the "Analytic" we can find a passage in which Kant claims that we represent the possibility of a natural object analogous to a causality that we "encounter in ourselves", and hence "we conceive of nature as *technical* through its own capacity (*CPJ* V 360.32–3). These statements from the "First Introduction" and the "Analytic" clearly show that Kant poses an analogy or similarity between natural and artificial practices and that he uses craftsmanship as a pattern for the generation of organic nature.

In the "Dialectic" Kant uses the phrase "technique of nature" repeatedly—twice in the headings of §§74 and 78 (*CPJ* V 395.23 – 4, *CPJ* V 410.15) and in several other passages.[9] In *CPJ* V 393.5 – 6, the term "technique" refers to the "correspondence of generated products with our concepts of ends". Likewise, in *CPJ* V 404.18 – 9 the "technique of nature" describes "a connection to ends in it". In the passage *CPJ* V 411.18, Kant explains "technique of nature" objectively as a "productive capacity" in nature, in *CPJ* V 413.17 and *CPJ* V 414.9 subjectively as an intentional mode of explanation that is superordinate to the mechanical mode of explanation of the possibility of a natural product. Kant nowhere articulates doubts about the analogy as he does in the "Analytic".

(2) At the end of the "Analytic" Kant states that teleological explanations should not be confused with theological ones (*CPJ* V 381.31– 382.1). In the "Dialectic", however, he claims that "teleology cannot find a complete answer for its inquiries except in a theology" (*CPJ* V 399.3 – 5), and he explicitly gives a version of the argument from design. He shows that both kinds of principles of nature, the mechanical as well as the teleological laws, can be unified in one and the same common ground: in the idea of "God" (*CPJ* V 399.37), the supernatural designer of nature. In the "Dialectic" Kant explicitly identifies the supersensible with theological content.

However, Kant does not claim God's objective existence, but only stresses the importance of the idea of God as a subjective principle of reflective judgment, which enables us to understand and to explain the systematic order of the two kinds of laws of nature. So he says at the end of §74 that an "original ground of nature" can be "thought without contradiction" but "is not good for any dogmatic determinations". An objective reality of the divine being cannot be guaranteed (*CPJ* V 397.18 – 9). And in §75, Kant points out that the "*intentionally acting supreme cause*" (*CPJ* V 399.11– 2) can only be established subjectively, not objectively. We cannot claim that God exists. Instead

> [a]ll that is allowed to us humans is the restricted formula: We cannot conceive of the purposiveness which must be made the basis even of our cognition of the internal possibility of many things in nature and make it comprehensible except by representing them and the world in general as a product of an intelligent cause (a God) (*CPJ* V 399.37–400.6, see *CPJ* V 400.28 – 401.2).

In §75, Kant points out that the self-generation of nature does not explain why certain features of nature are directed towards the idea of an end, i.e., towards the purpose of a natural product. The explanation of this specific constitution of

---

9 I leave out a difficult passage here that would require separate attention: *CPJ* V 390.21– 391.23.

organized beings requires an argument from design: God has created them in such a way that these features are directed toward their end. This God is an intelligent being outside of the world (*CPJ* V 399.2–3). It is a cause that "act[s] in accordance with intentions" and is productive analogous to "the causality of an understanding" (*CPJ* V 397.31–398.3).

Although Kant himself does not explicitly connect his remarks about the intelligent God in §§74–5 of the "Dialectic" with his notion of an intuitive understanding in §§76–7, the reconstruction of an argument from design in the "Dialectic" can include the §§76–7, based on Kant's lectures on rational theology in the 1780s, where he identifies the divine kind of consciousness with an intuitive understanding.[10]

II.3 Limits of the Notion of an Intelligent Divine Architect: the "Methodology of the Teleological Power of Judgment" (§§79–91)

Kant gives the argument from design[11] in the "Dialectic" without any major criticisms. This will change in the "Methodology" (§§79–91). In this part of the text, Kant gives a version of the argument from design; however, he criticizes it and describes its limitations and shortcomings. In §§85–7, Kant discusses the physicotheological[12] (§85) and the moral argument (§86) for the existence of God in the context of the unification of the realms of nature and freedom. The argument from design introduces a reduced concept of God, but does not lead to an identification of the designer with a God having the traditional divine attributes and moral perfection. This is because only the idea of an intentional and intelligent cause outside of the world is required for us to understand organisms. In order to conceive of natural products as purposes, we do need an intentional and intelligent but not an omniscient, omnibenevolent, and morally virtuous being.

---

10 I keep this point very brief here. For a detailed analysis see my paper "The Antinomy of Teleological Judgment" (unpublished manuscript).

11 In §75 Kant not only gives a physicotheological but also a cosmological argument for a supernatural being (as a regulative principle for the reflecting power of judgment), since his characterization of organized beings is also concerned with the ultimate necessity of the contingent empirical features of nature. The relationship between the contingent features of the world and their grounding in a necessary final being is discussed in the cosmological proof for the existence of God rather than in the physicotheological proof (see *CPJ* V 398.32–399.5).

12 The physicotheological argument is the traditional name for the teleological or design argument as a proof for the existence of God. Although Kant presents a version of the argument already in §§74 and 75 he does not use this term before §85.

But it is those classical divine attributes that would be required for a conception of a divine standpoint that unifies the realms of nature and freedom.

In §85, Kant explicitly takes the physicotheological argument for the existence of God into consideration. Physicotheology "is the attempt of reason to infer from the *ends* of nature (which can be cognized only empirically) to the supreme cause of nature and its properties" (*CPJ* V 436.5–7). A physicotheologian aims to draw the conclusion that a being outside the realm of nature exists (*CPJ* V 437.28–9). In this section, Kant emphasizes the empirical starting point of the argument, and he immediately rejects the possibility of drawing non-empirical conclusions (for instance the non-empirical properties of the supersensible architect) from empirical data. Physicotheology, "no matter how far it might be pushed, can reveal to us nothing about a *final end* of creation; for it does not even reach the question about such an end" (*CPJ* V 437.18–20). Since

> the data and hence the principles for *determining* that concept of an intelligent world-cause (as the highest artist) are merely empirical, they do not allow us to infer any properties beyond what experience reveals to us in its effects (*CPJ* V 438.5–9).

Empirical data cannot lead us to the presupposition of non-empirical properties such as the omniscience and intelligence, omnipotence, and omnibenevolence of a supernatural designer. According to Kant, physicotheology can never be more than a physical teleology because the relation to ends in it must be considered only as conditioned within nature. Physicotheology

> reduces the concept of a *deity* to that of an intelligent being that can be conceived by us, which may have one or more, or even many of the important properties that are requisite for the establishment of a nature corresponding to the greatest possible ends, but not all of them (*CPJ* V 438.16–22).

Beside his criticism of the empirical starting point of the argument, Kant now describes the physicotheological God as an intelligent and intentional agent that includes the theoretical use of reason only (*CPJ* V 440.26–442.5). God as an intelligence is a legislator for nature (*CPJ* V 444.13). An ethicotheological God, however, as Kant says in §86, would be *"omniscient"*, *"omnipotent"*, and *"omnibenevolent"*, a being that not only has theoretical reason, but practical wisdom also such as "justice" and "goodness" (*CPJ* V 444.15–28). An ethicotheological divine being could not only be a legislator for nature, but also a "sovereign in a moral realm of ends" (*CPJ* V 444.14–5).

III Kant's Account of the Argument from Design in the *CPJ*

III.1 Is Kant's Account of the Argument from Design in the *CPJ* Consistent?

Now I will clarify the systematic line in Kant's different perspectives on the argument from design. I will defend the view that Kant's remarks on this matter are consistent but entail significant shifts motivated by the different contextual questions that Kant treats in the "Analytic", the "Dialectic", and the "Methodology of the Teleological Power of Judgment".

(1) In the "Analytic", Kant tries to determine the significant features of an organism and establishes the two kinds of laws of nature: mechanical and teleological laws. Within this investigation, Kant adopts premises of the argument from design; however, he makes no use of the notion of a divine architect. He gives an account of biology as a science that is independent from external theological groundings: although organized beings that are subject to biological scientific investigation are among those beings that are subject to theological scientific investigation, the science of theology raises different questions about those objects than the science of biology. For this reason he has to keep two things in balance.

First, he has to demonstrate the internal principles of biology as a science and to secure that an organism and organic nature as a whole can be explained by these internal principles alone. The internal principles of biology as a science consist in a combined use of mechanical and final causal explanations of nature. Second, if biology as a science could be embedded into a broader metaphysical or theological context, Kant would have to secure that his account of biology does not contradict the premises of such a metaphysical or theological framework. I think this is what Kant is trying to do.

In the "Analytic", Kant develops an account of mechanical and final causation that suffices to explain the self-generation of the species, the individual, and the parts of an individual organism. But Kant does not go back to the first pair of the species, a question that might require theological explanations (as in the notion of generic preformism in §81). In addition, Kant approves the analogy between art and nature, although it remains subject to restrictions. The analogy does not include the hypothesis of a divine architect who produces organic beings. Kant uses the notion of a supersensible as a unifying teleological ground of the forms of nature; however, at this point he does not identify it with theological content. Instead, he interprets it as an idea of the judging human mind. And at the end of the "Analytic", Kant does not reject God as a possible object *per se*, but as an object that belongs to the scientific field of organized nature.

(2) In the "Dialectic", Kant raises the question of whether the application of two kinds of natural laws might cause an antinomy in our consciousness when we judge organized beings. Kant now searches for a unifying ground between the mechanical and teleological laws of nature to avoid a conflict between those laws in our consciousness. He finds this unifying ground in an intuitive understanding, i.e., the intentional and intelligent consciousness of a physicotheological God, based on an argument from design. In intuiting both kinds of laws, the divine architect creates and implants them into nature and determines the finalistic hierarchy between them: the subordination of mechanism under teleology. And in doing so it fulfills the idea of such a unifying ground of both laws outside of nature. Like an artist who subdues matter to his ideas of a purpose (the systematic form of an artificial object), the divine architect subdues natural matter to his idea of a purpose (the systematic form and order of the natural world).[13]

(3) In the "Methodology", Kant leaves the immediate context of biological investigations. He now discusses the systematic position of his theory of biology within the critical philosophy, and raises the question of whether there is a unifying ground between the realms of natural and moral laws. The answer to this question can no longer be adequately given by the idea of a physicotheological notion of God based on an argument from design. It requires a higher-ordered unifying ground than the idea of an intelligent, intentional supernatural being, namely an ethicotheological God based on an ethicotheological argument for the existence of God. In §85, Kant offers the argument from design as the penultimate step in the critical system that can bring the world's natural features in a finalistic order, but not its natural and moral ones. The intentional and intelligent designer of the physicotheological argument is an intelligent but not necessarily an omniscient, omnipotent, and it is not a morally good being. These properties are required for the unification of the realms of nature and morality.

Furthermore, the intelligence of an intelligent divine architect who brings the two types of natural laws into order involves only a "theoretical use of reason" (*CPJ* V 438.29–30, 439.20), for the insight into natural laws is based only on theoretical knowledge. In contrast, the moral perfection and practical wisdom of a divine being that unifies the realms of nature and morality also requires the "practical use" of reason (*CPJ* V 438.35). Therefore, a physicotheological God

---

13 For a detailed discussion of the notion of an intuitive understanding see my paper "The Antinomy of Teleological Judgment" (unpublished manuscript).

can only govern a kingdom of theoretical (natural), but not a kingdom of theoretical and practical (natural and moral) ends.

III.2 Further Considerations

In this final section of my paper, I will outline a difficulty for this reconstruction of Kant's account. It arises regarding the relation and necessity of the two notions of God involved in Kant's account. (a) Their relationship could be interpreted such that they are two irreducibly different ideas of supernatural divine beings at the end of the critical system—the idea of a physicotheological God that unifies the theoretical natural laws, and the idea of an ethicotheological God that unifies the theoretical natural and the practical moral laws. But is then not a third unifying ground required that unifies the physicotheological and the ethicotheological concepts of Gods to reach the highest *unity* of the critical system? Is Kant at fault for not having spelled out such a third notion of God? This is unlikely.

(b) One could instead take Kant's account to be that there are two notions of God at the end of the critical system, but that the notion of an ethicotheological God covers the functions of the physicotheological concept of God. The physicotheological concept of God would then be reducible to an ethicotheological one. Some of Kant's remarks seem to suggest such a reductive reading. For instance, Kant says that the moral proof of the existence of God does not "properly merely *supplement* the physico-teleological proof, thereby making it into a complete proof" but rather it "is a special proof that *makes good* the lack of conviction" in the physicotheological argument (*CPJ* V 478.13–6). Some remarks even indicate that the teleological proof is dispensable: the moral proof would "always remain in force even if we found in the world no material for physical teleology at all or only ambiguous material for it" (*CPJ* V 478.30–2). But why then do we need the concept of a physicotheological concept at all? Why did Kant give so much time to the discussion of the physicotheological argument and its function if it is entailed by the concept of the ethicotheological God? Why did Kant reintroduce the argument in the third *Critique* after examining and rejecting it in the first and abandoning it in the second?

As an answer, one could suggest a model of division of labor between both ideas of God that is similar to the division of labor in the government of a state. Although a king knows the function of those subordinate persons who are responsible for the order in the different realms within his kingdom, he nevertheless does not himself develop the detailed specific knowledge and capacities that are required to govern these different subordinate realms of the kingdom.

Analogously, one could say that the ethicotheological God as the supreme sovereign of his kingdom of ends knows the functions and potentially has the capacities of the physicotheological God, however he himself does not develop the specific facilities and particular knowledge that is required to unify the realm of nature. For instance, he himself cannot directly administer the endless manifold of empirical features of the natural world that the physicotheological God has to unify under teleological laws. This reading seems to be supported in passage *CPJ* V 447.16–448.13, where Kant says that the (subordinate) physicotheological God orders and organizes the outer natural world in accordance with the legislation of the inner moral world, which is given by the (superordinate) ethicotheological God. This process is analogous to legislation in a state or kingdom where the executive power transfers the inner order of laws, which are normatively legislated by the leader of the state or king, into the outer natural world (for a more extended discussion of this point see Cunico 2008, 317).

Reading (b) still argues for two notions of God that stand in the relation of a hierarchy to one another. (c) But one could also read Kant as holding that there is only one God at the end of the critical system whose different aspects can be proven in different ways from the perspective of human judgment. This one God (due to his theoretical reason) could be considered the creator of the more limited unity of the natural laws. It could then be proved based on the physicotheological proof from the human point of view. Or this one God (due to his theoretical and practical reason) could be considered as the creator of the unity of the natural and moral laws. It could then be proved based on the ethicotheological proof from the human point of view. In this reading Kant's remarks do not introduce two ideas of God, but only two aspects of one and the same God and two proofs of these aspects from the human point of view. It then suffices to give a reasonable explanation for the reintroduction of the physicotheological argument and its value for our human judgment. This could be the persuasive power of the physictheological argument for common sence. As already mentioned, Kant, from his early writings on, emphasizes the simplicity and easy empirical accessibility of the argument, which make it especially attractive for non-philosophical discourse and every day life. Thus Wood has pointed to physicotheology and its "unique value for morality and moral religion". He claims convincingly, that "the moral man believes in the governance of the world by a wise plan" and, for this reason,

> it is only natural that he should be on the lookout for signs of his wisdom, and that he should find in the purposive arrangements he observes in the natural world an apparent confirmation of his orally grounded convictions. The physicotheological proof, therefore,

is in common thinking very closely allied to moral faith (Wood 1978, 133, see also Wood 1970, 171).

Kant would then argue for one notion of God at the end of the critical system, but for two different modes of access to two aspects of this notion from a human point of view. This is the reading I favor.[14]

References

Ameriks, Karl 2008a, Status des Glaubens und Allgemeine Anmerkung zur Teleologie, in: *Immanuel Kant. Kritik der Urteilskraft*, ed. by Otfried Höffe, Berlin: Akademie Verlag, 331–49.

Aquinatis, Sancti Thomae 1266–72, *Summa theologiae. Prima Pars*, ed. by Fernando Sebastián Aguilar, Madrid: Biblioteca de Autores Cristianos ⁵1994.

Beck, Lewis White 1960, *A Commentary on Kant's Critique of Practical Reason*, Chicago: The University of Chicago Press.

Cunico, Gerardo 2008, Erklärungen für das Übersinnliche: physikotheologischer und moralischer Gottesbeweis, in: *Immanuel Kant. Kritik der Urteilskraft*, ed. by Otfried Höffe, Berlin: Akademie Verlag, 309–29.

Goy, Ina, The Antinomy of Teleological Judgment (unpublished manuscript).

Goy, Ina 2012, Kant on Formative Power, *Lebenswelt* 2, 26–49.

Himma, Kenneth Einar 2009, Design Arguments for the Existence of God, in: *Internet Encyclopedia of Philosophy*, http://www.iep.utm.edu/design/ [last visited: November 23, 2013].

Hume, David 1779, *Dialogues Concerning Natural Religion*, ed. by David Fate Norton and Mary J. Norton, Oxford: Clarendon Press 1976.

Mackie, John 1982, Arguments for Design, in: John Mackie, *The Miracle of Theism. Arguments for and Against the Existence of God*, Oxford: Oxford University Press, 133–49.

McFarland, John 1970, *Kant's Concept of Teleology*, Edinburgh: University of Edinburgh Press.

Paley, William 1802/⁶1819, *Natural Theology; Or, Evidences of the Existence and Attributes of the Deity, Collected from the Appearances of Nature*, London: S. Hamilton.

Ratzsch, Del 2001, *Nature, Design, and Science. The Status of Design in Natural Science*, New York: State University of New York Press.

14 I would like to thank the German Research Foundation for a four-year fellowship that enabled me to conduct the research for this essay. I presented earlier versions of this paper at the symposion on Kant's theory of biology in Tübingen (December 2010), at the History of Philosophy Roundtable in San Diego (April 2011), and at the Università degli Studi di Milano (May 2011). I am grateful to Julius Alves, Michael Demo, Piero Giordanetti, Otfried Höffe, Clinton Tolley, Eva Oggionni, Eric Watkins, Christoph Wehle, and Allen Wood for comments and discussions.

Sala, Giovanni B. 1990, Teleologie und moralischer Gottesbeweis in der Kritik der Urteilskraft, in: Giovanni B. Sala, *Kant und die Frage nach Gott. Gottesbeweise und Gottesbeweiskritik in den Schriften Kants*, Berlin: Walter de Gruyter, 426–50.

Schmucker, Josef 1983, *Kants vorkritische Kritik der Gottesbeweise: ein Schlüssel zur Interpretation des theologischen Hauptstücks der transzendentalen Dialektik der* Kritik der reinen Vernunft, Wiesbaden: Steiner.

Sober, Elliott 2004, The Design Argument, in: *The Blackwell Guide to the Philosophy of Religion*, ed. by William E. Mann, Oxford: Blackwell Publishing, 117–47.

Swinburne, Richard 1979/²2004, Teleological Arguments, in: Richard Swinburne, *The Existence of God*, Oxford: Clarendon Press, 153–91.

Theis, Robert 1994, Das Problem der Physikotheologie im *Beweisgrund*, in: Robert Theis, *Gott. Untersuchung zur Entwicklung des theologischen Diskurses in Kants Schriften zur theoretischen Philosophie bis hin zum Erscheinen der* Kritik der reinen Vernunft, Stuttgart Bad-Cannstatt: Fromann Holzboog, 111–43.

Wood, Allen W. 1970, *Kant's Moral Religion*, Ithaca: Cornell University Press.

Wood, Allen. W. 1978, The Physicotheological Proof, in: Allen W. Wood, *Kant's Rational Theology*, Ithaca: Cornell University Press, 130–45.

Paul Guyer
Freedom, Happiness, and Nature: Kant's Moral Teleology (*CPJ* §§83–4, 86–7)

1 Introduction: The Paradoxes in Kant's Moral Teleology

In the "Methodology of the Teleological Power of Judgment", the culminating part of the *CPJ* and thus of all three of his *Critiques*, Kant enlists what we can call his reflective teleology in service of his overarching vision of nature as the arena for the realization of human morality. Once our experience of the peculiar properties of organisms, described in Kant's philosophy of biology in the "Analytic", has led us to conceive of them and in turn the whole of nature as if they were the product of design, we are inevitably led by the very nature of human reason to conceive of the designer as unconditioned and as its end in designing nature as an unconditioned final end. As an end, the latter must be unconditioned in a normative sense, that is, something of unconditional value, even though its ground, the designer, is unconditioned in a purely ontological sense. Kant then argues on both *a priori* and empirical grounds that human happiness could not be the final end of nature: on *a priori* grounds, because happiness is not unconditionally valuable; on empirical grounds, because nature does not appear to particularly favor human happiness. The only alternative to happiness as the final end of nature is the realization of human freedom, which is, according to Kant's moral philosophy the essential end of humanity that makes humanity an end in itself.[1]

The idea that human freedom could be the final end of nature seems doubly paradoxical. First, on Kant's transcendental idealist account of freedom, it could not possibly be the *product* of nature, but is always conceived of as *external* to nature, although perhaps the *ground* of events within nature. So it is not clear in what way freedom could actually be the end of nature itself. Second, Kant's account of the highest good, particularly in the work that immediately preceded the *CPJ* and seems to have led directly to Kant's work on the latter, namely the *CprR*, holds that we must be able to conceive of nature as if it *were* hospitable to human happiness and that we must postulate the existence of God as the author of laws of nature that will make the realization of human happiness possible as

---

1 See *Lect. Moral Phil. Collins* XXVII/1 345.12.

the consequence of human morality. But this insistence that we must be able to conceive of nature as hospitable to the realization of human happiness as the condition for the possibility of the highest good seems to contradict Kant's insistence in the third *Critique* that happiness could not be the final end of nature. Thus Kant's moral teleology in the third *Critique* seems to contradict both his theory of human freedom and his theory of the highest good in the second *Critique*, two theories that are the heart of that work. This would be both disturbing in its own right and surprising, given the mere two years that separate the publication of the two books.

I argue, however, Kant's moral teleology in the third *Critique* does not in fact contradict the central doctrines of the second. The third *Critique*'s conception of human freedom as the final end of nature does not contradict the conception of freedom as external to nature in the second because what Kant actually argues is that we should conceive of the development of *conditions conducive to the successful exercise* of human freedom as the final, or more precisely ultimate end within nature, not the realization of human freedom itself. This position is consistent with Kant's application of his teleological conception of nature in the historical and political writings surrounding the third *Critique* as well as with Kant's treatment of free will in his other major work of the early 1790s, *Religion*. Second, Kant's treatment of the highest good in the third *Critique* is consistent with his treatment of it in the second *Critique* as well as later writings like the 1793 essay *On the Common Saying:* in the third *Critique* as well as in those other works Kant argues that happiness cannot be the *incentive* for morality even though it is the *object* of morality, the state of affairs that morality strives to bring about, and because happiness is only possible within nature the highest good must also be conceived of as possible within nature if our moral motivation is not to be undermined by the thought of its own futility. This means that the hostility of nature to the realization of human happiness that Kant emphasizes in the third *Critique* must ultimately be able to be interpreted in some way or other as an *appearance* that can be overcome from the proper point of view. But such a move is also necessary to make sense of Kant's argument about the highest good within the second *Critique*, where Kant argues that morality and happiness must be causally connected but do not appear to be so connected by human efforts, but then resolves this problem by postulating the existence of God as the author of *nature*, or highest original good, thus that it must be possible that morality and happiness be connected by *natural* means, as the highest derived good, which in the end can be nothing other than by human actions. So on this point also there is no contradiction between the two *Critiques*.

2 From the Purposiveness of Nature to its Final End

I will not review the preliminary stages of Kant's argument for the necessity of our conceiving of a final end of nature here.[2] Suffice it to say that Kant holds that it is only possible for us human beings to conceive of organic phenomena such as reproduction, growth, and self-preservation as processes directed by an antecedent conception of the organic individual and species as goals of intelligent creation, "not thinkable and explicable even through an exact analogy with human art" (*CPJ* V 375.15 – 6) but also not thinkable and explicable by us except by such an analogy, inexact as it may be. Moreover, once we have been led to conceive of "matter insofar as it is organized" as if it were the goal of design, "this concept necessarily leads us to the idea of the whole of nature as a system in accordance with the rule of ends, to which idea all of the mechanism of nature in accordance with principles of reason must now be subordinated", although, to be sure, only regulatively, "in order to test natural appearance by this idea" (*CPJ* V 378.37–379.4). But the only way in which we can reconcile the idea of nature as a system mechanically governed by uniform laws with the idea of it as designed is on the model of "theism", that is, the idea of the "world-whole" as the product of "an intentionally productive (originally living) intelligent being" (*CPJ* V 392.11– 2); thus even though theism is "incapable of dogmatically establishing the possibility of natural ends as a key to teleology [...] for us there remains no other way of judging the generation of [nature's] products as natural ends than through a supreme understanding as the cause of the world" (*CPJ* V 395.3 – 5, 17– 9). Kant warns us repeatedly that this "is only a ground for the reflecting, not for the determining ground of judgment, and absolutely cannot justify any objective assertion", but the non-objective or non-cognitive standpoint of reflecting judgment will be compatible with the practical rather than theoretical character of postulates of pure practical reason, and so the argument can continue to the morally significant point that Kant wants to reach, namely that if "we cannot form any concept at all of the possibility of" nature as a systematic and organized whole "except by conceiving of such an *intentionally acting* supreme cause" (*CPJ* V 399.10 – 2), then we also cannot conceive of such a supreme cause as having designed nature and realized this design without a sufficient reason for its action. Thus,

---

2 For my fuller treatment of this argument, see Guyer (2000, 2001, 2004, and 2007). See also the contribution by Watkins in this volume.

> if we assume that the connection to ends in the world is real and assume that there is a special kind of causality for it, namely that of an *intentionally acting* cause, then we cannot stop at the question why things in the world (organized beings) have this or that form, or are placed by nature in relation to this or that other thing; rather, once an understanding has been conceived that must be regarded as the cause of the possibility of such forms as they are really found in things, then we must also raise the question of the objective ground that could have determined this productive understanding to an effect of this sort, which is then the final end for which such things exist (*CPJ* V 434.16–435.3).

After all, *we* do not produce plans and try to realize them without having in mind some goal, some agreeable or good state of affairs that is to be brought into existence through the realization of those plans, and so since we cannot conceive of this productive understanding even greater than our own except on analogy with our own rational agency—"given the nature of our understanding and our reason we cannot conceive of [it] except in accordance with final causes" (*CPJ* V 429.7–9)—we also cannot conceive of this agent as acting without a goal in mind. But since we are conceiving of this cause as outside of nature and as greater than our own rational agency, indeed, given the nature of our reason, as unconditioned,[3] we certainly cannot conceive of its goal as in any way dependent "upon conditions which can be expected only from nature" (*CPJ* V 431.16–8); rather, we can only conceive of its goal as in some appropriate sense "unconditioned":

> the final end cannot be an end that nature would be sufficient to produce in accordance with its idea, because it is unconditioned [...]. A thing, however, which is to exist as the final end of an intelligent cause necessarily, on account of its objective constitution, must be such that in the order of ends it is dependent on no further condition (*CPJ* V 435.4–14).

In this passage, Kant emphasizes that the final end of an unconditioned cause must itself be unconditioned *in the order of ends*, which is presumably meant to be contrasted to unconditioned *in the order of causes*. It would be patently paradoxical if the final end of nature were supposed to be unconditioned in the order of causes, because if it were to be in any way produced by nature it would then be conditioned by nature after all. So presumably by saying that the final end of creation must be unconditioned in the order of ends Kant means to state that it must be something of unconditional *value*, and that only something of unconditional value could give a sufficient reason for creation

---

3 On this point, see again the contribution to this volume by Watkins.

by that which is unconditioned in the order of causes, the designer and creator of the world.

Such a Leibnizian thought would be natural for Kant, as would its transposition from a principle of speculative metaphysics into a principle of reflective judgment. But be that as it may, the candidate that Kant is now about to introduce as the only possible final end of nature, namely human freedom, is in fact both ontologically and normatively unconditional: freedom is normatively unconditional as the sole end in itself, that which should always be treated as an end and never merely as a means, but is also ontologically unconditioned as a spontaneous cause that can never be the effect of anything else. The unconditional value of freedom is the heart of Kant's foundation of morality, for according to Kant's lectures on ethics freedom is the "essential end of humanity" (*Lect. Moral Phil. Collins* XXVII/1 345.12) to which reason is merely the means (*Lect. Nat. Law Feyerabend* XXVII/2.2 1321.43), thus the defining feature of humanity as an end in itself that is the ground of any possible categorical imperative (*Groundwork* IV 428.5 – 6), while the spontaneity or unconditional causality of freedom is the idea of reason for which transcendental idealism makes room in the resolution of the third "Antinomy of Pure Reason" in the first *Critique* (*CPR* A 444 – 51/B 472 – 9, A 432 – 58/B 560 – 86). So what we will have to worry about is the two paradoxes I have already identified, the ontological paradox that non-natural freedom could be the final end of nature and the normative paradox that happiness is both excluded from being the final end of nature and yet included in it, as part of the highest good. So let us now look at Kant's arguments against happiness and for freedom as the final end that we must impute to nature, since it is these claims that give rise to these paradoxes.

To continue, then: having argued that we can conceive of nature only as the product of a supremely intelligent cause acting for the sake of a final end that is itself unconditioned in the order of ends, Kant now makes his argument that happiness is not a candidate for this unconditional end but that only human freedom is (which is, however, also unconditioned in the order of causes). Kant offers two different kinds of argument against the possibility that happiness could be the final end of nature.[4] The first is a conceptual argument that invokes the argument against happiness as the basis of the fundamental principle of morality in the *Groundwork* and the "Analytic" of the second *Critique* (see *Groundwork* IV 417.27, 419.11 and *CprR* V 25.12 – 26), and the second is a pair of empirical arguments that we could connect with the argument that happiness

---

4 Höffe finds four different arguments against happiness as the final end; see Höffe (2008, 298 – 300). His four arguments fall into the two types that I distinguish here.

does not appear to be naturally connected with the worthiness to be happy, which is the starting point of Kant's discussion of the highest good in the "Dialectic" of the second *Critique* (V 113.27–114.12). The conceptual argument is that our conception of happiness is a mere idea that is too indeterminate to furnish any determinate goal for nature and thus any determinate laws by means of which nature might be thought to reach this goal:

> it is a mere *idea* of a state to which [the human being] would make his instincts adequate [...]. He outlines this idea himself, and indeed, thanks to the involvement of his understanding with his imagination and his senses, in so many ways and with such frequent changes that even if nature were to be completely subjected to his will it could still assume no determinate universal and fixed law at all by means of which to correspond with this unstable concept and thus with the end that each arbitrarily sets for himself (*CPJ* V 430.8–16).

Another way of putting this point is to say that the word 'happiness' merely masquerades as the name of a determinate condition that could possibly be the goal of nature; in fact, it really refers to the indeterminate and potentially intra- and/ or interpersonally contradictory variety of ends the realization of which human beings think would bring them pleasure, and does not refer to anything determinate enough to be the unique and well-defined goal of nature.

This argument then segues into what we may think of as the first of Kant's empirical, or more empirical, arguments against the supposition that happiness could be the goal of nature. This argument concerns specifically human nature, or so to speak the nature within our own skins: no matter what satisfaction of our desires nature might actually provide for us, it is our own nature never to be satisfied with the happiness that we have and always to form some new desire that is not yet satisfied. Thus Kant follows the last sentence I quote with this:

> But even if we sought either to reduce this concept [happiness] to the genuine natural need concerning which our species is in thoroughgoing self-consensus, or, alternatively, to increase as much as possible the skill for fulfilling ends that have been thought up, what the human being understands by happiness and what is in fact his own ultimate natural end (not an end of freedom) would still never be attained by him; for his nature is not of the sort to call a halt anywhere in possession and enjoyment and to be satisfied (*CPJ* V 430.16–23).

This cannot be considered a conceptual argument, for human beings could have been otherwise; it is empirical, because all the evidence is that they are not. This first empirical argument is then followed by a second empirical argument, which is that there is no evidence that nature outside of our own skins has any concern for human happiness, and plenty of evidence that it does not:

> And further, it is so far from being the case that nature has made the human being its spe-
> cial favorite and favored him with beneficence above all other animals, that it has rather
> spared him just as little as any other animal from its destructive effects, whether of pesti-
> lence, hunger, danger of flood, cold, attacks by other animals great and small, etc. (*CPJ* V
> 430.24 – 7).

And Kant then completes his series of empirical arguments with a second argument that is explicitly about "the nature inside of us", namely that we seem to be
very good at making problems for each other and thus even thwarting any ar-
rangements that nature might seem to have made for our happiness:

> even more, the conflict in the *natural predispositions* of the human being, reduces himself
> and others of his own species, by means of plagues that he invents for himself, such as the
> oppression of domination, the barbarism of war, etc., to such need, and he works so hard
> for the destruction of his own species, that even if the most beneficent nature outside of us
> had made the happiness of our species its end, that end would not be attained in a system
> of nature upon the earth, because the nature inside of us is not receptive to that (*CPJ* V
> 430.23 – 36).

In sum, then, the concept of happiness is not determinate enough to furnish a
determinate end for nature; and while nature outside of us does not seem particularly friendly, or even seems hostile, to our happiness, our own nature
seems both inherently unsatisfiable and prone to make life miserable rather
than happy for ourselves and each other. So for both *a priori* and empirical reasons it hardly seems that happiness could be the final end of nature.

Kant follows his arguments against happiness as the final end of nature with
his argument for human freedom as the final end of nature. His formulation of
this argument makes it hard to overlook the first paradox I have described. Kant
begins the argument with an assertion of what I have called the ontological rather than normative conception of a final end: "A *final end* is that end which needs
no other as the condition of its possibility" (*CPJ* V 434.7 – 8). Kant then makes the
inference already quoted that once we have conceived of the idea of a productive
understanding as the cause of nature (the ontologically unconditioned in the
order of causes), we must also ask what "objective ground" could have been
its reason for the creation, and equates this objective ground with a final end
—thereby preparing for the switch to the normatively unconditional value in
the order of ends. He then states that the final end cannot actually be anything
in nature itself, because a final end is unconditioned but everything within nature is conditioned, which is still an ontological point:

> I have said above that the final end cannot be an end that nature would be sufficient to
> produce in accordance with its idea, because it is unconditioned. For there is nothing in

nature (as a sensible being) the determining ground of which, itself found in nature, is not always in turn conditioned; and this holds not merely for nature outside of us (material nature), but also for nature inside of us (thinking nature)—as long as it is clearly understood that I am considering only that within me which is nature (*CPJ* V 435.4–10).

This metaphysical argument actually renders both the conceptual and empirical arguments against happiness as the final end of nature otiose, although no doubt rhetorically effective, because even if the concept of happiness were determinate and there were every evidence rather than none that nature both within and without us favors the realization of happiness, happiness would still be a natural condition, thus conditioned and as such ineligible for the status of a final end. Kant does not himself emphasize this point, perhaps because he now makes the shift from an ontological conception of what is unconditioned in the final end to the normative conception, the move justified by the implicit assumption that an unconditioned ground of nature must have an unconditionally valuable end for its reason for the creation of nature: Kant now introduces freedom as the only possible final end of nature because the law in accordance with which human beings must make their own freedom the unconditioned condition of their pursuit of all other ends is itself necessary and unconditioned by any natural considerations. Thus Kant writes:

> Now we have in the world only a single sort of beings whose causality is teleological, i.e., aimed at ends and yet at the same time so constituted that the law in accordance with which they have to determine ends is represented by themselves as unconditioned and independent of natural conditions but yet as necessary in itself. The being of this sort is the human being, though considered as noumenon: the only natural being in which we can nevertheless cognize, on the basis of its own constitution, a supersensible faculty (*freedom*) and even the law of [this] causality together with the object that it can set for itself as the highest end (the highest good in the world).
>
> Now of the human being (and thus of every rational being in the world), as a moral being, it cannot be further asked why (*quem in finem*) it exists. His existence contains the highest end itself, to which, as far as he is capable, he can subject the whole of nature, or against which at least he need not hold himself to be subjected by any influence from nature [...] only in the human being, although in him only as a subject of morality, is unconditional legislation with regard to ends to be found, which therefore makes him alone capable of being a final end, to which the whole of nature is teleologically subordinated (*CPJ* V 435.15–436.2).

Three things must be noted about this crucial passage. First, it clearly treats the human being as moral being as unconditioned and therefore the only candidate for the final end of nature because of a normative ground, namely, the necessary and unconditional validity of the moral law to which the human being can and must submit his purely natural inclinations. The human being may be able to submit all influences from nature to this law because of the ontologically uncon-ditioned character of his freedom as spontaneity, but he ought to submit all his inclinations to this law because of the unconditional value of his freedom.[5] Sec-ond, the passage makes evident the problem about freedom as the end of nature that I have pointed out in its own wavering characterization of human beings as the final end: it starts out by calling them the only kind of being *in* the world that could be the final end of nature, but then describes them as the final end only "considered as noumenon", which would seem to imply that they are final ends only *outside* of nature (*CPJ* V 435.20). Kant immediately attempts to make his characterization more consistent by describing the human being as a "natural being" that nevertheless has a "supersensible faculty", namely freedom (*CPJ* V 435.21). This might take one step toward solving the problem of how something that is non-natural could be the final end of nature, namely that it is only partly non-natural, which also explains why it must be the human being and its free-dom which is the final end of nature, because it is at least partly within nature, but not God, who is wholly outside nature—but it still leaves the problem that the very faculty in virtue of which the human being is supposed to be the final end of nature is also supposed to be its non-natural faculty. Finally, we should also note that the passage suggests that it is not in fact just the freedom of human beings that is the final end of nature, but rather this "causality together with the object that it can set for itself as the highest end", namely "the highest good in the world" (*CPJ* V 435.23 – 4). But of course happiness is part of the highest good in the world, so this statement then raises our second paradox, namely how can Kant start by denying that happiness could possibly be the final end of na-ture but conclude by saying that something that includes happiness, namely the highest good, is itself part of the final end of nature?

---

5 In her comment on this paper, Goy points out that Kant actually calls at least five different things the final end of nature, namely the human being "as a moral being" (*CPJ* V 443.15 and 444.1), freedom (*CPJ* V 435.14 – 5, 21), its "law", i.e., the moral law (*CPJ* V 435.17, 22, 35), the good will (*CPJ* V 443.11), and the highest good (ibid.). However, all of these are closely related through the concept of the unconditional value of freedom: the human being is the final end because of its freedom, which is the final end; the moral law is the final end because the human being must act in accordance with it to preserve and promote its freedom and that of others; the good will is the final end because it is the human being's commitment to acting in accordance with the moral law and thereby preserving and promoting its freedom and the freedom of others; and finally the highest good is what the good human being wills because freedom is exercised in the selection of particular ends and the realization of particular ends constitutes happiness. Arguing these connections in detail is beyond the scope of the present paper, but I will discuss the last point further in section 4 of this paper.

In fact, there can be no doubt that Kant himself recognizes the first paradox, that is, that he cannot make human freedom itself the end of nature in the sense of something to be produced by nature. In §87, "On the moral proof of the existence of God", he states that "if reason is to provide a *final end a priori* at all, this can be nothing other than *the human being* (each rational being in the world) *under moral laws*", but then adds in a note:

> I deliberately say '*under* moral laws.' The final end of creation is not the human being *in accordance with* moral laws, i.e., one who behaves in accordance with them. For with the latter expression we would say more than we know, namely, that it is in the power of an author of the world to make it the case that the human being always behaves in accordance with moral laws, which would presuppose a concept of freedom and of nature (for the latter of which alone one can conceive of an external author) that would have to contain an insight far exceeding the insight of our reason into the supersensible substratum of nature and its identity with that which makes causality through freedom possible in the world [...]. According to our concepts of free causality, good or evil conduct depends upon ourselves (*CPJ* V 448.35 – 449.30).

Making human beings free is precisely what mere nature cannot do. More precisely, since human beings are by their very essence noumenally free but need to exercise their freedom in a way that preserves and promotes the possibility of the phenomenal exercise of their own freedom and that of others, that is what nature cannot make them do. Indeed, as Kant makes clear here, even God cannot make them do that; Kant assumes the ontologically unconditioned character of freedom here and infers that the morally correct exercise of human freedom can be due to nothing but human freedom itself. As free beings, humans are actually on a par with God. (There is thus of course a fundamental tension in Kant's idea that human freedom can be the final end of the unconditioned ground of creation; perhaps recognizing this is the first step to Kant's later, unpublished doctrine that God is nothing but an idea, namely the idea of our own freedom.)[6] Human freedom can be the normative final end of nature and its author, which must be respected by anything capable of respect, but it cannot be the ultimate product of anything natural.[7]

So Kant clearly recognized the first paradox I have identified, and even though he does not explicitly describe the conflict between his initial denial that human happiness can be the final end of nature and his subsequent asser-

---

6 See Guyer (2000, 43 – 50, and 2005, 305 – 12).

7 Höffe has formulated the first paradox thus: "The being, that belongs throughout to nature, namely the human being, is only the final end of nature if it transcends [*übersteigt*] nature" (Höffe 2008, 302; my translation).

tion that the highest good is at least part of that final end, that paradox is clear enough. Now how can Kant solve these paradoxes?

3 Nature's Preparation for Freedom

The solution to the first of our paradoxes is that, quite consistently with his radical view of freedom as non-natural, Kant does not actually say that nature itself can produce human freedom, but only that it can *prepare* us for the exercise of our freedom by producing *conditions favorable for the effective use* of that freedom. This is clear both within the "Critique of the Teleological Power of Judgment" and in the historical, political, and religious writings where Kant applies his critical teleology.

Kant's solution has both a conceptual and a substantive aspect. The conceptual component is the introduction of a distinction that I have so far overlooked between what nature itself can accomplish and what stands outside of nature and cannot be accomplished by nature; this is the distinction between the ultimate end (*der letzte Zweck*) of nature, which is so to speak the most important thing that nature itself can accomplish, *within* nature, and the actual final end (*Endzweck*) of nature, which stands *outside* of nature and cannot be accomplished by nature but is what gives nature its point and value. Kant thus says that

> [i]n order [...] to discover where in the human being we are at least to posit that *ultimate end* of nature, we must seek out that which nature is capable of doing in order to prepare him for what he must himself do in order to be a final end, and separate this from all those ends the possibility of which depends upon conditions which can be expected only from nature (*CPJ* V 431.12 – 7).

This makes it clear that human freedom itself cannot be produced by nature, thus that nature itself cannot produce human freedom, and that we must identify the ultimate end that can be produced by nature within nature with something related to freedom but not with freedom itself. Kant then implies that although happiness could take place within nature and thus is in that way at least a possible ultimate end of nature, it is incapable of being properly related to a final end, because if the human being makes happiness "into his whole end" that would "make him incapable of setting a final end for his own existence and of agreeing with that end" (perhaps this is yet another argument about happiness, that it can be freely chosen but not chosen as unconditionally valuable) (*CPJ* V 431.20 – 2). Kant next states that among all the ends of mankind "in nature there remains only the formal, subjective condition"—namely that of—"using nature as a means appropriate to the maxims of his free ends in general, as that which nature can accomplish with a view to the final end that lies outside of it and which can therefore be regarded as its", that is, nature's, "ultimate end" (*CPJ* V 431.23–8). In other words, what nature itself can produce can only be the means to freedom, not freedom itself; but since, as ontologically unconditioned, freedom, or at least noumenal freedom, cannot have and does not need any causal conditions, what Kant can mean is only that nature can develop in human beings means for the effective realization *in* nature of their genuinely free moral choices, which are outside of nature. And this is precisely what Kant goes on to identify as the ultimate end of nature: "*culture*", "[t]he production of the aptitude of a rational being for any ends in general (thus those of his freedom) [...] [b]ut not every kind of culture", in particular not the mere culture of "*skill* [...] for the promotion of ends in general", but rather "the culture of training (discipline)" for "the liberation of the will from the despotism of desires" (*CPJ* V 431.35–432.6). Nature can train us to control our desires, not to act immediately upon whatever desire occurs to us, and this is certainly a necessary condition for the effective phenomenal exercise of our genuine freedom to act in accordance with the moral law, which of course often requires us to resist desires. But it is certainly not a *sufficient* condition for our choice to act in accordance with the moral law, and thus to make freedom in ourselves and others our final end, because there are certainly many forms of evil-doing that also require concentration, delayed gratification, and so on, and thus control over at least some desires, even though in the end evil consists in allowing *some* desire that should not be so allowed to determine one's action. Thus, the discipline to which nature can train us can be a means for the effective use of our freedom, but to make use of this means only in accordance with the moral law requires a free choice that nature can never make for us. In this way, nature can prepare us to exercise our status as its final end, but it can never make us its final end or make us treat ourselves and others only as ends, never as means.[8]

---

8 Höffe states that the "mediation" between nature and freedom that Kant seeks "is accomplished by elements in nature that prepare the human being to be the final end through its own doing" (Höffe 2008, 302; my translation). I am suggesting that this is imprecise, for all that nature can do is to prepare us for the discipline that we need to be moral and thus fully realize our freedom, but it cannot in any way prepare us for the free choice that we need to make in order to make the freedom of ourselves and others always our end and never merely a means.

4 Nature, Happiness, and the Highest Good

The resolution of the first paradox for Kant's theory of the final end is thus clear. Let us now turn to the second paradox. Kant does not himself state the resolution of this paradox as plainly as he does his solution to the first, so here my proposal will be both more speculative and less precise.

There are two parts to this resolution. On the one hand, we must understand why happiness should be included as part of the moral end of mankind and therefore of nature at all, particularly given the hostile things that Kant says about happiness as a possible foundation for the moral law in both the *Groundwork* and the *CprR*. On the other hand, since happiness, unlike noumenally free choice, cannot be conceived of as something that takes place outside of nature, but can only be conceived of as taking place within nature, we must understand how Kant could conceive of happiness as an end of nature after all following his argument that nature favors human beings with beneficence no more than any of its other creatures. This is hardly the place for a detailed treatment of the highest good, a topic of such importance to Kant that he discusses it at length in each of the three *Critiques* (in each of which it is in fact the culminating topic), in the *Religion*, and yet again in the 1793 essay *Theory and Practice*. But let me address these two questions briefly.[9]

First, why does Kant include the highest good and thus happiness as part of the final end when, as he has argued, the final end can only be the human being in virtue of the supersensible capacity of freedom? To understand Kant's move, we must realize that, at least in his mature conception of the highest good, he does not think of happiness as a merely *natural* end of human beings, or an end of human beings as merely natural creatures, that is to be *constrained* by commitment to the moral law—in which case the happiness concerned would merely be one's own happiness, or the happiness of those near and dear to one with whom one happens naturally to identify; rather, he conceives of the relevant happiness as the happiness of *all*, collective or "unselfish" happiness, a concern for which can only be the *product* of commitment to the moral law (*Theory and Practice* VIII 280.40).

This conception depends upon Kant's recognition that to treat humanity both in one's own person and that of others as an end and never merely as a means includes a commitment to promoting the realization of the *particular ends* that all, oneself and others, freely choose in the exercise of their status as ends in themselves, combined with a definition of collective happiness as pre-

---

9 I have discussed the highest good at length in a series of papers, most recently Guyer (2011).

cisely the state of affairs that would result from the realization of the consistent and collective set of ends that would thus be promoted. As Kant argues in the *Groundwork*, "there is still only a negative and not a positive agreement with *humanity as an end in itself* unless everyone also tries, as far as he can, to further the ends of others" (*Groundwork* IV 430.21–4); as he finally admits in the "Doctrine of Virtue" of the *MM*, "since all *others* with the exception of myself would not be *all* [...] the law making benevolence a duty will include myself" (*MM* VI 451.6–9); and thus, as he argues in *Theory and Practice*, the "concept of duty [...] *introduces* another end for the human being's will, namely to work to the best of one's ability toward the *highest good* possible in the world (universal happiness combined with and in conformity with the purest morality throughout the world)", a "final end assigned by pure reason and comprehending the whole of all ends under one principle (a world as the highest good and possible through our cooperation" (*Theory and Practice* VIII 280.17–21). In *Theory and Practice* (VIII 279.28–9) Kant stresses against his critic Christian Garve that this does not make even universal happiness any part of the "incentive" of morality, but that "only in that ideal of pure reason does it also get an *object*". And this is precisely what Kant is assuming in the third *Critique* when he writes that

> [t]he moral law, as the formal rational condition of the use of our freedom, obligates us by itself alone, without depending on any sort of end as a material condition; yet it also determines for us, and indeed does so *a priori*, a final end, to strive after which it makes obligatory for us, and this is the *highest good in the world* possible through freedom [...] [which includes] the highest physical good that is possible in the world and which can be promoted, as far as it is up to us, as a final end, [...] [namely] *happiness* (*CPJ* V 450.4–14).

—just not only our own happiness.

In the first *Critique*, Kant's conception of the locus for the realization of the highest good had been ambiguous: he had held that the natural world must be transformed into a moral world (*CPR* A 808/B 836), but also conceived of the happiness included in the highest good as something that obviously cannot be counted upon in the natural world and must therefore be deferred to a life in a "world that is future for us", thereby employing the condition of the possibility of happiness as the ground for the postulate of immortality (*CPR* A 811/B 839). In the *CprR* and all subsequent works, however, Kant is adamant that happiness, as the fulfillment of ends that begin as desires, can only be achieved within nature or our natural lives, and thus uses it as the ground for the postulate of the existence of a "supreme cause of nature having a causality in keeping with the moral disposition" (*CprR* V 125.13–4) (although in the second *Critique* Kant also held that *virtue* could only be perfected in an immortal life, thereby inconsistently implying that the happiness of which we are made worthy might precede the perfection of virtue; after the second *Critique* Kant still mentions immortality as one of the postulates of pure practical reason but gives no further argument for it). So this brings us back to the second question that we must resolve in order to solve the second paradox in Kant's theory of the final end, namely how can he make happiness even part of the final end of nature when he has been so adamant that nature shows no special regard for our happiness? To this question Kant himself does not offer us an explicit answer, but there are two points that we might mention. First, it should be noted that in *Theory and Practice* Kant is careful to say that the object of morality is to work *to the best of our ability* toward universal happiness combined with the purest morality throughout the world; he is thus not committed to the claim that nature *guarantees* human happiness, but only to the claim that nature makes it *possible* for human beings to work toward happiness and even prepares them to work for it, but consistently with the other facts about nature. Thus Kant need not deny that nature produces floods and earthquakes that damage or destroy the happiness of many human beings, but need only suppose that nature produces dispositions within human beings that will enable them, if accompanied by the right moral choice, to work for as much human happiness as is possible in the face of such disasters. And nothing Kant has said about nature's indifference to human happiness is, strictly speaking, incompatible with that.

The second point to make here is that Kant had emphasized as early as the essay on *Universal History* that "[i]n the human being [...] those predispositions whose goal is the use of his reason were to develop completely only in the species, but not in the use of the individual", and that reason "knows no boundaries to its projects" (*Universal History* VIII 18.29 – 33, 19.1). Kant defines reason here as "a faculty of extending the rules and aims of the use of all its powers far beyond natural instinct" (*Universal History* VIII 18.32 – 4). This can include both technical reasoning, which figures out how to solve problems human beings encounter in nature, and pure practical reason, which informs us of the moral law. Kant can see the former as developing by entirely natural means, although of course he must see the latter as non-natural and, as I have argued, must conceive of nature as merely developing the means that it can use to its own ends if the pure will so chooses. But he can see both sorts of reason as being put to use for the complete development, including the happiness, of the species, not the individual, and even if he does not defer that happiness to a non-natural life, he could presumably defer it far into the future of the natural life of the species. Thus he could argue that even though nature *presently* seems indifferent or even hostile to human happiness, human ingenuity combined with human morality might *eventually* overcome many of nature's obstacles to human happiness—and that would suffice, for remember that our duty is only to work toward human happiness as far as we are capable of it. And we could then think of the development of human technical reason and of means necessary for the effective use of human pure practical reason, both of which together will ultimately produce the highest good insofar as we are capable of it, as the ultimate end of nature, or the final end of nature within nature, without outright contradiction.

Here I am no doubt helping Kant out; he could have made things easier for us had he explicitly said that nature's indifference or hostility to human happiness, which he so graphically described, is in fact only apparent or initial. But he did not. However, he did explicitly explain in what sense the non-natural end of human freedom can be an end of nature, and thus an end for the human as a biological organism living with and using other organisms, namely that nature can develop the means that human beings need to make their freedom effective in realizing the goals that it sets for their lives in the natural world, and he did explain clearly enough why happiness should be part of those goals even though it cannot be the foundation of the moral law or the incentive for our compliance with the moral law. So perhaps we should be content with that.[10]

References

Guyer, Paul 1995, Nature, Morality, and the Possibility of Peace, in: *Proceedings of the Eighth International Kant Congress*, ed. by Hoke Robinson, Milwaukee: Marquette University Press 1995, vol. I.1, 51–69; reprinted in: Paul Guyer 2000, *Kant on Freedom, Law, and Happiness*, Cambridge: Cambridge University Press, 408–34.

Guyer, Paul 2000, The Unity of Nature and Freedom: Kant's Conception of the System of Philosophy, in: *The Reception of Kant's Critical Philosophy*, ed. by Sally Sedgwick, Cambridge: Cambridge University Press, 19–53; reprinted in: Guyer 2005, 277–313.

Guyer, Paul 2001a, From Nature to Morality: Kant's New Argument in the 'Critique of Teleological Judgment', in: *Architektonik und System in der Philosophie Kants*, ed. by Hans Friedrich Fulda and Jürgen Stolzenberg, Hamburg: Meiner, 375–404; reprinted in: Guyer 2005, 314–42.

Guyer, Paul 2004, Zweck in der Natur: Was ist lebendig und was ist tot in Kants Teleologie?, in: *Warum Kant heute?*, ed. by Dietmar Heidemann and Kristina Engelhard, Berlin/New York: Walter de Gruyter, 383–412; English in: Guyer 2005, 343–72.

Guyer, Paul 2005, *Kant's System of Nature and Freedom*, Oxford: Oxford University Press.

Guyer, Paul 2006, The Possibility of Perpetual Peace, in: *Kant's Perpetual Peace: New Interpretative Essays*, ed. by Luigi Caranti, Rome: LUISS University Press, 161–82.

---

10 In preparing the final version of this paper I have benefited from the comments presented by Ina Goy as well as by insightful suggestions by Eric Watkins and Günter Zöller during the discussion of the previous version.

Guyer, Paul 2007, Natural Ends and the End of Nature, in: *Hans Christian Ørsted and the Romantic Legacy in Sciences*, ed. by Robert M. Brain, Robert S. Cohen, and Ole Knudsen, Dordrecht: Springer, 75 – 96.
Guyer, Paul 2011, Kantian Communities, in: *Kantian Communities: The Realm of Ends, the Ethical Community, and the Highest Good*, ed. by Lucas Thorpe, Rochester: University of Rochester Press.
Höffe, Ottfried 2008, Der Mensch als Endzweck, in: *Immanuel Kant: Kritik der Urteilskraft*, ed. by Otfried Höffe, Berlin: Akademie Verlag, 289 – 308.

Ernst-Otto Onnasch

The Role of the Organism in the Transcendental Philosophy of Kant's *Opus postumum*

In the late eighteenth century, there was perhaps no group of scientists that studied the critical philosophy of Immanuel Kant as intensively as the biologists, like Johann Friedrich Blumenbach (1752–1840), Johann Christian Reil (1759–1813) and Gottfried Reinhold Treviranus (1776–1837). Therefore, an impartial observer might come to think that Kant puts modern biology on the scientific map of his time. In fact, however, one can hardly claim that Kant took biology seriously as a science. In the *Metaphysical Foundations*, he leaves his readers with no illusions about the scientific status of what we would now call biology: biology is not a proper natural science and will as a matter of principle never gain such status. The reason for this is that biological determining grounds "in no way belong to representations of the outer senses, and so neither to the determinations of matter as matter" (*MAN* IV 544.15–6). Moreover, in contrast to physical sciences dealing with non-living objects (e. g., mechanics), biological disciplines do not allow of a metaphysical, i.e., a priori foundation.[1] If organisms do not allow for a presentation according to a priori principles in intuition, they cannot be intuited as organized. Already in the 1770s, Kant defined life as a "movement in the transcendental sense" (*Notes and Fragments* XVII 728.4), and thus not as a movement in the physical sense, which stands under physical conditions. In the first *Critique*, Kant points out that the principle of life is of immaterial nature and therefore "only intelligible, not subject to temporal alterations at all" (*CPR* A 779–80/B 807–8).

As a consequence of the impossibility of constructing biological (and chemical) phenomena in intuition under the conditions of space and time, a further problem emerges, namely whether these phenomena can as such be intuited at all and thus obtain the attribute of real existence within the Kantian framework (note that space and time are the conditions for the existence of things as appearances, see *CPR* B xxv). That at least organic phenomena have to have a real existence seems to be indisputable, at least because of the bare *distinction* Kant allows between organic and inorganic nature (which, of course, is

---

1 For a similar account of Kantian biology not being a proper science, see Friedman (2006, 56).

not a distinction between really existing and not existing things). In the *OP* Kant even notes that

> the classification of bodies in physics and for physics cannot be the division in organic and inorganic bodies for the very concept itself is problematic, i.e., it is doubtful whether bodies exist at all (namely corporal formations in figure, texture and energy) which contain a principle of life in itself (*OP* XXIII 483.29–34).

Consequently, the question arises as to how, within the Kantian framework, it is actually possible to distinguish organic from inorganic bodies.

In the following, I will first lay out the historical sources of Kant's distinction between organic and inorganic nature. In the second and third sections, I will develop an exposition of the systematic problems of this distinction mainly with respect to the solutions provided in the leaves of the *OP*. I will then elaborate the role of the organism within Kant's system of knowledge. The problem is that organisms belong to physics but are not constituted by the moving forces of matter; as a consequence, the organization of organisms cannot be experienced as something that is actual. In the final section, I will make a proposal—on the basis of Kant's account of the organism in the *OP*—as to how we ought to understand the actuality of organisms as being stated *through* experience *for the sake of experience*. I will argue that the actuality of the organism results from the fact of experience.

1 Historical Exposition of Kant's Understanding of the Organism

Without doubt, Georg Ernst Stahl (1659–1734) was the most influential eighteenth-century scientist who distinguished strictly between a physical or lifeless and an organic or vital realm of nature. For the organic realm he claimed distinct laws, aims, benefits, and effects.[2] It requires no explanation that Stahl's distinction disputed the mechanistic understanding of organisms that had been dominant since the time of Descartes. However, the criterion Stahl used to distinguish both realms of nature invokes an intelligent principle that acts within organisms according to purposes. Life consists of actions not *within* matter, but *regarding*

---

2 See Stahl (1714, in 1961, 52): "The *oeconomica vitae* has its own laws, its own goals, its usefulness and effects" ("Die Oeconomica vitae hat ihre eigenen Gesetze, ihre eigenen Ziele, ihren Nutzen und ihre Wirkungen"; my translation).

matter, or operating on it. By its actions it aggregates the heterogeneous particles of corporeal matter into an organism. Consequently, it is impossible to deduce life from the physical parts of which the organism consists, or from their relations and functions. On the contrary, the immaterial forces that constitute the organism—i.e., the vital forces[3]—are located within the soul, which animates the body not in a pre-established way, but spontaneously and according to *teleological causes* (Stahl 1708, 32). In the beginning, Stahl's theory of organic phenomena met with fierce opposition in German discussions on physics. It took almost half a century before Caspar Friedrich Wolff (1734 – 1794) took up his ideas for his embryo-genetic studies that were of great influence on Blumenbach, whom Kant mentioned favorably in his third *Critique* (*CPJ* V 242.19 – 34).

Stahl's sharp distinction between organic and inorganic bodies was of great importance for Kant's thoughts on the organism as well. In *Dreams* (1766), Kant was deeply skeptical of the dominant mechanical mode of explaining the organism. He argued that we can have scientific knowledge only on the basis of "the laws of the motion of mere matter" (*Dreams* II 331.20), and that organisms cannot be reduced to such moving forces of matter. In other words, proper scientific knowledge is based on the mechanical mode of explanation and organisms cannot be mechanically explained. For this reason he criticized the biologists of his time like Herman Boerhaave (1668 – 1738), who together with his fellow Newtonian biologists allowed for "mechanical causes" alone in explaining organisms (*Dreams* II 331.28 – 33)[4]. Kant defends the Stahlian perspective that organisms presuppose immaterial forces for their constitution.

Kant acknowledged Stahl's argument for the uniqueness of the organic realm (*Dreams* II 331.28 – 33), and adopted the *general* Stahlian distinction between an organic and inorganic nature for a very specific purpose. (Kant did not, however, adopt the specific ontological consequences of this account). This purpose was obviously to reject hylozoism, i.e., the view (widely accepted at the time) that matter can be animated in order to explain organic phenomena. Kant rejected this view, for he considered it to be self-contradictory. Kant adopted Stahl's distinction to argue for organisms as a second type of natural object in order to reject hylozoism, for accepting it entails the end of proper natural sci-

---

3 The term 'vital force' is not used by Stahl but is rather introduced by French physicians in the 1770s on the basis of Stahl's account, see Stollberg (2013).

4 Kant's criticism of the Newtonians is remarkable given that he dedicated his first publication, *True Estimation* (1747), to Johann Christoph Bohlius (1703 – 1785), who was a disciple of Boerhaave.

ence.[5] The acceptance of animated matter forces one to give up inertia as the essential feature of matter. By doing this, of course, the keystone of Newtonian physics would be removed. So Kant's earliest reflections on the nature of organisms are clearly directed against those scientific approaches that attempt to explain organisms by means of moving forces of matter alone. Instead, Kant defended the Newtonian idea of a proper science strictly based on a mechanical mode of explanation, dismissing all forces that do not fit this model of science. Kant's reference to Stahl's strict distinction between organic and inorganic nature, therefore, does not imply that he takes a position in the dispute within physiology, but rather that he uses this distinction in order to purify science from false presuppositions. In virtually all cases, Kant's reflections on biology are meant to criticize the exuberant use of reason by his contemporaries, among who most prominently of course was his former disciple Herder. For Kant, such exuberant use cannot rely on scientific innovations in contemporary biology. Also his view that purposiveness itself is something that needs a critique is not caused by contemporary debates in physiology, but arises from systematic problems of his critical philosophy.

2 Systematic Exposition of Kant's Understanding of Organisms

If we look at Kant's writings from his critical period, his comments on teleology in nature do not differ significantly from those of his earlier period. According to the first *Critique*, it is impossible to give a proper proof that there is no purposiveness in nature (see *CPR* A 688/B 716). Nevertheless, proper natural science cannot use teleology to explain natural phenomena. In his *Teleological Principles* (1788) Kant holds that "no one has an a priori insight that there are purposes in nature" (*Teleological Principles* VIII 182.16–7). Here the negative argument of the first *Critique*—namely that it is *impossible* to prove that there is no purposiveness in nature—receives a counterpart, as Kant now claims that no a priori argument can be given for the existence of purposive beings in nature, like organisms. A fortiori, no a priori argument can be given for the real existence of organisms. An empirical argument for their existence is impossible too, for this would entail appearing objects that are qualified as purposive. Intuition,

---

5 This argument can be found explicitly in *MAN* IV 544.25–6: "the death of all natural philosophy would be *hylozoism*", or in his lectures on metaphysics: "hylozoism is the opinion that matter has life—this is the death of all physics" (*Lect. Met. Dohna* XXVIII/2.1 687.6–7).

however, does not provide a framework that is able to recognize the teleological structure of the organism and therefore the organism as organism. Consequently, according to Kant, organisms can be no existing objects.

The organism consists of matter, which can be intuited, but the organization that accounts for the specific difference of the organism with regard to inorganic matter cannot be intuited. Therefore, the criterion for distinguishing organic or teleological determined nature from inorganic nature must depend on Kant's distinction between the two entirely heterogeneous cognitive faculties, namely the understanding for concepts (possibility) and the intuition for objects (actuality). Since the actuality of organisms cannot be found in any intuited natural object, the problem arises how organisms can be known to obtain in actuality. Kant states in the third *Critique* that this problem affects transcendental philosophy in general. It is due to "the nature of our (human) cognitive faculty" that we are obliged to think things as actual "without asserting that the basis for such a judgment lies in the object" (*CPJ* V 401.21–6). And further:

> The propositions, therefore, that things can be possible without being actual, and thus that there can be no inference at all from mere possibility to actuality, quite rightly hold for the human understanding without that proving that this distinction lies in the things themselves (*CPJ* V 402.14–8; see also Förster [2008, 267–8]).

§76 of the third *Critique* addresses this problem, but clarifies right away that its considerations "would certainly deserve to be elaborated in detail in transcendental philosophy" (*CPJ* V 401.5–6). For this reason, the exposition in this section remains "a digression" (*CPJ* V 401.6) without offering any proof of what is expounded. Kant refers to an exposition that does not belong to critical philosophy, but to transcendental philosophy.[6] Is he hinting here at an extra work that is yet to be written and which is to solve the problem at issue? This extra work must be related to the *OP*. Indeed, Kant designates this work (mainly in its latest leaves) as the highest standpoint of transcendental philosophy. A closer look at the *OP* and its exposition of the functions of the organism can, therefore, shed some light on the direction in which Kant was thinking in order to solve the proposed problem. However, before turning to the *OP*, I will briefly treat Kant's 1788

---

6 That critique and transcendental philosophy do not have the same extension is clearly stated in the first *Critique*: "Transcendental philosophy is here the idea of a science, for which the critique of pure reason is to outline the entire plan architectonically" (*CPR* A 13/B 27). In student transcripts from his lectures on metaphysics, Kant briefly states that transcendental philosophy is prior to the critique of reason (see *Notes and Fragments* XVIII 285.28). According to transcripts from his lectures on metaphysics (from 1794/5) transcendental philosophy is "the product of the critique of pure reason" (*Lect. Met. Vigilantius* XXIX/1.2 949.14–24).

244 — Ernst-Otto Onnasch

essay. This essay addresses several problems of the organism beyond the scope
of the third *Critique*, which is why it is more than likely that it precedes certain
discussions on the organism in the *OP*.

3 Developing the Problem of the Actuality of the Organism

Kant's *Teleological Principles* (1788) is an important publication with regard to
the systematic problems that eventually necessitated a third *Critique* (though it
is devoted primarily to a discussion of Georg Forster's views on human
races).[7] In general, the essay addresses the problem of how to understand purposiveness in nature. Kant explains first that we cannot understand the purposiveness in the objects of nature. An organic being is defined as

> a material being[8] which is possible only through the relation of everything contained in it to
> each other as end and means [...] Therefore a basic power that is effectuated through an
> organization has to be thought as a cause effective according to *ends*, and this in such a
> manner that these ends have to be presupposed [zum Grunde gelegt werden] for the pos-
> sibility of the effect (*Teleological Principles* VIII 181.1–7).

With regard to organisms, a purposive cause is effective only if the purpose
grounds the possibility of the effect. Therefore, the actuality of a purposive
being or an organism, i.e., its basic power, depends on its possible effects. However, to base actuality on possibility is, of course, highly problematic and opens
the door wide to speculation, and eventually renders objects of thought actual.

In order to prevent the invalid inference from possibility to actuality, Kant
specifies that we know such basic powers that bring about purposive objects,
"with respect to their *determining ground* by experience only *in ourselves*, namely
in our understanding and will" (*Teleological Principles* VIII 181.8–10). Understanding and will are qualified as our basic powers, and the will, if determined
by the understanding, is a faculty by means of which we are able to "bring about
something *in accordance with an idea* called purpose" (*Teleological Principles*
VIII 181.13–4). Such purposive products are artworks. However, it is by *experi-*

---

7 The first pages of *Teleological Principles* (VIII 159–61) summarize the content of the second,
teleological part of the third *Critique*.
8 Note that the organism may not be taken as living matter, for matter is characterized by inertia
(see *CPJ* V 394.26–8).

ence in ourselves with respect to a will determined by the understanding that we are aware of the actuality of purposive objects.

Such objects are, of course, not appearances, but are our own products produced by thought. And it is experience, as Kant states, that must offer an "example" or an empirically ascertainable "intention" that causes a purposive object, for otherwise "the concept of the faculty of a purposive acting being" would be "totally fictitious and void" and therefore "without the slightest guarantee that any object could correspond to it at all" (*Teleological Principles* VIII 181.17– 182.1). Moreover, it is only by means of empirically accessible intentions that we are allowed to assume the actuality of purposive beings or organisms (will and understanding that constitute the fundamental teleological powers are objectively real). The actuality of organisms is consequently not based on a possibility by means of mere thought, but is based on experience offering an example of actual organic bodies. According to Kant experience presupposes sensible beings, so sensibility in combination with the mere fact of experience offers the example for the actuality of the organization of a being with experience. Experience, however, seems to be possible only for sensible beings, so the mere fact of experience entails the actuality of my own organization. Is organization thus something that belongs to the conditions for the possibility of experience? If this were the case, it cannot come from experience, but must be a priori. And this must be prevented, for an a priori account of organization is fictional and void. In order to have experience of our organization

> [i]t is in fact indispensable for us to subject nature to the concept of an intention if we would even merely conduct research among its organized products by means of continued observation; and this concept is thus already an absolutely necessary maxim for the use of our reason in experience (*CPJ* V 398.12– 6).

In other words, we can understand ourselves as organisms because we subject nature to the concept of an intention or of a will determined by understanding that we insert into nature. However, this does not imply that Kant retracts in any way his fundamental criticism that *knowledge* of the ground of purposive connections in nature is impossible in whatever way.

The implicit argument in Kant's *Teleological Principles* is that if experience is real, then those who are in possession of experience must be understood as purposive beings that are also capable of positing their own purposes into nature according to the basic powers of will and understanding with which they are equipped. This leads to the conclusion that experience presupposes purposive beings, i.e., organisms, for only such purposive beings are empirically accessible.[9] However, in this conclusion one aspect is not captured, namely that all experience is the result of an intention determined by understanding. In Kant's essay, this crucial argument is not yet explicitly present, but it is elaborated in the *OP* that deals with this essential aspect of transcendental philosophy. But let us first clarify what is meant by the intention exercised by experience.

According to the first *Critique*, objective knowledge requires the use of categories with respect to data perceived through sense perception, i.e., data given in intuition. What is given in intuition, however, is a vast manifold of data, and certainly not all of these data are brought under concepts and thus raised to the level of knowledge. The reason for this is that we obviously *direct* our understanding towards the given manifold of data. In other words, from the vast manifold of data, we make a selection that we bring to knowledge. Consequently, knowledge of natural objects results—next to the categories and forms of intuition—from an *intention* or a *purpose* according to which the immediately intuited manifold is limited and organized into those appearances that we construct as objects of knowledge (i.e., mediated appearances).[10] Hence, the possibility of our empirical knowledge depends on an intentionally driven selection from the manifold given in intuition. Moreover, this intentionally driven being that makes its own experience cannot be merely possible in thought, but must be actual too, even though it is not itself given in intuition. (Note that we are talking here about an organism having knowledge.)

Let me give an example of what I mean. My sensible intuition of a rose does not involve that its particular redness, smell, and all its other particularities appear to me as well. In other words, only some of the particular attributes of the manifold data related to that intuited rose need actually to appear to me so that I can claim to have knowledge about that rose. What appears to me of this rose is just what I *deliberately*, i.e., by a certain intention, pick from the given manifold and turn from an immediate appearance into a mediated appearance. We thus intentionally turn that which is immediately given into an appearance of which we are conscious.[11] The intention, by means of which a mediated or sec-

---

9 Förster (2000, 26–8), correctly locates these types of arguments in the *OP*, presenting them as a novelty of Kant's last work. However, it seems clear that these arguments are already anticipated—albeit more or less implicitly—in Kant's *Teleological Principles*.

10 In the first paragraph of section VI of the "Introduction" to the third *Critique*, Kant clarifies that the correspondence of nature with its universal principles is contingent but nevertheless indispensable for the needs of our understanding, because nature has to agree with our *intention* to acquire knowledge, what implies that nature itself must be purposive (see *CPJ* V 186.25–187.10).

11 See Hübner (1953) for this conception of the production of appearances.

ond-order appearance is produced, must indeed be understood as a will determined by the understanding.

In the *OP* Kant shows that we make the objects that we have knowledge of by
means of this double structure of receiving appearances. Illuminating in this respect is a passage where Kant argues, first, that in order that we have appearances an object needs to affect us empirically (*per receptivitatem*), second, that this
results in a subject affecting itself (*per spontaneitatem*) and, third, that by this it
brings about the object of physics as it is made by us (see *OP* XXII 405.14–20).
Kant strongly emphasizes in the *OP* the fact that transcendental philosophy cannot depend on the given object, but must *produce* this object, for it has to be *my*
object. This produced object is the object that we have knowledge of; it is related
to the affecting object through the organism that mediates between the latter and
the faculty of understanding.

The mediating organism in this case, however, does not receive its organization from the fact that we are affected by an object, but rather results from the
spontaneous self-affection by which the subject produces its own object of
knowledge. The production of such an object, therefore, also entails that its producing subject turns itself into an object of physics, i.e., that the subject becomes an organism:

> The subject (object in the appearance) affected by empirical intuition is, insofar as it affects
> itself according to concepts, an organic body intuiting according to the five senses (*OP* XXII
> 388.3–6).

The organic body results from the subject that produces experience, and it therefore takes part in the process by which experience is made. In this process, however, its organization is not intuited, and insofar as it is called a product of nature, it is a product of the subject that makes all the objects of nature itself. Still,
the organism must be actual in this process. The question therefore is where this
actuality derives from, as it is not given by sensible intuition. This problem is indeed crucial, for it can make or break the entire critical project. Kant's line of
argument is that the organism does not receive its actuality from the affecting
object, nor from sensual intuition, but from the fact that the organism is necessary *for the sake of experience.*

That the organism plays an essential role in the process of acquiring knowledge is a view that becomes most prominent in the unfinished transition project, the so-called *OP.* But already in the third *Critique*, a glimpse of this essential
role can be discerned when Kant mentions in §75 that it is "indispensable for us
to subject nature to the concept of intention" for it is an "absolutely necessary
maxim for the use of our reason in experience" (*CPJ* V 398.12–6). In his lectures

on metaphysics delivered in 1792 he even moves one step further by stating that experience structuring thought entails an organism: "We can thus contrive no experience through which we could become aware of the faculty of thinking without body", i.e., an organic body (*Lect. Met. Dohna* XXVIII/2.1 683.38 – 684.1). In the *OP*, Kant draws heavily on the necessity of an organic body for experience. He states that for the sake of a doctrine of natural science "in the subject an organic principle of the moving forces" is presupposed "in [the form of] universal principles of the possibility of experience" (*OP* XXII 373.1 – 4). It is according to these universal principles of possible experience that a doctrine for the investigation of nature can be established in the first place. Here Kant is very clear that the organic forces, performed by the organism, belong to the universal principles of the possibility of experience. It is by means of this organism, i.e., the immaterial principle that determines a body according to laws, that the subject affects itself and "makes itself an object of experience" (see *OP* XXII 373.30 – 3). This is also why Kant calls the object an "appearance of an appearance" (ibid.).

This important notion, "appearance of an appearance", can be used in the *OP* to explain that in the process of producing knowledge the organism mediates the immediate appearance by making it an appearance for the subject. The vast manifold of the immediate appearance (e.g., all the data related to an intuited rose) is processed, in accordance with a purpose, by the organism into a mediated appearance that becomes the object of knowledge (e.g., a rose to which we assign specific sets of attributes). By means of this mediation the mediated appearances are part of the unity of experience, and this is the condition for formulating the doctrinal system of the investigation of nature, by means of which we know a priori that every appearance is part of *one* and the same system of experience. It is, indeed, impossible to know merely from an object given in intuition that an appearance is part of the unity of experience.

Kant remarks: "The unconditioned unity of the manifold in intuition is not *given* to the subject by another object, but is *thought* through itself" (*OP* XXII 443.28 – 30). That the unity of experience is thought through the subject entails that the subject provides the form according to which it affects itself and produces the object that is part of the unity of experience. If the intuited object can be thought in such a way that it becomes part of the unity of experience, its *actuality* in the intuition must always have been part of this unity. The latter is fulfilled by means of the organic body. The form by which it mediates the intuited objects in order to become objects for the understanding is prescribed to it by the subject. Because this form is always linked with the actuality of intuited objects that it mediates, the organism is also actual, for it brings forth actual objects that appear for the understanding. The intention executed by the organism is the cause of the mediated appearance of any actual object, and this cause is empirically accessible.

Kant's new understanding and interpretation of our organization seems completely plausible, for his main point is that we cannot understand ourselves as having experience if we are not an organism equipped with senses. However, throughout the entire *OP* Kant makes clear that we can neither prove nor postulate the possibility of our organization (see *OP* XXII 481.8–9), and that we know ourselves "in experience as an organic body" (*OP* XXII 481.10). Moreover, it is because experience provides the real concept of our organization (and of organization in general) that this is not a fiction. The idea that experience provides objective reality to the concept of organization is made more explicit in the *OP* than in previous works, like the third *Critique*. However, in many other respects the views articulated in the *OP* are quite similar to those of the third *Critique*. For example, the *OP* does not advocate the self-organization of matter, i.e., hylozoism, nor does it presuppose a soul for such an organization.

The efficient cause of the organism is analogous to the intellect, as the organism results from the subject that produces experience. Because of this, the very activity of the organism cannot be sensed and is thus not a part of the sensual world. Kant characterizes the organism as a "cause without a place" (*OP* XXII 291.8). His remark quoted above from the 1770s that the organism performs a "movement in the transcendental sense" therefore is still pertinent. New to the third *Critique*, however, is the idea that the movement of the organism is a transcendental condition for experience. This movement is qualified as a reciprocity embedded in a network of actions and reactions, which constitutes the whole of experience:

> The understanding has the faculty for making an empirical representation of a sense-object for itself, and so, too, the perception of an object, by means of the fact that it [i.e, the understanding] stimulates a priori the moving forces of the object on which it acts to reciprocity (*OP* XXII 503.14–7).

According to this reciprocity, the understanding "can enumerate a priori these actions with their reactions which, since they are merely relations of differing quality, only belong to perception" (*OP* XXII 503.17–20). And this is why the moving forces of the organism belong to the transition to physics, and not, once more, to the given empirical objects, as an intellect is presupposed "containing a priori a principle of the composition of the moving forces of matter that influence the faculty of perception" (*OP* XXII 401.16–8). Seen in this way, it is not the case that the representations of sense-objects enter the subject, but rather that these sense-objects, together with the subjective principle of their composition, become knowledge in order to think objects as appearances.

The criterion for the classification of organic and inorganic bodies, according to Kant's reflections in the *OP*, lies in the immaterial intellect according to which intentions and thus actions according to a purpose are possible at all. These intentions or purposive actions must be a priori for they are necessary conditions for the possible composition of the moving forces of matter by the understanding:

> The idea of organic bodies is *indirectly* contained *a priori* in that of a composite of moving forces, in which the concept of a real *whole* necessarily precedes that of its parts—which can only be thought by the concept of a combination according to *purposes* (*OP* XXI 213.1–5).

The possibility of organized bodies, however, cannot be known a priori, "hence their concept can only enter physics through *experience*" (*OP* XXII 356.3–5). Physics must therefore include

> thought-entities (*entia rationis*), as problematic, for the division of possible moving forces of matter; these are thought as so constituted that they *cannot* be thought *otherwise* than through experience. Of this kind are organic bodies (*OP* XXII 406.28–31).

In other words, experience entails thought-entities that require experience in order to be accessible for thought.

4 The Function of the Organism within the System of Experience

The later leaves of the *OP* clearly attempt to unite in our experience the realm of nature with the realm of purposes. The consequence of speaking of "our" experience is that the human being itself is addressed.

In the latest leaves of the *OP*, the seventh and first fascicles, Kant evaluates this consequence rigorously. In the seventh fascicle, the so-called "*Selbstsetzungslehre*" plays a central role in constituting us as embodied beings producing and having experience, and thus knowledge.[12] This doctrine concerns the constitution of transcendental philosophy itself, or knowledge of myself as a person that constitutes itself as a principle and that is its own originator (see *OP* XXII

---

12 Förster (1989) offers a general account of the *Selbstsetzungslehre* in the *OP*.

54.3–4). To be one's own originator has implications for experience, because we do not only produce what we can possibly know, but also the means for that very same knowledge, i. e., the system of reason. As we have seen, the organism plays an essential role in this system. Kant does not reflect on the *spontaneity* of the organism, but on its *receptivity* (the former would threaten to turn Kant's views into hylozoism):

> There is no spontaneity in the organization of matter but only receptivity from an immaterial principle of the formation of matter into bodies, which indicates [*geht auf*] the universe, and contains a thoroughgoing relation of means to ends. An understanding (which, however, is not a world-soul) [is] the principle of the system, not a principle of aggregation (*OP* XXII 78.16–20).

This quotation is significant, as it shows that the receptivity of the organism establishes a thoroughgoing purposive nexus regarding the complete knowledge of the universe, i. e., the world. Kant also makes clear that this purposive relation is not established by a Platonic world-soul,[13] but by understanding. This indicates that the understanding is not a principle of aggregation and thus not responsible for a *real* thoroughgoing purposive determination of the objects in the world, for such an account would obviously result in a kind of Spinozism. Rather, the receptivity of the organism appears as a principle of the system of understanding by providing a thoroughgoing relation of means and ends of which every appearance is a part. The organism's "real character (causality), […] points in the direction of a world organization (to unknown ends) of the galaxy itself" (*OP* XXII 549.26–8).

Understanding and will are exercised by man, and the very existence of a being endowed with these capacities brings a world into existence that is bound together by means and ends—even though we do not have any knowledge about the origin of this highly complex world-system. What is important, however, is that this existence is not established by the object but by experience itself, and it is because of this that there is no insight into the source of purposive relations in the real world.

With respect to the borders marked by the third *Critique*, Kant seems to make a new point in the *OP*, namely that our knowledge of the world is connected with the purposiveness of that world via the organism. Next to the fact of experience

---

13 It is most likely that Kant also rejected the views on the world-soul developed by Salomon Maimon (1753–1800) in 1790. Maimon had sent his essay *Ueber die Weltseele* to Kant, as indicated by Maimon's letter to Kant from May 9th, 1790 (*Correspondence* XI 171.10–30). Johann Schulz (1739–1805) planned a refutation of Maimon's views (see Kant's memo in *Notes and Fragments* XIX 317.27–30).

there is another fact presupposed by experience, namely the existence of an organism, i. e., a body exercising an intention in order to have or to produce experience. In the *OP*, however, no determination of the organism as an object of knowledge takes place. This is consistent by all means, for the organism is not an entity given in intuition.

However, if an intentional being and thus an organism is presupposed for experience in order to make experience, the purposiveness of nature is not added by thought and a fortiori purposiveness can no longer be understood as a subjective necessity of the power of judgment. This new status of the organism seems to violate the very heart of the critical philosophy, because the organism is a condition for possible experience and thus for our knowledge of objects, but it is also the condition for its own actuality, namely if experience is made.

Kant tackles this problem by distinguishing two kinds of sense-objects: first, objects "that can be given in experience" and second such objects "as can *only*— if they actually do exist—be given by means of experience; that is, one would not even be able to assume them as possible, were experience not to prove their actuality" (*OP* XXII 457.2–19). Of the latter kind are organic bodies, in contrast to inorganic objects, which are of the first kind. However, organisms are *not* given in experience. The organism, as the *OP* stresses over and over again, is condition *for the sake of experience*, i. e., objects without which experience would be from the outset impossible. Organisms are thus not constructed in intuition and therefore lack a direct existence proof by means of intuition. It is due to the intentional practice performed by the subject in the process of producing knowledge that an intentional being, i. e., an organism, becomes actual, and, because of its intentional character, singles out what is an actual object for my knowledge with regard to the immediately given manifold.[14] This is why Kant never grows tired of emphasizing that experience is not just received by us, but that we must actively *make* experience.

5 How can Organisms be Actual?

It seems as if Kant needs to present two different modes of actuality. In this final section, I will specify how we should understand these two modes of actuality. On the one hand, actuality comes with the things given in intuition and, on the

---

14 Förster (2000, 28) seems to argue that it is the experience of our body that produces the actuality of organisms. I would claim, however, that the actuality of organisms follows from the mere fact of experience.

other hand, actuality results from the activity performed by the will in relation to the understanding. The latter, however, does not determine the objects as they are constituted, but rather determines the form of their being, i.e., the connection of the given manifold for the sake of the possibility of experience, "by specification of perceptions in the apprehension of appearances and their coordination according to a law" (*OP* XXII 355.8 – 10).

In the *OP*, Kant establishes that formation entails both bringing the manifold under unity and bringing unity into the manifold. In this process the actuality of the things given in intuition becomes what it is through the form that is exercised by reason in order to produce experience. Seen thus, there is no contradiction with regard to the two modes of actuality mentioned, for it is always the case that any true existence in intuition is possible existence in thought, i.e., in the system of experience. So it is due to this system of experience that the immediate and not yet conceptualized actuality in intuition can be turned into conceptualized actuality as the ground of transcendental ontology.

Now we have an account of the first mode of actuality: actuality is attributed to things relative to the form of experience. This form of experience necessarily presupposes the organism. Let us treat this idea briefly.

According to Kant, nothing given in intuition will ever become an object of the understanding, if what is given is not actively brought to understanding. And for this the organism is needed, i.e., something that is able to *cause* that understanding gains a content that is actual (e.g., on the basis of content that is merely possible, empirical knowledge would not be possible at all). This cause resides in the intention executed by the organism of a rational being and performs—as receptive—the intentional part of the subject turning immediate or first order appearances into mediated appearances. It is because of this mediation by means of the organism that we take appearances to be actual. The form of the intention performed by the organism of a rational being gives or makes the actuality of the things in our experience (*forma dat esse rei*). In this sense something is actual if it can be assigned a place in the coordinated whole of experience, "[f]or whatever we have experience of there is required a formal principle of thoroughgoing determination" (*OP* XXII 499.5 – 7).

However, how is the actuality of the organism itself preserved? This brings us to the second mode of actuality. Kant's general remark is clear, for it is *experience* that proves the actuality of organic bodies, not the objects in intuition. Hence, the "possibility of an organic body cannot be assumed, without knowledge of its actuality in experience. Thus an organic body is such as is not thinkable otherwise *than through experience alone*" (*OP* XXII 499.22 – 4). Organic bodies are thinkable only because of the fact of experience that performs a form according to which we know a priori that every perception belongs to a coordinated whole of possible perceptions. However organisms are actual in thought, this actuality is no figment *as long actual things are experienced*. Therefore, the actuality of the organism in mere thought is secured by the actuality according to which experience "is *made* for the purpose of empirical knowledge" (*OP* XXII 498.21–2). In other words, the organism is actual, because experience has to be made according to a formal principle of thoroughgoing experience, "[f]or only thoroughly determined [perception] that is, existence grounds experience" (*OP* XXII 498.10–1). It is due to the organism and its purposive structure that we are able to predetermine—by directing the understanding according to an intention—"what kind" of "*perceptions* [...] the *thoroughgoing determination* of the object of perception (that is the latter's existence) will require in the production of experience" (*OP* XXII 497.11–5). We need a formal principle a priori of the connection of the manifold of empirical intuition in order that an aggregate of perceptions of an object can count as an object that is founded in experience. The organism, however, is not an object of experience; its possibility "cannot be assumed, without knowledge of its actuality in experience" (*OP* XXII 499.21–2). And we have this knowledge because we know that we direct our understanding according to an intention towards what is given in experience in order to produce experience that contains the totality of possible perceptions. In the margin of a late leaf, Kant even states that "the organism is contained in the consciousness of oneself. The subject makes its own form in accordance with *a priori* purposes" (*OP* XXII 78.22–3).

We have knowledge of the actuality of our organism as soon as we *make* experience. The concept of an organic body "presupposes experience: For, without the latter, the very idea of organic bodies would be an empty concept (without example)" (*OP* XXII 481.14–5). In analogy with our own organization, we can presuppose the actuality of other organisms and finally the organization of the world-body according to which—as the *OP* argues, but here cannot be elucidated in more detail—every *nexus effectivus* is at the same time the *nexus finalis*. The concept of organized bodies is part of the progression in the system of perceptions of the subject that affects itself, and as such the criterion to distinguish inorganic bodies from organic bodies is one of reason. [15]

15 The Netherlands Organisation for Scientific Research (NWO) has subsidized this research. I thank the editors of this volume for their supportive comments on the drafts of this paper and Ina Goy for kindly inviting me to participate in the international symposion on Kant's theory of biology in Tübingen. I am also greatfull to the Fritz-Thyssen-Foundation for making this conference possible.

References

Carvalho, Sarah 2004, *La controverse entre Stahl et Leibniz sur la vie, l'organisme et le mixte. Doutes concernant la vraie théorie médicale du célèbre Stahl, avec les répliques de Leibniz aux observations stahliennes. Texte introduit, traduit et annoté*, Paris: Vrin.

Förster, Eckhart 1989, Kant's *Selbstsetzungslehre*, in: *Kant's Transcendental Deductions. The Three 'Critiques' and the Opus Postumum*, ed. by Eckhart Förster, Stanford: Stanford University Press, 217–38.

Förster, Eckhart 2000, The 'Gap' in Kant's Critical Philosophy, in: Eckhart Förster, *Kant's Final Synthesis. An Essay on the Opus postumum*, Cambridge (MA): Harvard University Press, 48–74.

Förster, Eckhart 2008, Von der Eigentümlichkeit unseres Verstandes in Ansehung der Urteilskraft (§§74–78), in: *Immanuel Kant. Kritik der Urteilskraft*, ed. by Otfried Höffe, Berlin: Akademie Verlag, 259–88.

Friedman, Michael 2006, Kant—Naturphilosophie—Electromagnetism, in: *The Kantian Legacy in Nineteenth-Century Science*, ed. by Michael Friedman and Alfred Nordmann, Cambridge (MA): MIT Press, 51–79.

Hartmann, Fritz 2000, Die Leibniz-Stahl-Korrespondenz als Dialog zwischen monadischer und dualistisch-'psycho-somatischer' Anthropologie, in: *Georg Ernst Stahl (1659–1734) in wissenschaftshistorischer Sicht* (Acta Historica Leopoldina 30), ed. by Dietrich v. Engelhard and Alfred Gierer, Halle: J. A. Barth in Georg Thieme Verlag, 97–124.

Hübner, Kurt 1953, Leib und Erfahrung im Opus postumum, *Zeitschrift für philosophische Forschung* 7, 204–19.

Stahl, Georg Ernst 1708, *Theoria Medica Vera. Physiologiam et Pathologiam*, Halle: Orphanotropheum.

Stahl, Georg Ernst 1714, Über den Unterschied zwischen Organismus und Mechanismus, in: *Sudhoffs Klassiker der Medizin*, ed. by Bernward Josef Gottlieb, Leipzig: Barth 1961, vol. 36, 48–53.

Stollberg, Gunnar, Vitalism and Vital Force in Life Sciences—The Demise and Life of a Scientific Conception (unpublished manuscript), http://www.uni-bielefeld. de/soz/pdf/Vitalism.pdf [last visited: November 24, 2013].

Part III. **Kant's Theory of Biology in the Present Time**

Hannah Ginsborg
Oughts without Intentions: A Kantian Approach to Biological Functions

I

Talk of functions is pervasive both in the biological sciences and in our everyday talk about the parts and behavior of organisms. We say that the function of the heart is to circulate the blood, that the function of ribosomes is to make proteins, that grooming in cats serves a hygienic function, and so on. Specifically, we use the notion of function to register a distinction between the kinds of performances and roles just mentioned, and other causal roles played by the elements of biological systems: for example, making a thumping noise in the case of the heart, increasing a cell's RNA content in the case of ribosomes, and expending energy in the case of cats' grooming behavior.

But while biologists are typically happy to speak of functions in these contexts, philosophers often find such talk to be problematic. It characterizes living things, at least on the face of it, as though they were artifacts: to say that the function of the heart is to circulate the blood rather than to make a thumping sound is like saying that the function of the fan in the computer is to keep it cool rather than to produce white noise. But in the case of artifacts, we know that they were designed by an intelligent agent, and our function ascriptions typically reflect our beliefs about the designer's intentions. By contrast, at least since the Darwinian revolution, biologists are in agreement that organisms are not the product of intentional design. So the question arises of whether we are entitled to ascribe functions in a biological context, or whether our talk of functions relies implicitly on an illicit conception of organisms as designed by an intelligent agent.

One familiar response to this problem is to try to explain the notion of function without appealing to the notion of design, intelligent or otherwise. Various versions of this response can be seen in the considerable literature on functions which has emerged since the 1970s. Surveys of this literature typically distinguish two approaches: the historical or etiological approach pioneered by Wright, and continued by Millikan and Neander among others, and the causal role approach suggested by Cummins. On Wright's original version of the historical approach, to say that Y is a function of X, rather than a mere side-effect, is to say, very roughly, that X is there because it does Y. More recent versions of the historical approach, aimed at accounting specifically for biological as opposed to artifactual functions, appeal explicitly to etiology through natural selection: to say that Y is a function of X is to say that it was "selected for" (Neander 1991, 173), or, somewhat more precisely, that Y contributed to the fitness of the ancestors of the organism to which X belongs (ibid., 174). The causal role approach, on the other hand, explains the notion of function by appeal to the current as opposed to the past effects of whatever has the function. The functions of entities or traits, on this approach, are, roughly, the causal contributions they make to the overall activities of the system to which they belong.

However, both approaches are subject to well-known difficulties. Problems for the etiological view stem from the fact that in many biological fields, such as physiology and molecular biology, functions are ascribed without reference to the history of the relevant organism. Biologists in these fields have no intention of making historical claims when they ascribe functions to the parts of an organism or cell: they are concerned only with how the part actually contributes to the workings of the containing system. So the proposed approach seems to fall short of an (even partial) explanation of what we mean by 'function', providing only a characterization which is, at best, coextensive with it.[1] Relatedly, versions which make specific reference to natural selection are unable to account for the apparent continuity either between pre- and post-Darwinian ascriptions of biological functions or between the ascription of biological functions and the ascription of functions to artifacts and their parts. And while this last difficulty does not apply to Wright's original version, a strong point of which is that it offers a unified account of biological and artifactual function, that version is systematically vulnerable to counterexamples (requiring us to ascribe functions to the gas leak which renders unconscious the scientist who would otherwise mend it, the obesity which prevents someone from exercise that would enable him to lose weight, and so on [Boorse 1976]).

Some of these difficulties are avoided by the causal role approach, which allows the ascription of biological functions without regard to historical fact, and

─────────

1 This is not an objection if the etiological view is understood, following Millikan (1989, 289–91), as providing a "theoretical definition" of function, on analogy with the definition of water as HOH, rather than an explanation of what the term 'function' means. (For a somewhat contrasting approach, see Neander 1991.) However, I regard the analogy with the definition of water is misplaced, because I do not take the term 'function' to pick out a natural biological kind, the way 'water' picks out a chemical kind. At least on the face of it, biologists use the term 'function' not as a technical term within biology, but rather in its intuitive sense, that is, in the same sense in which we ordinarily speak of the functions of artifacts. Worries about the legitimacy of the intuitive notion of function in a biological context might motivate us to reject this *prima facie* plausible understanding of biologists' use of the term 'function', but, as will become clear in section III, part of my aim in this essay is to undermine that motivation.

accommodates artifactual as well as biological functions. But the causal role approach does not allow for functions of artifacts which are not part of complex systems (for example paperweights and doorstops [see Wright 1973]). Conversely, it seems to commit us to ascribing functions to parts of complex systems which we intuitively do not regard as functional, either because they are neither biological nor the products of intentional design (clouds in weather systems [Millikan 1989, 294]), or because they are themselves evidence of dysfunction (mutant DNA sequences in the formation of tumors [Kitcher 1993, 272]). And, in contrast to most versions of the historical view, it does not allow us to describe Y as a function of X unless X actually does Y, which seems to rule out the possibility of X's having a function which it fails to perform (Millikan 1989, 294–5; Neander 1991, 181–2).

Some philosophers have responded to these difficulties by attempting to rehabilitate the connection between function and design. Kitcher, in particular, proposes to identify the function of a trait or entity as what it is designed to do. This is compatible, he says, with the ascription of biological functions since, as he sees it, we can see not only the intentions of agents, but also the action of natural selection, as instances of design. As he puts it: "design is not always to be understood in terms of background intentions [...] one of Darwin's important discoveries is that we can think of design without a designer" (Kitcher 1993, 2). While his account qualifies as a version of the etiological view, Kitcher addresses the initial difficulty I mentioned for that view—that many biologists do not take function-ascriptions to be responsible to facts about etiology—by allowing that the connection between design and function may be indirect. If an organism or machine was designed to perform in a certain way, then we can ascribe functions to the structures or traits contributing to that performance, whether or not these were specifically designed to perform that contribution. For example, if a screw which has accidentally fallen into a machine enables the machine to work as the designer intended, then we can describe its contribution to the machine's performance as a function of it. Similarly, in the biological case, we can ascribe functions to the elements which enable an organism to perform in ways attributable to natural selection, whether or not those elements can be explained as a result of selection.

Kitcher's account has the advantage of keeping close to our intuitions about when biological functions should and should not be ascribed. Intuitively, we are inclined to say that an organic entity has a function in just those contexts where we would be inclined to say, if it were an artifact, that there was something that it was designed to do. But, as Kitcher recognizes, these intuitions leave open various possible criteria for function ascriptions. His account aims to respect that variety, allowing us to identify functions by appeal, depending on our explanatory interests, to past selection pressures, to more recent selection pressures, or to an item's current contribution to the behavior of the system to which it belongs. And this is attractive given that, in actual biological practice, what counts as a function varies with discipline: ecologists will tend to identify as functions those features of an item that are produced by selection pressure, whereas physiologists and molecular biologists will typically identify features currently contributing to the workings of the system to which the item belongs.

The account, however, faces an obvious problem: what entitles us to think of natural selection as a process of design? The notion of design on which Kitcher relies is, as pointed out in Krohs (2009, 74), not a technical notion but rather our ordinary intuitive notion; and this is important for his view given that he aims to capture a correspondingly intuitive notion of function, one which applies to the parts of artifacts as well as to organisms.[2] But, as Krohs (ibid., 72–4) also points out, the ordinary meaning of 'design' carries with it a reference to intention, whereas no intention is involved in the operation of natural selection. Natural selection results in organisms which are relatively well-adapted to their environment in the sense that they have a relatively high probability of surviving and reproducing in it. But the "selection" of these organisms over less well-adapted alternatives is not the selection associated with design, in which one alternative is chosen over another because it is recognized as better meeting a criterion which the designer has in mind. Although one alternative will be more successful than another at responding to environmental pressures, and this success will result in its relatively greater proliferation, this is on the face of it a paradigm case of selection without design: the better-adapted variants proliferate not

---

2 Krohs (2009, 75–8) himself defends an analysis of function in terms of a more technical notion of design, proposing that we can think of the ontogeny of individual organisms as a case of production by design (roughly, we can think of an organism's DNA as the locus of the design, like the construction plan in the case of an artifact). But I find Krohs' account unconvincing, largely because he takes the distinctive feature of production by design to be that a designed entity—in contrast to an individual work of art—is produced as one of a series, or, at least, can in principle be multiplied. The DNA of an organism qualifies as a design plan because it "fixes" (ibid., 75–7) types of proteins, rather than individual (token) proteins. However, unless 'fixing' is understood intentionally, so that the DNA in effect tells the agents assembling the organism which types of protein to use (the way a construction plan tells the builders which types of component to use) an intuitively essential element of design seems to go missing. Conversely, it seems to me that the notion of design is not limited to the production of series (or possible series), and that a unique work of art is no less designed than an industrial product. It might be replied that Krohs is drawing on a technical notion of design, so that appeals to the intuitive notion are beside the point. But his account loses philosophical interest to the extent that the notion of design it employs is divorced from our ordinary notion.

through the execution of a plan but rather through the "blind" workings of nature.[3] While it is indeed tempting to think of natural selection as a process of design, this is not because of any intrinsic similarity between the operation of natural selection and that of an actual designer, but rather because of a similarity in the end-products viewed in abstraction from their history. Our use of functional language suggests that we do consider organisms as if they were designed, but if this is so, it is not because of, but in spite of, our understanding of the processes which give rise to them.

My aim here is to suggest an alternative approach which is in the general spirit of Kitcher's but which takes as fundamental, not the notion of design itself, but a notion which I take to be presupposed by the notion of design, that of normativity. Like Kitcher's, my view aims to hew very close to our ordinary intuitions about functions and, relatedly, to accommodate the diversity of function ascription in biological practice. But, unlike Kitcher's, it does not depend on any assumptions about the actual etiology of the entities to which functions are ascribed. The approach is based on Kant's account of organisms as natural purposes, and I take it to be, in essence, Kantian, although this is not something I will defend here.[4] Very roughly, I want to say that the function of a trait or entity is not what it was in fact designed to do, or what it contributes to what the organism was designed to do, but simply what it should, or ought to do. To say that the function of the heart is to circulate the blood, as opposed to making a thumping noise, is to say that, in so far as a heart circulates the blood, it is doing what it should or ought to do, or as we might also put it, doing what is appropriate to it. Here I am highlighting an aspect of the notion of biological function which is mentioned by a number of philosophers on different sides of the debate, for example Millikan (1989), Neander (1991), Hardcastle (2002), Krohs (2009) and McLaughlin (2009). But, unlike these philosophers, I am taking the normativity of function ascriptions not as a feature to be explained by some other account of the meaning or significance of function ascriptions, but as itself capturing the meaning or significance of function ascriptions.[5]

---

3 For more detailed criticism along these lines, see Davies (2001, 58–67).

4 I offer a corresponding interpretation of Kant in Ginsborg (2001) and (2006).

5 Note that I am not proposing this as a full-blown analysis of the notion of function, but only as a partial analysis intended to make sense of the contrast between functions and side-effects. There is more to the notion of a function than a trait which a thing must have to be as it ought, as the example of whole organisms shows: a squirrel which cannot climb trees is defective, but we do not say that climbing trees is its function (see Godfrey-Smith 1994, 349). Many proposed accounts of functions, including Kitcher's, are in the same position. While the point cannot be pursued here, I think that the goal of a full-blown analysis is unrealistic, since almost all

Now one of the intuitions motivating my approach is that our ascription of functions to biological entities reflects a conception of them as analogous to artifacts. So I might be characterized as holding that the function of a biological entity is what it is *as if* designed to do. That would relate my approach to that of Broad, for whom the notion of teleology (closely connected, for Broad, with that of function) "involves a hypothetical reference to design" (Broad 1925, 84), and also to that of Kant as he is commonly—although, on my view, wrongly—understood.[6] However, as I take Kant himself to have recognized, there is something puzzling about the idea of treating organisms as analogous to artifacts. How can we investigate organisms as if they were designed, while at the same time recognizing that they originate through natural processes, and thus that they are, precisely, *not* products of design? The idea that organisms can be characterized in normative terms is, as I see it, the answer to that puzzle, and I take it to be Kant's answer as well. We treat organisms as analogous to artifacts precisely by taking them to be governed by normative constraints, and it is these normative constraints which we express when we characterize their parts and behavior in functional terms, or, in Kant's terms, regard them as purposive. So while we can indeed say that the function of the heart is to circulate the blood because it is as if designed to circulate the blood, this is a circuitous way of expressing what I take to be the essential significance of the functional ascription, namely that a heart which circulates blood is not merely doing something, but doing what it ought to do.

II

An obvious objection is that talk of what a biological trait or entity 'ought' to do is just as problematic as talk of its function is, and for the same reason. How can we claim that the heart ought to circulate the blood[7] without supposing that it was consciously designed, or intended, to circulate the blood? Indeed this reference to intention, or at least conscious thought, seems to be built into the locutions 'is meant to' and 'is supposed to', which are typically interchangeable, in regard to organisms and artifacts, with 'ought to' and 'should'. When we say

---

concepts—not just philosophical ones—resist analysis (for a classic discussion, see Fodor 1981, 284–8).

6 A view of this kind is defended in Breitenbach (2009).

7 I use locutions of the form 'X ought to Y' as shorthand for the more longwinded: 'In so far as X Ys, it is doing as it ought', although the latter are more idiomatic, and, I suspect, philosophically preferable.

the fan is meant to, or supposed to, keep the computer cool, so that, if it fails to do so it is not doing as it ought, we seem on the face of it to imply that the designer intended the fan to keep the computer cool.

It will be helpful as a preliminary to distinguish our use of the term 'ought' in connection with artifacts and (perhaps illicitly) in biological contexts, from two other uses. The first applies exclusively to rational beings, and picks out, correspondingly, a kind of normativity which is directly associated with reasons. This use occurs in practical contexts when we speak of what a rational agent ought to do, or perhaps more precisely, what she ought to intend. The same use is manifested in theoretical contexts when we talk about what a rational thinker ought to believe. In these contexts the term 'ought to' can typically be paraphrased directly as an attribution of reasons: to say that someone ought to act in a certain way or to form a certain belief is to say that she has good or conclusive reasons for the action or belief. This use can be uncontroversially described as normative. The second applies to phenomena of all kinds, and is typically used to express predictions or to convey that something might be, or might have been predicted. We say for example that the weather ought to be sunny tomorrow, that the train ought to be here any minute, or that the poison ought to have taken effect by now. This use is typically not thought of as normative, although it might be indirectly paraphrased in terms of reasons for belief: to say, on this use, that something ought to have happened is to say that there was reason to predict that it would happen. The use of 'ought' with which we are concerned differs from both of these. To say that the fan or the heart ought to, or are meant to, do such-and-such, is not to say that they have reasons to do such-and-such; it is not to treat them as rational agents or subjects of belief. But nor is it to predict what they will do, or to claim that such predictions are or were reasonable. We might indeed say that the fan ought to make white noise, meaning that it is reasonable to predict that it will, but this does not imply that it ought to make white noise in the sense with which we are concerned: conversely, to say that it ought to cool the computer (in that sense) does not imply that there is reason to predict that it will.

What are we saying, then, when we use the 'ought' locution in connection with artifacts? Although we are not ascribing reasons to the artifacts, there still seems to be something normative about the locution: in particular, it seems to register a norm or standard to which the artifact, or the relevant part of the artifact, is subject. We can say, for example of a fan which does not cool the computer, that it is defective or malfunctioning. But it might be suggested—and this would support the objection—that the apparent normativity is illusory: that talk of 'ought' in the case of the fan can be reduced to talk of design or intention, where this in turn carries no normative implications. To say that the fan ought to cool the computer is simply to say that the fan was designed or intended to cool the computer, and a malfunctioning fan is one which does not perform as its designer intended. If this suggestion is correct, my proposal is clearly a non-starter, since it implies that the function of the heart is to circulate the blood only if a designer intended it to circulate the blood. However, this suggestion oversimplifies the relation between talk of 'ought' and talk of design or intention. We cannot simply reduce the relevant notion of 'ought' to the notion of design, since the notion of design already presupposes the corresponding notion of normative constraint. Roughly: it is not sufficient, in order for something to count as designed, that there be a conception of the thing in a designer's mind, and that that conception be causally operative in the thing's production. Rather, it is required in addition that the designer recognize the conception as normatively binding on the thing. It is not enough for her to think of the fan's cooling the computer, and for that thought to be responsible for her including a fan in the design for the computer. She must, in effect, think that the fan *is to*, that is that it *should* cool the computer. We cannot make sense of an artifact as designed unless we suppose that the designer thinks of it in normative terms, as something which should or ought to be this or that way, and which will count as defective if it is not.

If that thought is correct, then we cannot simply understand the 'ought' associated with artifacts as shorthand for talk of a designer's intentions. The 'ought' conveys a distinctive notion of normative constraint presupposed by, and thus not reducible to, the notion of design. But this is not sufficient to defuse the objection. For even if the notion of design presupposes that of how something ought to be, it might still be the case that this latter notion makes sense only in the context of a designer's intentions. If that is so, then the notions of 'ought' and design are mutually dependent. Although we cannot think of an object as designed unless we ascribe to the designer the thought that it ought to be this or that way, we can, conversely, speak of how something ought to be only on the assumption of a designer who intends that it be this or that way. And that seems to rule out the possibility of regarding natural phenomena in normative terms.

I want to address this objection by appealing to a notion which I have discussed elsewhere under the name of 'primitive normativity'.[8] If this notion makes sense, then we can intelligibly ascribe oughts without corresponding intentions: more specifically, we can intelligibly characterize natural phenomena as being or not being as they ought to be, where the ought is normative rather

---

8 See Ginsborg (2011).

than that associated with prediction. This does not directly warrant our characterizing organisms in normative terms. But it does remove what I take to be the major conceptual obstacle to our doing so. Once we recognize, from the kind of case which I am about to describe, that there can in principle be oughts without intentions, that is, natural oughts, then we need no longer have a bad conscience about regarding other natural phenomena—in particular biological phenomena —in normative terms as well.[9]

The notion of primitive normativity, while not itself Wittgensteinian, is most easily introduced in the context of a point which is emphasized by Wittgenstein, about the way in which our use of language, and relatedly our grasp of concepts, is conditioned by our proto-cognitive dispositions to react to the world, dispositions that are sometimes referred to as "ways of going on". The locus classicus of this point is *Philosophical Investigations* §185, where Wittgenstein considers the case of the pupil for whom it comes naturally to continue the series 2, 4, 6, 8 ... 1000 with 1004. Part of the moral of this example is that our ability to grasp basic arithmetical concepts like *add two* and addition more generally depends on our having the natural reactions which we, in contrast to the pupil in the example, in fact have—in other words, on our having a natural tendency to "go on" with 1002. Wittgenstein compares the case of the pupil to that of someone who naturally responds to a pointing hand by looking in the direction from fingertip to wrist. Our understanding of the hand as pointing in the direction that it does—an understanding that is crucial for the possibility of ostensive learning —depends on the contingent fact that we do not react to it in the aberrant way described in the example. The examples here can be multiplied indefinitely, in particular for all of the responses that are responsible for our sorting objects in the ways that we do, and which we tend to describe as "finding similarities" among things. A child who had no natural tendency to group together things of the same color or shape, but who instead was inclined to sort in "grue"-like ways (finding it natural, for example, to sort blue cubes with green spheres, and blue spheres with green cubes) would not be able to acquire concepts like *green* and *blue*, at least not in any normal way.

To this Wittgensteinian point I want to add a further claim: that, in exercising our natural tendencies to react as we do, we not only react, but also take our reactions to be appropriate to the contexts which engender them. We do not merely feel ourselves impelled to say '1002' after '1000', as if in the grip of a compulsion to produce that number and not some other, we take the numeral '1002' to fit the preceding sequence; that is, we take it that, in saying '1002', we are

---

9 Here again I take myself to be following Kant, see Ginsborg (2006, 464–6).

going on as we should or ought. Similarly, when we look in the direction of the pointing finger, we take ourselves to be reacting appropriately: we are conscious of our reaction as "called for" by the hand, and not as merely elicited by it. We are thus ascribing normativity to our natural reactions, taking them to be as they ought to be with respect to what we are reacting to. But—and this is a crucial point—I want to claim that this ascription of normativity does not derive from, but rather makes possible, our grasp of the corresponding concepts and hence our cognition of the objects or situations to which we are responding. The child's sense, when she puts the green cube with the other green cubes, that this is where the cube "belongs", or where she "should" or "ought to" put it, is not based on her having recognized that the cube is green and that she is sorting it with the other green cubes. That she has this sense of appropriateness is rather, like the natural tendency itself, part of the conditions for her grasping the concept *green* in the first place. So I take this awareness of normativity itself to be immediate, as opposed to being based on an inference from the applicability of a concept. Relatedly, the normativity she ascribes to her reaction is "primitive". The idea of responding as she ought which is implicit in her attitude cannot be reduced to, or explicated in terms of, that of responding in a way which conforms to an applicable rule.

What is the character of the 'ought' that figures in the attitude I have described, and, in particular how does it relate to the three senses of 'ought' I distinguished earlier? We can rule out, I think, that it is the non-normative ought which I associated with prediction. It might indeed be reasonable to predict that someone continuing the series of even numbers up to 1000 will go on with 1002, but this is not required for us to be able to take ourselves, when we say '1002', to be going on as we ought. Less obviously, I think we can also rule out that it is the ought through which we ascribe reasons to a rational agent or epistemic subject. The thought that 1002 is appropriate is not the thought that we have reason to say '1002', whether we conceive of the possible reasons here as practical or theoretical. Someone continuing the series might indeed take it that she ought to, in the sense of having a practical reason to, say '1002' after '1000'. Perhaps she knows that if she does so, she will please her teacher or pass a test. But she need not take herself to have such a reason in order to find '1002' the appropriate thing to say. She may indeed have a reason not to say it—perhaps she has reason to annoy, frustrate or tease her teacher—and she may think, as a result, that she ought, as a matter of practical rationality, to say '1004' instead. But that is compatible with her continuing to take it that, in a more fundamental sense, '1002' is appropriate and '1004' is not. Turning now to theoretical reasons, someone might be aware, in saying '1002', that she is expressing a belief which is rationally justified, for example that 1000 plus two is 1002. So she may ascribe to herself a theoretical reason for saying '1002', a reason which she could express by means of a very simple arithmetical proof. But her awareness of '1002' as appropriate to the preceding series does not depend on her appreciation of such a reason. Indeed, as in the practical case, she may take the belief that she would express by her utterance of '1002' to be false, and a fortiori unjustified, for example if she has been deliberately and explicitly following the rule to add two up to 1000 and add four thereafter. In that case, her saying '1002' would express the claim that 1000 + 4 = 1002, and there would be no theoretical reason supporting that claim. But she can recognize this and still take it that '1002' represents the appropriate continuation of the series of numbers as such. The point illustrated by these examples—which I can here do little more than gesture at—is that the primitively normative 'ought' applies not to the intentional actions or beliefs of a rational agent or epistemic subject, but rather to the natural reactions of a human being responding to her environment. These natural reactions can indeed amount to, or express, actions and beliefs, and, correspondingly, the human being responding to her environment can also be viewed, and can view herself, as a rational agent and epistemic subject governed by rational norms. But this is possible only in virtue of the fact that, so to speak as a human being rather than a rational agent or subject, she takes her natural responses to be appropriate in the pre-rational sense I have described. It is only in virtue of her taking this more primitively normative attitude to them that her responses can come to have intentional content and thus to be the kinds of items to which the 'ought' of reasons is applicable.

We now have a basis for responding to the objection raised at the beginning of this section. According to that objection, the claim that something is (in a normative sense) as it ought to be to be depends on the assumption that the thing was or is intended to be that way. This rules out, according to the objection, the ascription of oughts to natural phenomena, and in particular to organisms and their parts and traits. But according to the line of thought I have sketched, the possibility of intentional content, and a fortiori of cognition, depends on our being able to take a normative attitude to at least some natural phenomena— specifically, our own precognitive responses to the world. So, if cognition is to be possible, the ascription of (normative, non-rational) oughts cannot presuppose the assumption of design. What I am suggesting here is something like a transcendental argument for the intelligibility of natural normativity in general, which can then be deployed to defuse an objection to the application of norms to biological phenomena. The argument shows that we must regard one particular class of natural phenomena—what I have called our pre-cognitive responses to the world—in normative terms. From this it emerges that the kinds of oughts that we are tempted to invoke when we try to understand biological phenomena—when we say, for example, that a heart which does not circulate the blood is not functioning as it ought—cannot be ruled out simply on the grounds that the phenomena to which they apply are natural rather than the product of design. There may still be objections to the use of 'ought' in a specifically biological context, but they do not include the objection that the relevant notion of 'ought' implies that of design or intention.[10]

III

The argument presented in the last section might be challenged on the grounds that it licenses the ascription of norms—and thus, according to the account in section I, of functions—not only to the parts and traits of organisms, but everywhere in nature. If the indispensability of our ascriptions of primitive normativity to our own proto-cognitive natural reactions entitles us to apply normative notions to hearts, ribosomes and grooming behavior, then why not also to clouds in weather systems, mutant DNA sequences in tumors, and, for that matter, falling rocks?

This challenge mistakes the aim of the argument, which was to remove a conceptual obstacle in the way of normative ascriptions in biology—namely, the thought that we cannot ascribe oughts without intentions—and not to provide a positive argument for their legitimacy. The argument indeed implies that we can legitimately ascribe appropriateness to the kinds of natural reactions which figured in our examples, but not that we are entitled to any other specific claims to appropriateness, for example, that a heart which circulates the blood is functioning as it ought. It aims to conclude only that such a claim cannot be rejected, as a matter of principle, on the grounds that it presupposes that the heart

---

10 Banham (2008, 438) objects to my (2006) ascription of this line of argument to Kant that it "requires it to be the case that we are in some sense designed to view organisms as designed, a view that appears dangerously circular". This characterization strikes me as inaccurate: the view does indeed require that we are "as if" designed to respond to, and thus view, the world around us in certain characteristic ways rather than others, but the point is not that we are "as if" designed to view organisms as "as if" designed; rather it is that the requirement to view our own responses to the world in terms of "as if" design licenses, or at least removes an obstacle to, our viewing biological phenomena in the same way. There would be a genuine threat of circularity if the awareness of normativity in our own responses depended on the kind of biological ought in question: if, for example, we took our response of '1002' to be appropriate on the grounds that it is characteristic of a healthy human being with faculties in good working order. But it is a part of the "primitive" character of the awareness of normativity that it does not rest on conceiving oneself as an organism subject to biological norms.

was designed. Now it might be objected that oughts without intentions are intelligible only in the special case of our own natural reactions: although their intelligibility must be assumed in order to make sense of the possibility of cognition, no inference is available to their intelligibility in other contexts. But here the burden of proof is on the objector, to show that ascriptions of natural oughts outside the cognitive context are not only false or unjustified, but also unintelligible or self-contradictory.

Another way to make the same point is to note that the argument of the previous section, while aiming to establish the intelligibility of ought-, and thus, on my account, function-ascriptions to natural phenomena, does not commit us to doing so in any particular circumstances (beyond the core case of our own natural reactions). My approach to function ascriptions thus differs from the etiological and causal role approaches considered earlier, both of which do require us to ascribe functions to phenomena which satisfy a certain set of naturalistic criteria. On a causal role view, it is sufficient for something's having a function that it contribute in a certain kind of way to the activities of a certain kind of system. Such a view, depending on how it is articulated, can commit us to ascribing functions, say, to clouds in weather systems; and that is one of the problems of the causal role approach. Similarly, etiological views require us to ascribe functions to items which have a particular causal history. This again, as we saw in section I, can lead to over-generous function ascription (for example in the case of the gas leak which renders the scientist unconscious). But these problems do not arise for my account because it does not offer specific empirical criteria for function ascription.

This might provoke a challenge from another angle: given the limited commitments of my account, how can it shed any light on the notion of a function? Shouldn't we expect, from an account of biological function, that it offer criteria for function ascription by appeal to which our actual applications of the term 'function' can be justified? I have offered one criterion, namely that if X does Y we are entitled to regard Y as a function of X only if, in doing Y, X is doing as it ought. But, leaving aside the difficulty (shared with other theories) that this offers at most a necessary condition for function ascription,[11] it can be objected that we are no less in need of criteria for determining when Y is what X ought to do, than for determining when Y is the function of X. What is needed, roughly speaking, are naturalistic criteria, such that it is a matter of empirically discoverable fact whether we are entitled to apply the notion of a function in a given context.

---

11 See footnote 5.

Here I want to claim that it is an advantage rather than a limitation of my analysis that it does not provide such criteria. For this allows it, like Kitcher's view, to respect the diversity of criteria which biologists in different areas invoke to justify ascriptions of function. It is true that it also leaves open—and this is something which Kitcher does not intend—that the notion of function could conceivably be applied to natural phenomena that are not biological. This again, however, seems to me to speak in favor of the analysis. I do not think that a philosophical analysis of the notion of function, even one intended to do justice specifically to the use of the term 'function' in biology, should rule out in advance that we might encounter non-biological phenomena for which functional characterizations turn out to be scientifically indispensable.

This is not to deny that there can be any philosophical account of the circumstances in which function ascriptions are or are not justified. One such account, compatible with the analysis I have suggested, would link the justification of function ascriptions with the demands of understanding.[12] We are justified in ascribing functions, on this account, when this is required for a satisfactory understanding of the relevant phenomena. The behavior of a weather system can be adequately understood (although, notoriously, not predicted) through a grasp of physical regularities concerning the effect of temperature on the movement of air masses, the conditions under which water evaporates and condenses, and so forth. We are not required to conceive of any of the elements of the system in functional or normative terms: while we might have to recognize that clouds have a tendency to produce rain under certain circumstances, we do not need to suppose, further, that this is one of their functions. But it is at least arguable that a full understanding of biological phenomena requires us to invoke normative notions. We might indeed regard an individual organism as an assemblage of inorganic molecules governed by physical laws, and, with enough information about the arrangement of the molecules we might even be able to predict its behavior. But we would be missing something about it if we did not understand it also as a system of organic parts (a brain, lungs, heart, arteries, and so forth), and, again at least arguably, this requires understanding of these parts not only as having tendencies to do various things (pump blood, make a noise, add to the total body weight) but also as meant to do them, or as malfunctioning when they fail to do them.

Furthermore, philosophers might be able to offer substantive answers to the question why—if the account just sketched is correct—our understanding of biological phenomena is unlike our understanding of meteorological phenomena in

---

12 I ascribe such an account to Kant in Ginsborg (2001).

presupposing normative notions. It might be suggested, for example, that this is due to their incomparably greater complexity. Perhaps organisms present us with such a diversity of empirical regularities that we can comprehend them only by demarcating, as our proper object of study, those regularities in fact tending towards the maintenance of the organism. And perhaps we can treat those regularities as privileged, for the purpose of scientific enquiry, only by thinking of them in normative terms. It might also be argued, that where we discover such complexity in nature, it could only be as a result of Darwinian processes, so that, as a matter of fact, we are entitled to ascribe functions only in domains where natural selection is operative.

However, I think it not only undesirable, but also unnecessary to build any of these substantive proposals into a philosophical analysis of the notion of a function. We might indeed, as philosophers, argue that functional ascriptions are justified only when required for an understanding of the phenomena and, more specifically, that this requirement holds, in the case of natural phenomena, only for highly complex systems, or, indeed, for products of natural selection. But we do not need to incorporate such criteria into an account of what functions are. As I see it, philosophers have tried to offer specific theoretical accounts of the notion of a function— for example in terms of the history or causal role of a trait—only because of the worry that, because of its apparent dependence on the idea of intentional design, the intuitive notion of function cannot legitimately be applied to natural phenomena. In the absence of this worry, there is no obvious motivation for offering the kind of naturalistic criterion for function-ascription which my analysis can be accused of failing to provide. For it can simply be assumed that biologists are using the term 'function' in the sense familiar from our talk of artifacts.

Now if my analysis were meant to compete directly with these accounts, so to speak at the same level, then the objection would be well-taken. But my aim is not to offer an alternative to these accounts, but rather to counter the worry which motivates them in the first place. As we have seen, that worry derives from the assumption that the intuitive notion of function depends on that of intentional design. But, on my analysis, the crucial distinction between functions and side-effects is made out not in terms of intentional design, but rather in terms of the notion of 'ought'. If, as I argued in section II, we can in turn make sense of oughts without intention, then the worry about the intuitive notion of function is addressed. This does not put an end to questions about the justification for any particular functional claim or set of claims, whether in biological science or elsewhere, and exploration of such questions might throw up a new reason for denying that biologists are using 'function' in the intuitive sense. But, I have tried to argue, the supposed dependence of functional ascriptions on the assumption of intentional design is an illusion. Functional ascriptions in biology depend on the ascription of norms to natural phenomena, but there is nothing intrinsically objectionable about regarding natural phenomena in normative terms.[13]

References

Banham, Gary 2008, New Work on Kant, *British Journal for the History of Philosophy* 16 (2), 431–9.

Boorse, Christopher 1976, Wright on Functions, *Philosophical Review* 85, 70–86.

Breitenbach, Angela 2009b, Teleology in Biology: A Kantian Approach, in: *Kant Yearbook 1*, ed. by Dietmar Heidemann, Berlin/New York: Walter de Gruyter, 31–56.

Broad, Charles Dunbar 1925, *Mind and its Place in Nature*, London: Kegan Paul.

Cummins, Robert 1975, Functional Analysis, *The Journal of Philosophy* 72, 741–60.

Davies, Paul Sheldon 2001, *Norms of Nature*, Cambridge (MA): MIT Press.

Fodor, Jerry A. 1981, The Present Status of the Innateness Controversy, in: Jerry A. Fodor, *Representations*, Cambridge (MA): MIT Press, 257–333.

Ginsborg, Hannah 2001, Kant on Understanding Organisms as Natural Purposes, in: *Kant and the Sciences*, ed. by Eric Watkins, Oxford: Oxford University Press, 231–58.

Ginsborg, Hannah 2006, Kant's Biological Teleology and its Philosophical Significance, in: *A Companion to Kant*, ed. by Graham Bird, Oxford: Blackwell, 455–69.

Ginsborg, Hannah 2011, Primitive Normativity and Skepticism about Rules, *Journal of Philosophy* 108 (5), 227–54.

Godfrey-Smith, Peter 1994, A Modern History Theory of Functions, *Noûs* 28 (3), 344–62.

Hardcastle, Valerie 2002, On the Normativity of Functions, in: *Functions: New Essays in the Philosophy of Psychology and Biology*, ed. by André Ariew, Robert Cummins, and Mark Perlman, Cambridge: Cambridge University Press, 144–56.

Kitcher, Philip 1993, Function and Design, *Midwest Studies in Philosophy* 18, 379–97.

Krohs, Ulrich 2009, Functions as Based on a Concept of General Design, *Synthese* 166, 69–89.

McLaughlin, Peter 2009, Functions and Norms, in: *Functions in Biological and Artificial Worlds*, ed. by Ulrich Krohs and Peter Kroes, Cambridge (MA): MIT Press, 93–102.

Millikan, Ruth Garrett 1989, In Defense of Proper Functions, *Philosophy of Science* 56 (2), 288–302.

Neander, Karen 1991, Functions as Selected Effects, *Philosophy of Science* 58 (2), 168–84.

Wittgenstein, Ludwig 1953, *Philosophical Investigations,* trans. by G.E.M. Anscombe, ed. by G.E.M. Anscombe and Rush Rhees, New York: Macmillan.

Wright, Larry 1973, Functions, *Philosophical Review* 82, 139–68.

---

13 I thank Frank Hofmann, André Wunder, Günter Zöller, and the editors of this volume for helpful comments.

Siegfried Roth

Kant, Polanyi, and Molecular Biology

The most salient feature of modern biology is the attempt to provide molecular explanations at each level of life's hierarchical organization, ranging from basic metabolism and core functions of the cell to higher-level phenomena like growth and development, functions of complex organs including the brain, and aspects of behavior like learning and memory. Modern biology clearly follows a reductionist approach: phenotypic features of the whole organism are correlated with and frequently causally explained by particular molecular changes. Nevertheless, Kant's statement that there will never be "a Newton who could make comprehensible even the generation of a blade of grass according to natural laws" (*CJP* V 400.18–9) is relevant, as I will argue, even to today's molecular biologist. Indeed, the strongest support for this claim is rooted in a deeper understanding of organisms at the molecular level. This view was anticipated by the physico-chemist Michael Polanyi, whose ideas on reductionisms will be presented in the middle part of this paper. Its first part provides a reconstruction of Kant's views of the organism, which highlights two aspects: (i) Organisms are objects of nature that integrate two levels of lawfulness and (ii) they possess an inner self-representation. My claim is that Kant's conceptual account captures essential aspects of modern molecular definitions of life.

1 Kant's Theory of the Organism

Already in his pre-critical writings, Kant's statements about organisms always presume that there is a clear distinction between the organic and the inorganic world (*Theory of Heavens* I 230.14–26; Roth 2011b). But what leads Kant in the first place to sharply separate the living from the non-living? Self-directed movement, which characterizes animal life, is ruled out as Kant repeatedly refers to plants and chooses the tree as the key example when introducing his definition of the organism in §64 of the *CPJ*. In a later passage, though, he admits that regarding plants the difference is not always obvious and that it requires close attention to recognize the "indescribably wise organization" which distinguishes them from the "mineral kingdom" (*CPJ* V 426.26–7). An obvious connotation of "indescribably wise organization" (ibid.) might be structural intricacy, which in the seventeenth and eighteenth centuries had become apparent through the introduction of the microscope (Wilson 1997, 70–102). Organisms seemed to possess an indefinitely complex inner organization, which lacked the geometric regularity represented by crystals. This was an important argument for Haller to reject explanations of organismic form based on the concept of force, even if such a force, like Wolff's *vis essentialis*, was construed as being specific to living beings (Haller 1766, 118).

Kant, however, never explicitly uses irregularity of form or internal structural complexity as an argument distinguishing the organic from the inorganic world. Indeed, he is quite sensitive in recognizing the power of the blind mechanism to produce a large variety of patterns and forms in nature. He states "nature displays everywhere in its free formations [...] much mechanical tendency to the generation of forms" (*CPJ* V 348.5 – 6). Besides crystals, such as the vast variety of snowflakes, he mentions a number of inorganic processes that lead to irregularly shaped objects (*CPJ* V 349.1 – 19). The general mechanisms of nature might even account for structurally complex organic forms like "skin, hair and bone" (*CPJ* V 377.19) or for aspects of outer shape and surface coloration of animals and plants which "can be ascribed to nature and its faculty for forming itself [...] in accordance with chemical laws" (*CPJ* V 349.36 – 7). Thus, mechanical and chemical processes can produce geometrically irregular patterns and shapes and with this capacity they even contribute to organismic form. Given this situation how could Kant be so certain in distinguishing inorganic from organic beings?

It is instructive in this regard to compare two examples of material objects having the same degree of complexity although they come about in completely different ways: the hexagonal structure of a snowflake that according to Kant results from the blind mechanism of nature, and the hexagon drawn in the sand that Kant uses in §64 to introduce his preliminary discussion of organisms as natural ends (*CPJ* V 370.16 – 32). The hexagonal geometry of the ice crystal is an intrinsic property of water vapor given the right temperature and composition of the air (*CPJ* V 348.21 – 35). However, the arrangement of sand grains to form the shape of a hexagon is external to their material properties. As a materially realized figure, the hexagon drawn in the sand is defined by the positions in space of small material objects. However, these positions are brought about by an external force guided by the concept of a hexagon present in the mind of the person who has drawn the figure. The completed figure unites two levels of organization: the lower level of the sand grains held in their positions by the law of gravity and the higher level which determines the macroscopic distribution. There is no lawful connection, as there is in the ice crystal, that can explain the higher-level organization from processes at the lower level.

From the textual context it is clear that for Kant the hexagon drawn in the sand provides an example of how we initially identify natural ends, i. e., organisms, and distinguish them from inorganic objects. The characteristic features of organisms are neither their material complexity nor the absence of the laws that govern the inorganic world, but rather the employment of the latter according to higher-level principles. Kant's frequent characterization of natural ends or organisms as lawfulness of the contingent (*First Introduction* XX 240.27– 8, *CPJ* V 360.8 – 10, 370.5 – 6, 393.19 – 20, 396.26 – 7, 398.32 – 5, 404.24 – 30, 407.5 – 7) represents an abbreviated formula describing a structure in which two levels of lawfulness are at work. Contingency refers to the uncoupling of the two levels. The higher-level laws represent particular constellations of lower-level laws chosen from an almost indefinite set of possibilities. Referring to the anatomy of a bird Kant says: "nature, considered as mere mechanism, could have formed itself in a thousand different ways without hitting precisely upon the unity in accordance with such a rule" (*CPJ* V 360.16 – 8).

After presenting the example of the hexagon in §64 Kant immediately moves on to distinguish artifacts from organisms. However, as Ginsborg (2004) has shown convincingly, much of what Kant says about organisms in the "Critique of the Teleological Power of Judgment" can be understood on the basis of their artifact- or machine-like character.[1] In particular, artifacts and organisms pose the same problem for reductive theories: their form and function appear to be inexplicable on the mere basis of the universal laws of nature. Since the higher-level laws cannot be derived from those at the lower level, and since they account for the unique structure and functional organization of the organism (or artifact), the relation between both levels is highly asymmetric. This point is reflected in Kant's resolution of the antinomy of teleological judgment presented in §78, where he concludes that there is only one way to unify both levels of lawfulness: "the one (mechanism) can only be subordinated to the other" (*CPJ* V 414.9 – 10). Subordination implies that the "original organization [...] uses that mechanism itself in order to produce [...] its [...] configurations" (*CPJ* V 418.14 – 6).

The methodological advice Kant gives to the biologist in the "Critique of the Teleological Power of Judgment" follows directly from the characterization of the organism as artifact-like. Since the higher level cannot be deduced from the lower level, the existence of the former cannot be denied when studying organisms, otherwise the object of research is lost (*CPJ* V 376.34 – 6, 411.4 – 6, 418.11– 4). On the other hand, the mechanism of nature is still at work within the organism, and while teleological reasoning provides the heuristics, a deeper scientific un-

---

1 It is, however, interesting to note that artifacts can and often do possess a very simple material organisation, e. g., a table, a fork, or a match. Organisms, on the other hand, are always highly complex with regard to their material composition, which greatly reinforces the improbability and the apparent contingency of their structure. The complexity argument will become important for my discussion of modern theories of life.

derstanding requires the mechanistic approach (*CPJ* V 410.16–9). In particular, there is no *a priori* definition of the limits of mechanistic research: "the authorization to seek for a merely mechanical explanation of all natural products is in itself entirely unrestricted" (*CPJ* V 417.27–8).

Despite the fact that much of what Kant says about organisms can be understood on the basis of their artifact- or machine-like organization, the central project of the "Critique of the Teleological Power of Judgment", namely, finding a satisfactory account of a natural purpose, depends on characterizing organisms as true products of nature. Thus, Kant has to answer the questions (1) where the design-like, purposive structure of organisms comes from, and (2) what this entails for the inner constitution of the organism.

(1) The problem of the origin of the design-like nature of organisms comprises two aspects: the transition from the inorganic to the organic world (the origin of life) and the emergence of particular features of design given the existence of living beings. Given the prevalent belief in spontaneous generation in the eighteenth and early nineteenth centuries, it is remarkable that Kant clearly distinguished these two points (Roth 2011b). For example, the assumption that simple forms of life constantly arise from inorganic matter is essential for Lamarck's influential theory of evolution published almost two decades after the *CPJ*. In general, Kant rejects the idea of spontaneous generation (*generatio aequivoca*) (*CPJ* V 419.28–30, 421.4–8, 424.16–8). Only once within the *CPJ* does he consider a transition from the non-living to the living (*CPJ* V 419.4–8). Yet even in this passage he clearly points out the special character that such a transition would have, distinguishing it from all other potential evolutionary processes (*generationes heteronymae*). However, given organized matter to start with, Kant provides explanations for design-like organismic features using normative language and the concepts of appropriateness or adaptedness. For example, the structure of the eye[2] is such "that it ought to have been suitable for seeing" (*First Introduction* XX 240.25–6). Organisms reproduce themselves "with appropriate deviations, which are required in the circumstances for self-preservation" (*CPJ* V 374.32–3). Finally, an increase in adaptedness may have provided direction to life's evolution, which Kant imagines to have started with "creatures of less purposive form, which in turn bear others that are formed more suitably for their place of origin and their relationships to one another" (*CPJ* V 419.16–8). Thus, although Kant lacks a mechanistic theory of evolution, he applies a rudimentary

---

[2] The design of the eye as an optical instrument played a crucial role in the eighteenth-century debate between mechanists and teleologists (Schramm 1985, 34–5, 53–4, 58–60, 158–60).

concept of adaptation to abiotic and biotic environments to explain the emergence of purposive or design-like organismic features.

(2) However, these functional considerations only address the content, viz., particular features of the higher-level design-like organization. They do not explain how this organization in general is compatible with viewing organisms as products of nature. The obvious difference between a machine and an organism is the ability of the latter to produce itself. Thus, a deeper understanding of organisms as natural products should be linked to a mechanistic account of self-reproduction. A central aspect of any account of self-reproduction is a theory of generation (development or ontogeny) explaining how the visible, design-like complexity of the organism arises from the seed or the egg. Kant treats this topic in §81 of the appendix of the "Critique of the Teleological Power of Judgment". He defends a synthetic theory of generation which combines elements of epigenesis and preformation, the two prevailing models for generation in the mid-eighteenth century.[3] Preformationists assumed that the structural complexity of an organism arises from a preformed miniature copy present in germ cells. The formation of the organism is essentially the growth of preexisting structures which go back to the initial creation of the world. The defenders of epigenesis, on the other hand, assume that organismic complexity arises anew in each generation cycle.

Kant discusses the conceptual problems of the theories of generation with amazing clarity and, although he defends epigenesis, he immediately qualifies this statement. If there is a *de novo* generation of complexity in each reproduction cycle, there must be a principle at work that guides this process such that the outcome is an organism that is almost identical to its parents. Thus, the outcome has to be predetermined or preformed, though not in the manner of miniature material copy, i.e., not as a representation in space. Kant characterizes this alternative way of representation as "preformed virtualiter" (*CPJ* V 423.8) and calls the corresponding version of epigenesis "*generic preformation*" (*CPJ* V 423.5). Unfortunately, Kant does not provide more explanation of what he meant by "preformed virtualiter ". From the context, however, one has to assume that he is alluding to an internal self-representation of the whole organism, since typically preformation is associated with a copy of the organism within itself. Since in the case of generic preformation the self-representation does not correspond to a copy in the form of spatial relations, it ought to be, I suggest, a concept-like copy of the organism representing, as Kant says, "the internally purposive predispositions that were imparted to its stock, and thus the specific form"

---

3 See Fisher in this volume.

(*CPJ* V 423.7–8). 'Concept-like', however, cannot have the meaning of 'immaterial', because this would imply hylozoism or animism, which Kant explicitly rejects (*CPJ* V 374.33–375.5, 394.26–395.2). Moreover, Kant clearly points out that his theory of epigenesis refers to a naturally occurring process (*CPJ* V 424.11–6).[4]

This interpretation is in agreement with Kant's account of Blumenbach's "formative drive" (*CPJ* V 424.34). Kant agrees with Blumenbach that we cannot explain the origin of life from inorganic matter. However, as soon as an original principle of organization is in place, this principle provides "guidance and direction" (*CPJ* V 424.33) to a "capacity of matter" (*CPJ* V 424.31) that already governs processes of inorganic form production, like crystallization.[5] Kant believes that it is this capacity under the "guidance and direction" of higher-order organismic principles which Blumenbach calls "formative drive". Irrespective of the question of whether he misunderstood Blumenbach (Lenoir 1980, 87–96; Richards 2000, 30–2), Kant clearly asserts that the formative drive is a "natural mechanism" (*CPJ* V 424.28), which gains its particularity only in the organismic context.[6]

Taken together, according to Kant's theory of generic preformation, a concept-like, inner self-representation of the organism controls the epigenetic process by which organismic complexity arises and thus accounts for the machine-like structure of the organism. The representation of the whole exists (in the seed) prior to its realization, i.e., the goal of the development is predetermined, characterizing generic preformation as a future-directed process comparable to an intentional act.[7] Kant's ideas about generic preformation represent probably his most elaborate theory of life. Therefore, it ought to be informative to revisit Kant's conceptual analysis of organisms as natural ends in the "Analytic" in the light of these ideas.

Introducing the tree as an example in §64, Kant had characterized living beings in a threefold way. The first two, reproduction and development, correspond

---

4 Further evidence for this interpretation is found in the title of §81, which announces a treatment of "the association [Beigesellung] of mechanism with the teleological principle" (*CPJ* V 421.31–3).

5 Unfortunately he uses here the term "*Bildungskraft*" (*CPJ* V 424.32) for the general formative capacity of matter, which sounds very similar to "*bildende Kraft*" (*CPJ* V 374.23) used in §65 with an organism-specific meaning.

6 Therefore, I do not fully agree with Ginsborg's (2004, 46–54) claim that self-production represents a second type of mechanical inexplicability. It seems that Kant is strict in denying mechanistic explanations only with regard to the first emergence of self-producing systems (origin of life).

7 The concept of future-directedness plays an important role in Zuckert's (2007, 86, 135–9) reconstruction of Kant's idea of a natural end.

to the future-directed aspects of the organism and are covered by the theory of generic preformation. However, the third aspect analyses organismic structure at a given point in time. The organism is viewed as being composed of parts and the instantaneous relations of the parts to each other as well as that of the parts to the whole are discussed. The parts of the organism are characterized as self-causing in two respects. Firstly, they are indirectly self-preserving: "the preservation of the one is reciprocally dependent on the preservation of the others" (*CPJ* V 371.31–2). Reciprocal self-preservation also characterizes the relation between the parts and the whole: the whole (the tree) preserves itself through the functioning of its parts (leaves), the existence of which depends on the whole. Secondly, the parts are also self-generating. To explain this, Kant refers again to developmental phenomena and formulates an interesting thought experiment. After grafting a piece of a tree onto another tree, the grafted piece will give rise to an almost complete tree of the first type. Thus, a particular part can generate other parts and in the extreme case the entire organism. Since grafting is possible with many different parts, Kant envisages the normal tree as composed of many potential trees: "[h]ence one can regard every twig or leaf of one tree as merely grafted [...] into it, hence as a tree existing in itself" (*CPJ* V 371.35–7).[8] Together with the idea of generic preformation, this implies that the parts of the organism also contain self-representations of the whole.[9]

The final conceptual analysis of the organism as a natural end in §65[10] refers exclusively to the third point of the tree example in §64. However, in §65 Kant states that the whole that results from the reciprocal interactions of the parts does not exist "as a cause [...] but [*only*] as a ground for the cognition" (*CPJ* V 373.22–3). This subjective-regulative account of the concept of a natural end seems to be in conflict with Kant's statements in §64[11] and in particular with his theory of generation presented in §81. The organism seen at a given moment of time could be viewed as a system of reciprocal interactions maintaining its

---

8 This thought experiment anticipates the famous distinction of Driesch (1894, 11–2, 75–87) between the prospective fate and the prospective potential of a given part of the organism during its development.

9 The spatial preformists had developed a similar theory. They postulated the so-called "*Ergänzungskeime*" (literally, "germ of supplementation") to account for the specifity of the regeneration process (Roth 2008).

10 Beisbart (2009) has provided an excellent reconstruction of Kant's argument in §65, stressing the non-realist, regulative aspects of the idea of a natural end. A shortcoming of his interpretation is the focus on part-whole relations. It seems to me that circular causality between parts plays an important role for Kant because it provides a model for a structure whose unity results from causal closure.

11 This has been pointed out by Goy (2008, 230).

structure, the unity of which can only be represented by an outside observer. However, the organism viewed as a dynamic structure that can reproduce itself requires the idea of a causal self-representation. It seems to me that Kant was aware of this fact and falls short of his own rich analysis in §64 or the appendix when claiming that the whole cannot be causal. Moreover, the later passages in §65, which frequently have provided difficulties to commentators, can be more easily understood on the basis of the account in §81. Kant tries to find analogues for the causal structure of the organism and arrives at the conclusion that organisms are "not thinkable and explicable in accordance with any analogy to any physical, i.e., natural capacity that is known to us" (*CPJ* V 375.13–4). Comparisons can only be made to systems that are not strictly natural. For example, Kant compares the organism with the way people are organized in a state (*CPJ* V 375.31–7). A state is characterized by institutions that are inner representations of the whole or its parts. At the end of §65 he speaks of "a remote analogy with our own causality in accordance with ends" (*CPJ* V 375.20).[12] If organisms were characterized by an internal concept-like self-representation, as Kant's theory of generic preformation suggests, they would indeed have an inner organization with a remote structural resemblance to that of human minds or human institutions.

2 Michael Polanyi: Life's Irreducible Structure

After a successful research career in physical chemistry lasting from the 1910s to the 1930s, Michael Polanyi (1891–1976) turned his interests to socioeconomic and philosophical problems (Jha 2002, 3–47). In 1958 he published his chief work *Personal Knowledge: Towards a Post-critical Philosophy*, which already contains a basic outline of his antireductionist theory of technical artifacts and organisms. The most comprehensive treatment of his ideas on biology is found, however, in the influential 1968 *Science* paper "Life's Irreducible Structure" (reprinted in Polanyi 1969).

Like Kant, Polanyi starts his account of organisms with an analysis of artifacts. In particular he discusses the causal structure of machines, which he claims are objects controlled by two principles:

> The higher one is the principle of the machine's design, and this harnesses the lower one, which consists in the physical chemical processes on which the machine relies [...]. The

---

12 Breitenbach (2009, 84–108) has analyzed this analogy in great detail.

construction of the machine imposes boundary conditions on the laws of physics and chemistry (Polanyi 1969, 226).

These boundary conditions cannot be explained in terms of physics and chemistry. If one would just provide a total description of the molecular topography of, e. g., a typewriter, specifying the position and movement of all its molecules, this would not provide any insight into its function (Polanyi 1969, 227). In *Personal Knowledge* normative aspects had provided the strongest arguments that a description of machines in physical and chemical terms is inadequate: "The operational principle of the machine [...] functions as an ideal: the ideal of the machine in good working order. [...]. A physical and chemical investigation [...] can say nothing at all about the way the machine works or ought to work" (Polanyi 1958, 329). In *Life's Irreducible Structure* normative aspects are not explicitly mentioned; the focus of the antireductionist argument is the non-deducibility of higher-level from lower-level laws.

Living beings are classed with machines and their irreducibility is mainly explained by their machine-like character. Polanyi is aware that this is a reversal of a view that dominated biology for centuries (Polanyi 1967 in 1997, 295). Like a machine, the organism can be called a system under dual control: the shape of the organism can be compared to the boundary conditions of the machine. "Morphogenesis, the process by which the structure of living beings develops, can then be likened to the shaping of a machine which will act as boundary for the laws of inanimate nature" (Polanyi 1969, 227). Like the structure of a machine, the process of morphogenesis has a normative aspect that, according to Polanyi, precludes a mathematical treatment: "normal shapes—as distinct from abnormal, malformed [...] would have to be identified [...]. Mathematical relations are, like the processes of physics and chemistry, neutral in respect to morphogenetic success or failure" (Polanyi 1958, 358).

However, morphogenetic processes can be traced back, at least in principle, to the genetic material (DNA), the chemical nature of which is known. Does this not imply that morphogenesis, and thus the machine-like boundary conditions for organismic function, can be reduced to the laws of chemistry and physics? In general, molecular structure is defined by maximal stability, corresponding to a minimum of free energy. If alternative sequences of base pairs of a given DNA molecule would lead to different stabilities of the molecule, certain sequences would be favored just by minimal free energy principles. However, this is not the case. Structural features of the DNA molecule allow alternate sequences of a given length to have very similar stability:

the four organic bases have equal probability of forming any particular item of the series [...]. As the arrangement of (words on) a printed page is extraneous to the chemistry of the printed page, so is the base sequence in a DNA molecule extraneous to the chemical forces at work in the DNA molecule (Polanyi 1969, 229).

Polanyi summarizes this idea by claiming: "The physical indeterminacy of the sequence that produces the improbability of the occurrence of any particular sequence thereby enables it to have a meaning" (ibid.). Of course, improbability provides only the potential to carry information. The particular sequence of a DNA molecule is only meaningful if we assume that it generates the structure of an organism.

Taken together, the organism according to Polanyi is a system under twofold dual control: the structure or shape of the organism provides boundary conditions for its function; this structure emerges from a process of morphogenesis which itself depends on the genetic information stored in the DNA sequence. However, the DNA sequence cannot be described in purely chemical terms, but requires additional information determining its particular composition and is thus itself a structure under dual control. Polanyi avoids an explicit discussion of the higher level controlling principles, which account for a particular DNA sequence. However, he speculates about the origin of life in general terms. A system under dual control allows for continuity since higher-level principles can be realized in different degrees: "The fact that the effect of a higher principle over a system under dual control can have any values down to zero may allow us also to conceive of the continuous emergence of irreducible principles within the origin of life" (Polanyi 1969, 231).

What are the methodological consequences of Polanyi's concept of life? First, the control by boundary conditions does not disempower the laws of physics and chemistry. In fact, the operation of higher-level principles relies on the laws of physics and chemistry. Therefore, research on lower-level principles is essential for understanding the system. However, explaining the system just starting with the lower level "may be beyond our powers" (Polanyi 1969, 235).

It is obvious that there are many parallels between Polanyi and Kant:
1) For both, machines and organisms are not reducible to the general laws of nature because they are objects uniting two levels of lawfulness.
2) Both, however, avoid vitalistic conclusions. The general laws of nature remain valid within the organism.
3) An understanding of organisms, therefore, requires both research at the lower and higher levels of control.
4) Both use normative language to describe organisms.

5) Kant postulates a concept-like inner representation to explain the self-production of the organism. Polanyi points out that the genetic material containing the instructions for self-production itself is a system under dual control.

3 Molecular Biology

The most prevalent kind of functional analysis in modern biology is the production of genetically altered (mutant) organisms (Alberts et al. 2007, 553–5). This has two advantages. First, the mutant organism is not the product of external manipulations (ablations, transplantations, or injections), which frequently cause damage resulting in unwanted side effects. It rather results from a breeding scheme and is intact except for the loss or alteration of a particular gene. Second, despite the fact that genetic manipulations affect the molecular level of life's organization they do not necessarily require prior knowledge of molecular structure. In fact, among the most successful genetic experiments are large-scale mutagenesis screens, in which mutations are produced randomly, followed by subsequent phenotypic screening.[13] The phenotypic consequences of a given mutation, i.e., the deviations from the normal or 'wild type' phenotype, allow one to draw conclusions about the function of the mutated gene independent of knowledge of its molecular nature. However, even if experiments aim for the functional analysis of a given molecular structure, genetic approaches are usually unavoidable. For example, a protein might show structural similarity to other proteins known from prior experiments, suggesting a similar function. Genetic loss-of-function experiments are nevertheless required to validate this suggestion, since predictions of function based on structural similarity have turned out to be very limited. Paradoxically the language of physics and chemistry describing the structural feature of macromolecules plays only a minor explanatory role in modern molecular biology. Instead, the use of functional and normative terms is as prevalent as in other areas of biological research (e.g., physiology, behavior, ecology).

Given this background it is surprising that discussions about reductionism in the philosophy of biology have been dominated by the assumption that molecular biology, regarded as a branch of macromolecular chemistry, is the reducing theory. Studies defending antireductionist (e.g., Kitcher 1999) or reductionist

---

13 The Nobel Prize (1995) winning screen of Nüsslein-Volhard and Wieschaus represents a famous example, which provided the breakthrough for the molecular analysis of multicellular development.

(e. g., Rosenberg 1997) views accepted as common ground that a successful reduction of biological phenomena to the level of molecular biology implies a reduction to physics and chemistry. Only in recent years can an increasing awareness of the complex relation between molecular biology and the physical sciences be observed. In his newest account of the reductionism debate, Rosenberg writes:

> these irreducible biological facts are to be found at least at the level of molecular biology. So even if the rest of biology can be grounded in molecular biology, there will still be an unbridgeable gap between it and physical sciences (Rosenberg/McShea 2008, 114).

He illustrates this statement by, e. g., the chemical difference between RNA and DNA, which surprisingly cannot be explained on chemical grounds but requires functional and evolutionary considerations. This situation calls for attempts to acquire a deeper understanding of the relation between chemistry and biology. After all, molecular biology is dealing with macromolecules. What does it then mean that the sequence-based macromolecules of molecular biology cannot be entirely understood in chemical terms, and how is their structure related to the properties of the physical world?

Recall the arguments of Polanyi regarding the structure of DNA. A molecule whose composition (sequence of building blocks) would be largely determined by thermodynamic principles would have limited capacity to store information since only a small number of sequences would result in stable structures. However, the DNA structure is such that almost all possible sequences of base pairs in a molecule of a given length have similar stability.[14] This is a consequence of particular structural features of the DNA molecule: e. g., the geometry of base-pairing, the way the backbone is constructed by linking the nucleotides, etc. The fact that DNA can be a carrier of sequence information can be chemically explained, but not the particular sequence of a given DNA molecule. The latter is the result of a complex evolutionary process. Thus, it is paradoxically the particular chemistry of nucleic acids that allows the formation of a structure the composition of which is not determined on purely chemical grounds.

The employment of chemical structures or mechanisms to overcome the limits of chemical processes is one of the most general principles of molecular biology, in particular at the level of the other large class of sequence-based macromolecules, the proteins. Many proteins serve as catalysts (enzymes) to select and

---

14 This is only approximately correct. The stability of double helices varies depending on the base composition since Guanine-Cytosine base pairs are more stable than Adenine-Thymine base pairs.

facilitate particular chemical reactions so that they take place in an 'ideal' or undisturbed way (Alberts et al. 2007, 152–93). For example, nucleic acid replication is inherently prone to error. Proofreading enzymes are used to overcome this limitation. There is almost no free-running molecular process in the cell that is not controlled by specific proteins providing a certain type of idealized chemistry for a particular reaction or transport phenomenon. Proteins have the capacity to provide these highly specific conditions since their sequence-based structure is the result of evolutionary optimization processes. Thus, they owe their ability to provide conditions for "idealized, error-free chemistry" to the fact that their own structure cannot be explained just in terms of chemical reactions. In addition to the reactions connecting the building blocks (the amino acids) of proteins, instructions must be available that determine the linear order of these connections. Hence, proteins also are structures under dual control. The higher-order law accounting for their structure and function is provided by the genetic information.

We so far used the term 'genetic information' in a colloquial sense as it is common among working molecular biologists following Crick's (1958) introduction of the term. This practice has proven to be extremely fruitful. Today in the area of whole genome sequencing, the analysis of informational aspects of the genetic material has become a large and indispensable subdiscipline of molecular biology called bioinformatics. Despite the overwhelming success of this research agenda, philosophers of science have objected to the use of the term information in molecular biology (reviewed in Stegmann 2005 and Weber 2005, 251–3). In particular, they point out that the semantic aspect of information cannot be recovered from a system that is entirely determined by molecular interactions. Semantic features, they claim, require the concept of 'aboutness' and this in turn needs at least minimal forms of intentionality, which by definition are lacking in purely molecular systems. However, following Eigen's (1971) theory of the origin of life, Küppers (1986, 61–92) has suggested how the concept of information can be successfully applied to molecular biology. From a different angle Stegmann (2005) has arrived at similar conclusions. The elementary process underlying all three core mechanisms of molecular biology (replication, transcription, and translation [Alberts et al. 2007, 263–410]) is the process of template-directed synthesis. Küppers and Stegmann claim that template-directed synthesis, in short *templating*, represents a physicochemical process with qualitatively very special properties that can be explained independently from the context of the organism.

Template-directed synthesis is the process by which a molecular template P, e. g., the stretch of a nucleic acid molecule containing the sequence GUACG gives rise to Q, the complementary stretch CAUGC (G pairs with C and U pairs with A).

The overall process is a combination of chemical reactions. The building blocks, the single nucleotides A, U, C, G, are present in the solution, but become arranged in a unique sequence because of the prior presence of the template. Thus, the process possesses an elementary form of future- or goal-directedness. One could argue that in a trivial sense any causal sequence of events is future-directed. However, templating means that a particular structure already exists and ought to give rise to its duplicate (in the case of nucleic acids, via the complementary configuration). Importantly, 'ought to' has not only a metaphorical meaning. Templating defines a goal, which also can be missed. The synthesis of the product requires that a number of molecular recognition events (the pairing of the bases) and chemical reactions (the linking of the nucleotides) occur correctly. This process is error-prone, as pointed out earlier. Therefore, Stegmann interprets the semantic content or "aboutness" of a template as its "instructional content for the synthesis of the product. A template P is about the production of Q insofar as P provides the instructional content for synthesizing Q" (Stegman 2005, 435). Polanyi would express the same idea by saying: the template provides the higher-order law which guides the chemical events required to synthesize the product. Given the template as a starting structure, in Kantian language, one would probably say: it is the purpose of the template to produce a copy of itself employing the general laws of nature: "original organization [*the template*] uses that mechanism itself in order to produce [...] its [...] configurations [*the complementary strand*]" (*CPJ* V 418.13 – 5). Taken together one can argue that templating reactions have special organizational features that permit their classification as primitive forms of goal-directedness and intentionality in nature.

Although the templating reaction by itself represents a chemical process with qualitatively very special properties, it does not fully account for the molecular structure of life. The instructional content of the template that guides the production of a complementary strand is represented just by the base sequence of the template. This sequence does not require a particular meaning beyond its instructional content. Thus, any sequence can be replicated (Stegmann 2005, 426). The neutrality regarding sequence variation is indeed an important aspect of the process. However, in living systems the sequences being replicated indeed have a meaning that goes beyond the mere instructional content. In living systems the template also contains the instructions for producing macromolecules, mainly proteins which allow the templating reaction to occur in an idealized environment. Some of these products are very close to the actual process of templating. For example, in all existing organisms templating cannot occur without particular proteins (polymerases) that catalyze even the simple reaction of connecting the building blocks. The proofreading enzymes have already been mentioned. One could extend the list of processes controlled by proteins (i.e., by instructed catalysts) moving further and further from the actual process of templating. One may regard all cell metabolism and cell structure—even the physiology and anatomy of the whole organism—as being there for the sake of templating, resulting in the replication of the genetic information, an idea that has been forcefully defended by Dawkins in his famous book *The Selfish Gene* (1976). In a very broad and abstract sense, the molecular structure of life is that of a templating reaction, which is mediated directly and indirectly by self-instructed catalysts.

This abstract account of the molecular structure of life focusing on templating reactions also provides a framework for the origin of life. It has been shown experimentally and by mathematical modeling that repeated template-directed synthesis (replication) leads to Darwinian evolution at the molecular level (Küppers 1983, 37–94; Yarus 2010, 30–5). A replicating system of molecules constantly produces new molecular variants, since the process of replication is necessarily error-prone. The new variants differ with respect to stability, speed and fidelity of templating. This leads to selection processes. Mathematical equations describing templating reactions introduce a quality factor, the value of which provides information on how a particular molecule will behave in a selection experiment (Eigen 1971). Thus, even the formal treatment of templating requires the introduction of terms with normative meaning.

It is likely that the first evolutionary processes that occurred on our earth resulted from simple reactions of molecular templating. How the step towards life as we known it (involving protein synthesis) occurred is less clear. This step requires considerable genetic complexity, which in current organisms can only be maintained with the help of proteins. There is no room here to discuss theories on the origin of life. However, the high degree of genetic complexity required for protein synthesis has an important philosophical implication. Life as we know it presupposes a high degree of material complexity resulting from multiple evolutionary transitions in which, e. g., the genetic code emerged or a basic repertoire of protein folds was selected (Alva et al. 2010). Thus, even the simplest forms of life on our earth harbor myriads of features resulting from the stochasticity of evolutionary events and thus do not follow any particular chemical or physical logic. It is unlikely that we will ever find convincing chemical explanations for why a certain nucleotide triplet represents a particular amino acid (genetic code) or why particular protein folds exist in extant proteins. In this sense, essential features of the particular realization of life on our planet are contingent given the general laws of nature. Undoubtedly, only physical and chemical processes have been at work in generating life. However, due to their complexity, contingent nature, and most importantly their historical singularity, they are not accessible to us. Depending on the particular problem the modern molecular biologist attempts to understand, this evolutionary contingency of life may cause considerable limits for a mechanistic analysis (Roth 2011a). It is in this sense that the modern biologist would sympathize with Kant and rephrase one of his central statements by claiming that there will never be "a Newton who could make comprehensible even the generation [*of the simplest bacterium*] according to natural laws" (*CPJ* V 400.18 – 9).

However, the parallels between Kant and molecular biology run deeper. The prevailing functional approach in biology and the widespread use of normative language can be traced back to the basic molecular structure of life, which itself can be viewed as representing a primitive form of goal-directedness and intentionality. Even a researcher who feels uneasy with applying these terms cannot deny that experiments and theories on the origin of life have demonstrated that templating reactions lead to a qualitatively new behavior of matter (Darwinian molecular evolution). Thus, molecular biology, I suggest, has molecularized the idea of a natural end and thus provides a deep understanding of why organisms are unique among all physical objects in our world.[15]

References

Alberts, Bruce/Johnson, Alexander/Walter, Peter/Lewis, Julian/Raff, Martin/Roberts, Keith
2007, *The Molecular Biology of the Cell*, New York: Garland.

Alva, Vikram/Remmert, Michael/Biegert, Andreas/Lupas, Andrei N./Söding, Johannes 2010, A
Galaxy of Folds, *Protein Science* 19, 124 – 30.

Beisbart, Claus 2009, Kant's Characterisation of Natural Ends, in: *Kant Yearbook 1*, ed. by
Dietmar Heidemann, Berlin: Walter de Gruyter, 1 – 30.

Breitenbach, Angela 2009, *Die Analogie von Vernunft und Natur. Eine Umweltphilosophie
nach Kant*, Berlin/New York: Walter de Gruyter.

Crick, Francis 1958, On Protein Synthesis, *Symposia of the Society for Experimental Biology*
12, 138 – 67.

Dawkins, Richard 1976, *The Selfish Gene*, Oxford: Oxford University Press.

Driesch, Hans 1894, *Die analytische Theorie der organischen Entwicklung*, Leipzig:
Engelmann.

Eigen, Manfred 1971, Self-organisation of Matter and the Evolution of Biological
Macromolecules, *Naturwissenschaften* 58, 465 – 523.

15 I am grateful to Ute Deichmann and to Dietmar Heidemann for giving me the opportunity to present earlier versions of this paper at a conference on "Philosophies in Biology" (March 4 – 5, 2008) of the Jacques Loeb Centre (Ben-Gurion University) and on "Teleologie und Biologie im Kontext von Kants 'Kritik der Urteilskraft'" (December 10 – 11, 2009) of the University of Luxembourg, respectively. I am also thankful to Kristen Panfilio for critical reading and correcting the manuscript.

Ginsborg, Hannah 2004, Two Kinds of Mechanical Inexplicability in Kant and Aristotle, *Journal of the History of Philosophy* 42, 33–65.

Goy, Ina 2008, Die Teleologie in der organischen Natur (§§ 64–68), in: *Immanuel Kant. Kritik der Urteilskraft*, ed. by Otfried Höffe, Berlin: Akademie Verlag, 223–39.

Jha, Stefania Ruzsits 2002, *Reconsidering Michael Polanyi's Philosophy*, Pittsburgh: University of Pittsburgh Press.

Kitcher, Philip 1999, The Hegemony of Molecular Biology, *Biology and Philosophy* 14, 195–210.

Küppers, Bernd-Olaf 1983, *Molecular Theory of Evolution: Outline of a Physico-Chemical Theory of the Origin of Life*, Heidelberg: Springer.

Küppers, Bernd-Olaf 1986, *Der Ursprung der biologischen Information*, Munich: Piper.

Lenoir, Timothy 1980, Kant, Blumenbach, and Vital Materialism in German Biology, *Isis* 71, 77–108.

Polanyi, Michael 1958, *Personal Knowledge. Towards a Post-Critical Philosophy*, Chicago: The University of Chicago Press.

Polanyi, Michael 1967, Life Transcending Physics and Chemistry, in: *Society, Economics and Philosophy, Selected Papers by Michael Polanyi*, ed. by Richard Allen, New Brunswick: Transaction Publisher.

Polanyi, Michael 1969, Life's Irreducible Structure, in: *Knowing and Being, Essays by Michael Polanyi*, ed. by Marjorie Grene, Chicago: The University of Chicago Press, 225–39.

Richards, Robert J. 2000, Kant and Blumenbach on the Bildungstrieb: A Historical Misunderstanding, *Studies in the History and Philosophy of Biological and Biomedical Sciences* 31 (1), 11–32.

Rosenberg, Alex 1997, Reductionism Redux: Computing the Embryo, *Biology and Philosophy* 12, 445–70.

Rosenberg, Alex/McShea, Daniel W. 2008, *Philosophy of Biology. A Contemporary Introduction*, New York/London: Routledge.

Roth, Siegfried 2008, Kant und die Biologie seiner Zeit, in: *Immanuel Kant. Kritik der Urteilskraft*, ed. by Otfried Höffe, Berlin: Akademie Verlag, 275–88.

Roth, Siegfried 2011a, Mathematics and Biology. A Kantian View on the History of Pattern Formation Theory, *Development, Genes and Evolution* 221, 255–29.

Roth, Siegfried 2011b, Evolution und Fortschritt. Zum Problem der Höherentwicklung in der organischen Evolution, in: *Zweck und Natur*, ed. by Tobias Schlicht, Paderborn: Wilhelm Fink Verlag, 195–247.

Schramm, Matthias 1985, *Natur ohne Sinn? Das Ende des teleologischen Weltbilds*, Graz: Styria.

Stegmann, Ulrich E. 2005, Genetic Information as Instructional Content, *Philosophy of Science* 72, 425–43.

Weber, Marcel 2005, *The Philosophy of Experimental Biology*, Cambridge: Cambridge University Press.

Wilson, Catherine 1997, *The Invisible World: Early Modern Philosophy and the Invention of the Microscope*, Princeton: Princeton University Press.

Yarus, Michael 2010, *Life from an RNA World*, Cambridge: Harvard University Press.

Zuckert, Rachel 2007, *Kant on Beauty and Biology. An Interpretation of the* Critique of Judgment, Cambridge: Cambridge University Press.

Bibliography

1 Primary Sources

1.1 Original Edition of Kant's Works

Kant, Immanuel, *Kants gesammelte Schriften*, Akademie-Ausgabe, Berlin: Walter de Gruyter, 1902–.

1.2 Translations of Kant's Works

Unless otherwise noted, translations of Kant's works in this volume are from *The Cambridge Edition of the Works of Immanuel Kant*. But citations give an abbreviated English title of the relevant work, rather than referring to the titles of the volumes of the *Cambridge Edition*. The key to abbreviations found at the beginning of this volume also indicates the title and page numbers of the relevant *Cambridge Edition* volume for each work cited in translation.

The Cambridge Edition of the Works of Immanuel Kant, ed. by Paul Guyer and Allen W. Wood, Cambridge: Cambridge University Press 1993–.

Anthropology, History, and Education, ed. by Günter Zöller and Robert B. Louden, trans. by Mary Gregor, Paul Guyer, Robert B. Louden, Holly Wilson, Allen W. Wood, Günter Zöller, and Arnulf Zweig, Cambridge: Cambridge University Press 2007.
Correspondence, ed. and trans. by Arnulf Zweig, Cambridge: Cambridge University Press 1999.
Critique of Pure Reason, ed. and trans. by Paul Guyer and Allen W. Wood, Cambridge: Cambridge University Press 1997.
Critique of the Power of Judgment, ed. by Paul Guyer, trans. by Paul Guyer and Eric Matthews, Cambridge: Cambridge University Press 2000.
Lectures on Anthropology, ed. by Allen W. Wood and Robert B. Louden, trans. by Robert Clewis, Robert B. Louden, Felicitas Munzel, and Allen W. Wood, Cambridge: Cambridge University Press 2012.
Lectures on Ethics, ed. by Peter Heath and Jerome B. Schneewind, trans. by Peter Heath, Cambridge: Cambridge University Press 1997.
Lectures on Logic, ed. and trans. by J. Michael Young, Cambridge: Cambridge University Press 1992.
Lectures on Metaphysics, ed. and trans. by Karl Ameriks and Steve Naragon, Cambridge: Cambridge University Press 1997.
Natural Science, ed. by Eric Watkins, trans. by Lewis White Beck, Jeffery Edwards, Olaf Reinhardt, Martin Schönfeld, and Eric Watkins, Cambridge: Cambridge University Press 2012.

Notes and Fragments, ed. by Paul Guyer, trans. by Curtis Bowman, Paul Guyer, and Frederick Rauscher, Cambridge: Cambridge University Press 2005.

Opus Postumum, ed. by Eckart Förster, trans. by Eckart Förster and Michael Rosen, Cambridge: Cambridge University Press 1993.

Practical Philosophy, ed. and trans. by Mary Gregor, introd. by Allen W. Wood, Cambridge: Cambridge University Press 1996.

Religion and Rational Theology, ed. by Allen W. Wood and George di Giovanni, Cambridge: Cambridge University Press 1996.

Theoretical Philosophy 1755–1770, ed. and trans. by David Walford, in collaboration with Ralf Meerbote, Cambridge: Cambridge University Press 1992.

Theoretical Philosophy after 1781, ed. by Henry E. Allison and Peter Heath, trans. by Henry E. Allison, Michael Friedman, Gary Hatfield, and Peter Heath, Cambridge: Cambridge University Press 2002.

2 Historical and Primary Literature

anonymous 1794, Ueber die Kantische Teleologie, Philosophisches Archiv 2 (3), 1–16.

Aquinatis, Sancti Thomae 1266–72, Summa theologiae. Prima Pars, ed. by Fernando Sebastián Aguilar, Madrid: Biblioteca de Autores Cristianos ⁵1994.

Bernard, Claude 1878, Leçons sur les phénomènes de la vie communs aux végétaux et aux animaux, Paris: Baillière.

Blumenbach, Johann Friedrich 1781, Über den Bildungstrieb und das Zeugungsgeschäfte, Göttingen: Johann Christian Dieterich.

Bourguet, Louis 1762, Lettres philosophiques sur la formation des sels et de crystaux et sur la génération et le méchanisme organique des plantes et des animaux, Paris: François L'Honoré.

Buffon, Georges-Louis Leclerc comte de 1775, Histoire naturelle, générale et particulière, Supplément vol. 2, Paris: Imprimerie Royale.

Charleton, Walter 1652, The Darknes of Atheism Dispelled by the Light of Nature, London: Lee, http://eebo.chadwyck.com/home [last visited: November 18, 2013].

Cudworth, Ralph 1678, The True Intellectual System of the Universe, London: Royston, http://eebo.chadwyck.com/home [last visited: November 18, 2013].

Descartes, René 1964–76, Œuvres de Descartes, ed. by Charles Adam and Paul Tannery, Paris: Vrin.

Forster, Georg 1786, Noch etwas über Menschenrassen, Der Teutsche Merkur 56, 57–86, 150–66.

Hegel, Georg Wilhelm Friedrich 1807, Phänomenologie des Geistes, in: Georg Wilhelm Friedrich Hegel, Werke, Frankfurt/M.: Suhrkamp 1970, vol. 3.

Herder, Johann Gottfried 1785, Ideen zur Philosophie der Geschichte der Menschheit. Erster und Zweiter Teil, in: Johann Gottfried Herder, Sämmtliche Werke, ed. by Bernhard Suphan, Berlin: Weidmannsche Buchhandlung 1887, vol. 13, 1–484.

Herder, Johann Gottfried 1797, Briefe zu Beförderung der Humanität, in: Johann Gottfried Herder, Sämmtliche Werke, ed. by Bernhard Suphan, Berlin: Weidmannsche Buchhandlung 1883, vol. 18, 1–302.

Herder, Johann Gottfried 1799, Eine Metakritik zur Kritik der reinen Vernunft, in: Johann
  Gottfried Herder, *Sämmtliche Werke*, ed. by Bernhard Suphan, Berlin: Weidmannsche
  Buchhandlung 1881, vol. 21.
Herder, Johann Gottfried 1800, Kalligone, in: Johann Gottfried Herder, *Sämmtliche Werke*, ed.
  by Bernhard Suphan, Berlin: Weidmannsche Buchhandlung 1887, vol. 22, 1–332.
Hume, David 1779, *Dialogues Concerning Natural Religion*, ed. by David Fate Norton and
  Mary J. Norton, Oxford: Clarendon Press 1976.
Johann Georg Hamann. Briefwechsel, ed. by Walther Ziesemer and Arthur Henkel (vols. 1–3)
  and ed. by Arthur Henkel (vols. 4–7), Wiesbaden: Insel 1955–1979.
Leibniz, Gottfried Wilhelm von 1683–1685 [?], Genera terminorum substantiae, in: Gottfried
  Wilhelm von Leibniz, *Sämtliche Schriften und Briefe*, ed. by The Berlin-Brandenburgian
  Academy of Sciences and the Academy of Sciences Göttingen, Berlin: Akademie Verlag
  1999, series 6, vol. 4, part A, 566–9.
Malebranche, Nicolas 1674, Recherche de la vérité, in: Nicolas Malebranche, *Œuvres
  complètes*, ed. by Geneviève Rodis-Lewis, Paris: Vrin 1965, 20 vols., vol. 1.
Maupertuis, Pierre-Louis Moreau de 1744/²1745, *The Earthly Venus*, trans. by Simon Brangier
  Boas, New York & London: Johnson Reprint Corporation 1966.
Paley, William 1802/⁶1819, *Natural Theology; Or, Evidences of the Existence and Attributes of
  the Deity, Collected from the Appearances of Nature*, London: S. Hamilton.
Spinoza, Benedictus de 1677, *Ethics*, ed. and trans. by George Henry Radcliffe Parkinson,
  Oxford: Oxford University Press 2000.
Stahl, Georg Ernst 1708, *Theoria Medica Vera. Physiologiam et Pathologiam*, Halle:
  Orphanotropheum.
Stahl, Georg Ernst 1714, Über den Unterschied zwischen Organismus und Mechanismus, in:
  Sudhoffs Klassiker der Medizin, ed. by Bernward Josef Gottlieb, Leipzig: Barth 1961,
  vol. 36, 48–53.
Tetens, Johann Nicolaus 1776/7, Von der Entwicklung des menschlichen Körpers, in: Johann
  Nicolaus Tetens, *Philosophische Versuche über die menschliche Natur*, Hildesheim/New
  York: Georg Olms 1979, vol. 2, 459–64.
Trembley, Abraham 1744, *Mémoires pour servir à l'histoire d'un genre d'eau douce, à bras en
  forme de cornes*, Leiden: Jean & Herman Verbeek; English in: Lenhoff, Sylvia G./Lenhoff,
  Howard M. 1986, *Hydra and the Birth of Experimental Biology—1744. Abraham
  Trembley's Memoirs Concerning the Natural History of a Type of Freshwater Polyp with
  Armes Shaped like Horns*, with a translation of Trembley's *Memoirs* from the French,
  Pacific Crove: The Boxwood Press.
Wittgenstein, Ludwig 1953, *Philosophical Investigations*, trans. by G.E.M. Anscombe, ed. by
  G.E.M. Anscombe and Rush Rhees, New York: Macmillan.
Wolff, Caspar Friedrich 1759, *Theoria generationis*, Halle: Litteris Hendelianis; reprinted in:
  Caspar Friedrich Wolff, *Theorie von der Generation in zwei Abhandlungen erklärt und
  bewiesen/Theoria generationis*, Hildesheim: Georg Olms 1966.
Wolff, Caspar Friedrich 1764, *Die Theorie der Generationen, in zwo Abhandlungen erklärt und
  bewiesen*, Berlin: Friedrich Wilhelm Birnstiel; reprinted in: Caspar Friedrich Wolff,
  *Theorie von der Generation in zwei Abhandlungen erklärt und bewiesen/Theoria
  generationis*, Hildesheim: Georg Olms 1966.
Wolff, Caspar Friedrich 1768/9, De formatione intestinorum praecipue, tum et de amnio
  spurio aliisque partibus embryonis gallinacei, nondum visis, observationes, in ovis
  incubatis institutae, in: *Novi Commentarii Academiae Scientiarum Imperialis*

Petropolitanae, Sankt Petersburg: Petropoli Typis Academiae Scientiarum, Tom. XII
(1768), 403–507 [§§1–119 [sic!] = Pars I/II]; Tom. XIII (1769), 478–530 [§§119 [sic!]–155
= Pars III].

Wolff, Caspar Friedrich 1789, *Von der eigenthümlichen und wesentlichen Kraft der
vegetabilischen, sowohl als auch der animalischen Substanz*, St. Petersburg: Imperial
Academy of Sciences.

3 Contemporary and Secondary Literature

Adickes, Erich 1923, *Kant als Naturforscher*, Berlin: Walter de Gruyter.

Alberts, Bruce/Johnson, Alexander/Walter, Peter/Lewis, Julian/Raff, Martin/Roberts, Keith
2007, *The Molecular Biology of the Cell*, New York: Garland.

Allison, Henry E. 1983/²2004, *Kant's Transcendental Idealism*, New Haven: Yale University
Press.

Allison, Henry E. 1991, Kant's Antinomy of Teleological Judgment, *Southern Journal of
Philosophy* 30, Supplement, 25–42, reprinted in: *Kant's 'Critique of the Power of
Judgment': Critical Essays*, ed. by Paul Guyer, New York: Rowman & Littlefield 2003,
219–36.

Alva, Vikram/Remmert, Michael/Biegert, Andreas/Lupas, Andrei N./Söding, Johannes 2010, A
Galaxy of Folds, *Protein Science* 19, 124–30.

Ameriks, Karl 2008a, Status des Glaubens und Allgemeine Anmerkung zur Teleologie, in:
Immanuel Kant. Kritik der Urteilskraft, ed. by Otfried Höffe, Berlin: Akademie Verlag,
331–49.

Ameriks, Karl 2008b, The End of the *Critiques:* Kant's Moral 'Creationism' in: *Rethinking
Kant*, ed. by Pablo Muchnik, Newcastle: Cambridge Scholars Publishing, 165–90.

Ameriks, Karl 2009, The Purposive Development of Human Capacities, in: *Kant's Idea for a
Universal History with a Cosmopolitan Intent*, ed. by Amélie Oskenberg Rorty and James
Schmidt, Cambridge: Cambridge University Press, 46–67.

Ameriks, Karl 2011, Das Schicksal von Kants *Rezensionen* zu Herders *Ideen*, in: *Immanuel
Kant. Schriften zur Geschichtsphilosophie*, ed. by Otfried Höffe, Berlin: Akademie Verlag,
119–36.

Axelsson, Eva 2006, Magnesium Chelatase—a Key Enzyme in Chlorophyll Biosynthesis, Lund.

Banham, Gary 2008, New Work on Kant, *British Journal for the History of Philosophy* 16 (2),
431–9.

Beck, Lewis White 1960, *A Commentary on Kant's Critique of Practical Reason*, Chicago: The
University of Chicago Press.

Beisbart, Claus 2009, Kant's Characterisation of Natural Ends, in: *Kant Yearbook 1*, ed. by
Dietmar Heidemann, Berlin/New York: Walter de Gruyter, 1–30.

Boorse, Christopher 1976, Wright on Functions, *Philosophical Review* 85, 70–86.

Breidbach, Olaf 1999, Einleitung. Zur Mechanik der Ontogenese, in: Caspar Friedrich Wolff,
*Theoria Generationis. Ueber die Entwicklung der Pflanzen und Thiere. I., II. und III. Theil
(1759)*, Thun/Frankfurt/M.: Harry Deutsch, i–xxxiv.

Breitenbach, Angela 2006, Mechanical Explanation of Nature and its Limits in Kant's *Critique
of Judgment*, *Studies in History and Philosophy of Biological and Biomedical Sciences*
37, 694–711.

Breitenbach, Angela 2008, Two Views on Nature: A Solution to Kant's Antinomy of Mechanism and Teleology, *British Journal for the History of Philosophy* 16, 351–69.

Breitenbach, Angela 2009a, *Die Analogie von Vernunft und Natur. Eine Umweltphilosophie nach Kant*, Berlin/New York: Walter de Gruyter.

Breitenbach, Angela 2009b, Teleology in Biology: A Kantian Approach, in: *Kant Yearbook 1*, ed. by Dietmar Heidemann, Berlin/New York: Walter de Gruyter, 31–56.

Broad, Charles Dunbar 1925, *Mind and its Place in Nature*, London: Kegan Paul.

Carrier, Martin 2001, Kant's Theory of Matter and His Views on Chemistry, in: *Kant and the Sciences*, ed. by Eric Watkins, Oxford: Oxford University Press, 205–30.

Cartwright, Nancy 1999, *The Dappled World: A Study of the Boundaries of Science*, Cambridge: Cambridge University Press.

Carvalho, Sarah 2004, *La controverse entre Stahl et Leibniz sur la vie, l'organisme et le mixte. Doutes concernant la vraie théorie médicale du célèbre Stahl, avec les répliques de Leibniz aux observations stahliennes. Texte introduit, traduit et annoté*, Paris: Vrin.

Cheung, Tobias 2006, From the Organism of a Body to the Body of an Organism: Occurrence and Meaning of the Word 'Organism' from the Seventeenth to the Nineteenth Centuries, *The British Journal for the History of Science* 39, 319–39.

Cheung, Tobias 2009, Der Baum im Baum. Modellkörper, reproduktive Systeme und die Differenz zwischen Lebendigem und Unlebendigem bei Kant und Bonnet, in: *Kants Philosophie der Natur. Ihre Entwicklung im Opus Postumum und ihre Wirkung*, ed. by Ernst-Otto Onnasch, Berlin/New York: Walter de Gruyter, 25–49.

Chiereghin, Franco 1990, Finalità e idea della vita. La recezione hegeliana della teleologia di Kant, *Verifiche* 19, 127–229.

Churchill, Frederick B. 1970, The History of Embryology as Intellectual History, *Journal of the History of Biology* 3, 155–81.

Crick, Francis 1958, On Protein Synthesis, *Symposia of the Society for Experimental Biology* 12, 138–67.

Cummins, Robert 1975, Functional Analysis, *The Journal of Philosophy* 72, 741–64.

Cunico, Gerardo 2008, Erklärungen für das Übersinnliche: physikotheologischer und moralischer Gottesbeweis, in: *Immanuel Kant. Kritik der Urteilskraft*, ed. by Otfried Höffe, Berlin: Akademie Verlag, 309–29.

Davies, Paul Sheldon 2001, *Norms of Nature*, Cambridge (MA): MIT Press.

Dawkins, Richard 1976, *The Selfish Gene*, Oxford: Oxford University Press.

Detlefsen, Karen 2006, Explanation and Demonstration in the Haller-Wolff Debate, in: *The Problem of Animal Generation in Early Modern Philosophy*, ed. by Justin E. H. Smith, Cambridge: Cambridge University Press, 235–61.

Driesch, Hans 1894, *Die analytische Theorie der organischen Entwicklung*, Leipzig: Engelmann.

Duchesneau, François 1998, *Les modèles du vivant de Descartes à Leibniz*, Paris: Vrin.

Duchesneau, François 2006, Essential Force and Formative Force: Models for Epigenesis in the 18th Century, in: *Self-Organization and Emergence in Life Sciences*, ed. by Bernard Feltz, Marc Crommelinck, and Philippe Goujon, Dordrecht: Springer, 171–86.

Dupont, Jean-Claude 2007, Pre-Kantian Revival of Epigenesis: Caspar Friedrich Wolff's *De formatione intestinorum* (1768–69), in: *Understanding Purpose: Collected Essays on Kant and the Philosophy of Biology*, ed. by Philippe Huneman, Rochester: University of Rochester Press, 37–49.

Effertz, Dirk 1994, *Kants Metaphysik: Welt und Freiheit*, Freiburg: Alber.

Eigen, Manfred 1971, Self-organisation of Matter and the Evolution of Biological Macromolecules, *Naturwissenschaften* 58, 465–523.

Fisher, Mark 2007, Kant's Explanatory Natural History: Generation and Classification of Organisms in Kant's Natural Philosophy, in: *Understanding Purpose. Kant and the Philosophy of Biology*, ed. by Philippe Huneman, Rochester: Rochester University Press, 101–21.

Fisher, Mark 2008, Organisms and Teleology in Kant's Natural Philosophy, Emory University.

Fodor, Jerry A. 1981, The Present Status of the Innateness Controversy, in: Jerry A. Fodor, *Representations*, Cambridge (MA): MIT Press, 257–333.

Förster, Eckhart 1989, Kant's *Selbstsetzungslehre*, in: *Kant's Transcendental Deductions. The Three 'Critiques' and the Opus Postumum*, ed. by Eckhart Förster, Stanford: Stanford University Press, 217–38.

Förster, Eckhart 2000, The 'Gap' in Kant's Critical Philosophy, in: Eckhart Förster, *Kant's Final Synthesis. An Essay on the Opus postumum*, Cambridge (MA): Harvard University Press, 48–74.

Förster, Eckhart 2002a, Die Bedeutung von §§76, 77 der *Kritik der Urteilskraft* für die Entwicklung der nachkantischen Philosophie, *Zeitschrift für philosophische Forschung* 56 (2), 170–90.

Förster, Eckhart 2002b, Die Bedeutung von §§76, 77 der *Kritik der Urteilskraft* für die Entwicklung der nachkantischen Philosophie (Teil II), *Zeitschrift für philosophische Forschung* 56 (3), 321–45.

Förster, Eckhart 2008, Von der Eigentümlichkeit unseres Verstandes in Ansehung der Urteilskraft (§§74–78), in: *Immanuel Kant. Kritik der Urteilskraft*, ed. by Otfried Höffe, Berlin: Akademie Verlag, 259–88.

Förster, Eckhart 2011, *Die 25 Jahre der Philosophie*, Frankfurt/M.: Klostermann.

Frank, Manfred/Zanetti, Véronique (eds.) 2001, *Schriften zur Ästhetik und Naturphilosophie*, Frankfurt/M.: Suhrkamp, vol. 3.

Friedman, Michael 1992, *Kant and the Exact Sciences*, Cambridge (MA): Harvard University Press.

Friedman, Michael 2006, Kant—Naturphilosophic—Electromagnetism, in: *The Kantian Legacy in Nineteenth-Century Science*, ed. by Michael Friedman and Alfred Nordmann, Cambridge (MA): MIT Press, 51–79.

Frigo, Gian Franco 2009, Bildungskraft und Bildungstrieb bei Kant, in: *Kants Philosophie der Natur. Ihre Entwicklung im* Opus Postumum *und ihre Wirkung*, ed. by Ernst-Otto Onnasch, Berlin/New York: Walter de Gruyter, 9–23.

Garrett, Don 2003, Teleology in Spinoza and Early Modern Rationalism, in: *New Essays on the Rationalists*, ed. by Rocco Di Gennaro and Charles Huenemann, Oxford: Oxford University Press, 310–36.

Ginsborg, Hannah 1997, Kant on Aesthetic and Biological Purposiveness, in: *Reclaiming the History of Ethics: Essays for John Rawls*, ed. by Andrews Reath, Barbara Herman, and Christine Korsgaard, Cambridge: Cambridge University Press, 329–60.

Ginsborg, Hannah 2001, Kant on Understanding Organisms as Natural Purposes, in: *Kant and the Sciences*, ed. by Eric Watkins, Oxford: Oxford University Press, 231–58.

Ginsborg, Hannah 2004, Two Kinds of Mechanical Inexplicability in Kant and Aristotle, *Journal of the History of Philosophy* 42, 33–65.

Ginsborg, Hannah 2006, Kant's Biological Teleology and its Philosophical Significance, in: *A Companion to Kant*, ed. by Graham Bird, Oxford: Blackwell, 455–69.

Ginsborg, Hannah 2011, Primitive Normativity and Skepticism about Rules, *Journal of Philosophy* 108 (5), 227–54.

Godfrey-Smith, Peter 1994, A Modern History Theory of Functions, *Noûs* 28 (3), 344–362.

Godfrey-Smith, Peter 2007, Information in Biology, in: *The Cambridge Companion to the Philosophy of Biology*, ed. by David L. Hull and Michael Ruse, Cambridge: Cambridge University Press, 103–19.

Goy, Ina 2008, Die Teleologie in der organischen Natur (§§ 64–68), in: *Immanuel Kant. Kritik der Urteilskraft*, ed. by Otfried Höffe, Berlin: Akademie Verlag, 223–39.

Goy, Ina 2012, Kant on Formative Power, *Lebenswelt* 2, 26–49.

Goy, Ina 2013, On Judging Nature as a System of Ends. Exegetical Problems of §67 of the *Critique of the Power of Judgment*, in: *Akten des XI. Internationalen Kant-Kongresses, Pisa 2010*, ed. by Claudio LaRocca, Stefano Bacin, Alfredo Ferrarin, and Margit Ruffing, Berlin/Boston: Walter de Gruyter, vol. 5, 65–76.

Goy, Ina, The Antinomy of Teleological Judgment (unpublished manuscript).

Greene, Marjorie/Depew, David 2004, *The Philosophy of Biology: An Episodic History*, Cambridge: Cambridge University Press.

Grier, Michelle 2001, *Kant's Doctrine of Transcendental Illusion*, Cambridge: Cambridge University Press.

Guyer, Paul 1995, Nature, Morality, and the Possibility of Peace, in: *Proceedings of the Eighth International Kant Congress*, ed. by Hoke Robinson, Milwaukee: Marquette University Press 1995, vol. I.1, 51–69; reprinted in: Paul Guyer 2000, *Kant on Freedom, Law, and Happiness*, Cambridge: Cambridge University Press, 408–34.

Guyer, Paul 2000, The Unity of Nature and Freedom: Kant's Conception of the System of Philosophy, in: *The Reception of Kant's Critical Philosophy*, ed. by Sally Sedgwick, Cambridge: Cambridge University Press, 19–53; reprinted in: Guyer 2005, 277–313.

Guyer, Paul 2001a, From Nature to Morality: Kant's New Argument in the 'Critique of Teleological Judgment', in: *Architektonik und System in der Philosophie Kants*, ed. by Hans Friedrich Fulda and Jürgen Stolzenberg, Hamburg: Meiner, 375–404; reprinted in: Guyer 2005, 314–42.

Guyer, Paul 2001b, Organisms and the Unity of Science, in: *Kant and the Sciences*, ed. by Eric Watkins, Oxford: Oxford University Press, 259–81.

Guyer, Paul 2004, Zweck in der Natur: Was ist lebendig und was ist tot in Kants Teleologie?, in: *Warum Kant heute?*, ed. by Dietmar Heidemann and Kristina Engelhard, Berlin/New York: Walter de Gruyter, 383–412; English in: Guyer 2005, 343–72.

Guyer, Paul 2005, *Kant's System of Nature and Freedom*, Oxford: Oxford University Press.

Guyer, Paul 2006, The Possibility of Perpetual Peace, in: *Kant's Perpetual Peace: New Interpretative Essays*, ed. by Luigi Caranti, Rome: LUISS University Press, 161–82.

Guyer, Paul 2007, Natural Ends and the End of Nature, in: *Hans Christian Ørsted and the Romantic Legacy in Sciences*, ed. by Robert M. Brain, Robert S. Cohen, and Ole Knudsen, Dordrecht: Springer, 75–96.

Guyer, Paul 2011, Kantian Communities, in: *Kantian Communities: The Realm of Ends, the Ethical Community, and the Highest Good*, ed. by Lucas Thorpe, Rochester: University of Rochester Press.

Hall, Ned 2004, Two Concepts of Cause, in: *Causation and Counterfactuals*, ed. by John Collins, Ned Hall, and L.A. Paul, Cambridge (MA): MIT Press, 225–76.

Hardcastle, Valerie 2002, On the Normativity of Functions, in: *Functions: New Essays in the Philosophy of Psychology and Biology*, ed. by André Ariew, Robert Cummins, and Mark Perlman, Cambridge: Cambridge University Press, 144–56.

Hartmann, Fritz 2000, Die Leibniz-Stahl-Korrespondenz als Dialog zwischen monadischer und dualistisch-'psycho-somatischer' Anthropologie, in: *Georg Ernst Stahl (1659–1734) in wissenschaftshistorischer Sicht* (Acta Historica Leopoldina 30), ed. by Dietrich v. Engelhard and Alfred Gierer, Halle: J. A. Barth in Georg Thieme Verlag, 97–124.

Haym, Rudolf 1954, *Herder*, Berlin: Aufbau Verlag, vol. 2.

Heidegger, Martin 1976, Vom Wesen und Begriff der Physis. Aristoteles, Physik B, 1, in: Wegmarken, in: Martin Heidegger, *Gesamtausgabe*, ed. by Friedrich-Wilhelm von Hermann, Frankfurt/M.: Vittorio Klostermann 1975–, vol. 9, 239–301.

Himma, Kenneth Einar 2009, Design Arguments for the Existence of God, in: *Internet Encyclopedia of Philosophy*, http://www.iep.utm.edu/design/[last visited: November 23, 2013].

Höffe, Ottfried 2008, Der Mensch als Endzweck, in: *Immanuel Kant: Kritik der Urteilskraft*, ed. by Otfried Höffe, Berlin: Akademie Verlag, 289–308.

Hoffheimer, Michael 1982, Maupertuis and the Eighteenth-Century Critique of Preexistence, *Journal of the History of Biology* 15, 119–44.

Horn, Laurence R. 2010, Contradiction, in: *The Stanford Encyclopedia of Philosophy*, ed. by Edward Zalta, http://plato.stanford.edu/archives/win2010/entries/contradiction/[last visited: November 18, 2013].

Hübner, Kurt 1953, Leib und Erfahrung im Opus postumum, *Zeitschrift für philosophische Forschung* 7, 204–19.

Hull, David L. 1999, The Use and Abuse of Sir Karl Popper, *Biology and Philosophy* 14, 481–504.

Huneman, Philippe 2006, From Comparative Anatomy to the "Adventures of Reason", *Studies in History and Philosophy of Biological and Biomedical Sciences* 37 (4), 649–74.

Huneman, Philippe (ed.) 2007a, *Understanding Purpose: Kant and the Philosophy of Biology*, Rochester: University of Rochester Press.

Huneman, Philippe 2007b, Reflexive Judgment and Embryology: Kant's Shift Between the First and the Third Critique, in: *Understanding Purpose: Kant and the Philosophy of Biology*, ed. by Philippe Huneman, Rochester: University of Rochester Press, 75–100.

Huneman, Philippe 2008, *Métaphysique et biologie: Kant et la constitution du concept d'organisme*, Paris: Kimé.

Illetterati, Luca 2008, Being-for: Purposes and Functions in Artefacts and Living Beings, in: *Purposiveness: Teleology Between Nature and Mind*, ed. by Luca Illetterati and Francesca Michelini, Frankfurt/M.: Ontos, 135–62.

Illetterati, Luca/Michelini, Francesca (eds.) 2008, *Purposiveness. Teleology Between Nature and Mind*, Frankfurt/M.: Ontos.

Jha, Stefania Ruzsits 2002, *Reconsidering Michael Polanyi's Philosophy*, Pittsburgh: University of Pittsburgh Press.

Kauffmann, Stuart 1993, *Self-organisation and the Origins of Order*, New York: Oxford University Press.

Kitcher, Philip 1993, Function and Design, *Midwest Studies in Philosophy* 18, 379–97.

Kitcher, Philip 1999, The Hegemony of Molecular Biology, *Biology and Philosophy* 14, 195–210.

Krohs, Ulrich 2009, Functions as Based on a Concept of General Design, *Synthese* 166, 69–89.

Küppers, Bernd-Olaf 1983, *Molecular Theory of Evolution, Outline of a Physico-Chemical Theory of the Origin of Life*, Heidelberg: Springer.

Küppers, Bernd-Olaf 1986, *Der Ursprung der biologischen Information*, Munich: Piper.

Lenoir, Timothy 1980, Kant, Blumenbach, and Vital Materialism in German Biology, *Isis* 71, 77–108.

Lenoir, Timothy 1981, The Göttingen School and the Development of Transcendental Naturphilosophie in the Romantic Era, *Studies in History of Biology* 5, 111–205.

Lenoir, Timothy 1982a, Blumenbach, Kant, and the Teleomechanical Approach to Life, in: Timothy Lenoir, *The Strategy of Life. Teleology and Mechanics in Nineteenth-Century German Biology*, Dordrecht/Boston/London: Springer, 17–34.

Lenoir, Timothy 1982b, *The Strategy of Life: Teleology and Mechanism in Nineteenth-Century German Biology*, Dordrecht: Reidel.

Lewens, Tim 2004, *Organisms and Artefacts. Design in Nature and Elsewhere*, Cambridge (MA): MIT Press.

Löw, Reinhard 1980, *Die Philosophie des Lebendigen*, Frankfurt/M.: Suhrkamp.

Look, Brandon 2006, Blumenbach and Kant on Mechanism and Teleology in Nature. The Case of the Formative Drive, in: *The Problem of Animal Generation in Early Modern Philosophy*, ed. by Justin E. H. Smith, Cambridge: Cambridge University Press, 355–72.

Mackie, John 1982, *The Miracle of Theism. Arguments for and Against the Existence of God*, Oxford: Oxford University Press.

Marc-Wogau, Konrad 1938, *Vier Studien zu Kants Kritik der Urteilskraft*, Uppsala: A.-B. Lundequistska Bokhandeln.

Mayr, Ernst 1961, Cause and Effect in Biology, *Science* 134, 1501–6.

Mayr, Ernst 1982, *The Growth of Biological Thought: Diversity, Evolution, and Inheritance*, Cambridge (MA): The Belknap Press of Harvard University Press.

McFarland, John D. 1970, *Kant's Concept of Teleology*, Edinburgh: University of Edinburgh Press.

McLaughlin, Peter 1989a, *Kants Kritik der teleologischen Urteilskraft*, Bonn: Bouvier.

McLaughlin, Peter 1989b, What Is an Antinomy of Judgment?, in: *Proceedings: Sixth International Kant Congress*, ed. by Gerhard Funke and Thomas Seebohm, Washington, D.C.: University Press of America, vol. II/2, 357–67.

McLaughlin, Peter 1990, *Kant's Critique of Teleology in Biological Explanation. Antinomy and Teleology*, Lampeter: Edwin Mellen Press.

McLaughlin, Peter 2001, *What Functions Explain?* Cambridge: Cambridge University Press.

McLaughlin, Peter 2009, Functions and Norms, in: *Functions in Biological and Artificial Worlds*, ed. by Ulrich Krohs and Peter Kroes, Cambridge (MA): MIT Press, 93–102.

Millikan, Ruth Garrett 1989, In Defense of Proper Functions, *Philosophy of Science* 56 (2), 288–302.

Müller-Sievers, Helmut 2000, From Preformation to Epigenesis/Self-Generation in Philosophy: Kant, in: Helmut Müller-Sievers, *Self-generation: Biology, Philosophy, and Literature around 1800*, Stanford: Stanford University Press, 26–64.

Mumford, Stephen 2009, Causal Powers and Capacities, in: *The Oxford Handbook of Causation*, ed. by Helen Beebee, Christopher Hitchcock, and Peter Menzies, Oxford: Oxford University Press, 265–78.

Mumford, Stephen/Anjum, Rani Lill 2011, *Getting Causes from Powers*, Oxford: Oxford University Press.

Neander, Karen 1991, Functions as Selected Effects, *Philosophy of Science* 58 (2), 168–84.

Nunziante, Antonio 2004, "Corpus vivens est automaton sui perpetuativum ex naturae istituto". Some Remarks on Leibniz's Distinction between "Machina naturalis" and "Organica artificialia", *Studia Leibnitiana*, Sonderheft 32, 203–16.

Polanyi, Michael 1958, *Personal Knowledge. Towards a Post-Critical Philosophy*, Chicago: The University of Chicago Press.

Polanyi, Michael 1967, Life Transcending Physics and Chemistry, in: *Society, Economics and Philosophy, Selected Papers by Michael Polanyi*, ed. by Richard Allen, New Brunswick: Transaction Publisher.

Polanyi, Michael 1969, Life's Irreducible Structure, in: *Knowing and Being, Essays by Michael Polanyi*, ed. by Marjorie Grene, Chicago: The University of Chicago Press, 225–39.

Pyle, Andrew 2006, Malebranche on Animal Generation: Preexistence and the Microscope, in: *The Problem of Animal Generation in Early Modern Philosophy*, ed. by Justin E. H. Smith, Cambridge: Cambridge University Press, 194–214.

Quarfood, Marcel 2004, *Transcendental Idealism and the Organism: Essays on Kant*, Stockholm: Almquist & Wiksell.

Quarfood, Marcel 2006, Kant on Biological Teleology: Towards a Two-Level Interpretation, *Studies in History and Philosophy of Biological and Biomedical Sciences* 37, 735–47.

Ratzsch, Del 2001, *Nature, Design, and Science. The Status of Design in Natural Science*, New York: State University of New York Press.

Richards, Robert J. 2000, Kant and Blumenbach on the *Bildungstrieb:* A Historical Misunderstanding, *Studies in the History and Philosophy of Biological and Biomedical Sciences* 31 (1), 11–32.

Richards, Robert J. 2002, Early Theories of Development: Kant und Blumenbach, in: Robert J. Richards, *The Romantic Conception of Life: Science and Philosophy in the Age of Goethe*, Chicago: University of Chicago Press, 207–37.

Roe, Shirley A. 1979, Rationalism and Embryology: Caspar Friedrich Wolff's Theory of Epigenesis, *Journal of the History of Biology* 12, 1–43.

Roe, Shirley A. 1981, *Matter, Life, and Generation: Eighteenth-Century Embryology and the Haller-Wolff Debate*, Cambridge: Cambridge University Press.

Roger, Jaques 1963/²1971, *Les sciences de la vie dans la pensée fraçaise au XVIIIe siècle*, Paris: A. Colin.

Rosenberg, Alex 1997, Reductionism Redux: Computing the Embryo, *Biology and Philosophy* 12, 445–70.

Rosenberg, Alex/McShea, Daniel W. 2008, *Philosophy of Biology. A Contemporary Introduction*, New York/London: Routledge.

Roth, Siegfried 2008, Kant und die Biologie seiner Zeit, in: *Immanuel Kant. Kritik der Urteilskraft*, ed. by Otfried Höffe, Berlin: Akademie Verlag, 275–88.

Roth, Siegfried 2011a, Mathematics and Biology. A Kantian View on the History of Pattern Formation Theory, *Development, Genes and Evolution* 221, 255–29.

Roth, Siegfried 2011b, Evolution und Fortschritt. Zum Problem der Höherentwicklung in der organischen Evolution, in: *Zweck und Natur*, ed. by Tobias Schlicht, Paderborn: Wilhelm Fink Verlag, 195–247.

Runes, Dagobert D. (ed.) 1942, *The Dictionary of Philosophy*, New York: Philosophical Library.

Sala, Giovanni B. 1990, Teleologie und moralischer Gottesbeweis in der Kritik der
    Urteilskraft, in: Giovanni B. Sala, *Kant und die Frage nach Gott. Gottesbeweise und
    Gottesbeweiskritik in den Schriften Kants*, Berlin/New York: Walter de Gruyter, 426–50.
Schmucker, Josef 1983, *Kants vorkritische Kritik der Gottesbeweise: ein Schlüssel zur
    Interpretation des theologischen Hauptstücks der transzendentalen Dialektik der* Kritik
    der reinen Vernunft, Wiesbaden: Steiner.
Schramm, Matthias 1985, *Natur ohne Sinn? Das Ende des teleologischen Weltbilds*, Graz:
    Styria.
Sloan, Phillip R. 2002, Preforming the Categories: Eighteenth-Century Generation Theory and
    the Biological Roots of Kant's A Priori, *Journal of the History of Philosophy* 40 (2),
    229–53.
Sloan, Phillip R. 2006, Kant on the History of Nature: The Ambiguous Heritage of the Critical
    Philosophy for Natural History, *Studies in History and Philosophy of Biology and
    Biomedical Sciences* 37 (4), 627–48.
Sober, Elliott 2004, The Design Argument, in: *The Blackwell Guide to the Philosophy of
    Religion*, ed. by William E. Mann, Oxford: Blackwell Publishing, 117–47.
Stegmann, Ulrich E. 2005, Genetic Information as Instructional Content, *Philosophy of
    Science* 72, 425–43.
Steigerwald, Joan (ed.) 2006, Kantian Teleology and the Biological Sciences, *Studies in
    History and Philosophy of Biological and Biomedical Sciences* 37, 621–792.
Stollberg, Gunnar, Vitalism and Vital Force in Life Sciences—The Demise and Life of a
    Scientific Conception (unpublished manuscript), http://www.uni-bielefeld.de/soz/pdf/Vi
    talism.pdf [last visited: November 24, 2013].
Strawson, Peter F. 1995, *The Bounds of Sense. An Essay on Kant's* Critique of Pure Reason,
    London: Routledge.
Šustar, Predrag 2008, The Organism Concept: Kant's Methodological Turn, in: *Purposiveness.
    Teleology between Nature and Mind*, ed. by Luca Illetterati and Francesca Michelini,
    Frankfurt/M.: Ontos, 33–57.
Šustar, Predrag 2013, Normativity and Biological Lawlikeness: Three Variants, in: *Akten des
    XI. Internationalen Kant-Kongresses, Pisa 2010*, ed. by Claudio LaRocca, Stefano Bacin,
    Alfredo Ferrarin, and Margit Ruffing, Berlin/Boston: Walter de Gruyter.
Swinburne, Richard 1979/²2004, Teleological Arguments, in: Richard Swinburne, *The
    Existence of God*, Oxford: Clarendon Press, 153–91.
Teufel, Thomas 2011a, Wholes that Cause their Parts: Organic Self-reproduction and the
    Reality of Biological Teleology, *Studies in History and Philosophy of Biological and
    Biomedical Sciences* 42, 252–60.
Teufel, Thomas 2011b, Kant's *Non*-Teleological Conception of Purposiveness, *Kant-Studien* 102
    (2), 232–52.
Theis, Robert 1994, Das Problem der Physikotheologie im *Beweisgrund*, in: Robert Theis,
    *Gott. Untersuchung zur Entwickung des theologischen Diskurses in Kants Schriften zur
    theoretischen Philosophie bis hin zum Erscheinen der* Kritik der reinen Vernunft,
    Stuttgart Bad-Cannstatt: Fromann Holzboog, 111–43.
Toepfer, Georg 2004, *Zweckbegriff und Organismus: Über die teleologische Beurteilung
    biologischer Systeme*, Würzburg: Königshausen und Neumann.
Toepfer, Georg 2008, Teleology in Natural Organised Systems and Artefacts: Interdependence
    of Processes versus External Design, in: *Purposiveness: Teleology Between Nature and
    Mind*, ed. by Luca Illetterati and Francesca Michelini, Frankfurt/M.: Ontos, 163–81.

Tonelli, Giorgio 1959, La nécessité des lois de la nature au 18ème siècle et chez Kant en 1762, *Revue d'histoire des sciences* 12, 225–41.

Ungerer, Emil 1922, *Die Teleologie Kants und ihre Bedeutung für die Logik der Biologie*, Berlin: Bornträger.

van den Berg, Hein 2009, Kant on Vital Forces. Metaphysical Concerns versus Scientific Practice, in: *Kants Philosophie der Natur. Ihre Entwicklung im* Opus Postumum *und ihre Wirkung*, ed. by Ernst-Otto Onnasch, Berlin/New York: Walter de Gruyter, 115–35.

van den Berg, Hein 2011, *Kant on Proper Science. Biology in the Critical Philosophy and the Opus Postumum*, Graz: Styria.

Walker, Caroline J./Willows, Robert D. 1997, Mechanism and Regulation of Mg-chelatase, *Biochemical Journal* 327, 321–33.

Warda, Arthur 1922, *Immanuel Kants Bücher*, Berlin: Breslauer.

Watkins, Eric 2003, Forces and Causes in Kant's Early Pre-Critical Writings, *Studies in History and Philosophy of Science* 34, 5–27.

Watkins, Eric 2005, *Kant and the Metaphysics of Causality*, Cambridge/New York: Cambridge University Press.

Watkins, Eric 2008, Die Antinomie der teleologischen Urteilskraft und Kants Ablehnung alternativer Teleologien (§§69–71 und §§72–73), in: *Immanuel Kant. Kritik der Urteilskraft*, ed. by Otfried Höffe, Berlin: Akademie Verlag, 241–58.

Watkins, Eric 2009, The Antinomy of Teleological Judgment, in: *Kant Yearbook 1*, ed. by Dietmar Heidemann, Berlin/New York: Walter de Gruyter, 197–221.

Watkins, Eric 2010, The Antinomy of Practical Reason: Reason, the Unconditioned, and the Highest Good, in: *Kant's 'Critique of Practical Reason': A Critical Guide*, ed. by Andrews Reath and Jens Timmerman, Cambridge/New York: Cambridge University Press, 145–67.

Weber, Marcel 2005, *The Philosophy of Experimental Biology*, Cambridge: Cambridge University Press.

White, David A. 1997, Kant's Notion of a Purpose, in: Special Issue: Final Causality in Nature and Human Affairs, ed. by Richard F. Hassing, *Studies in Philosophy and the History of Philosophy* 30, 125–50.

Wilson, Catherine 1997, *The Invisible World: Early Modern Philosophy and the Invention of the Microscope*, Princeton: Princeton University Press.

Wilson, Robert A. 2005, *Genes and the Agents of Life. The Individual in the Fragile Sciences: Biology*, Cambridge: Cambridge University Press.

Wood, Allen W. 1970, *Kant's Moral Religion*, Ithaca: Cornell University Press.

Wood, Allen W. 1978, The Physicotheological Proof, in: Allen W. Wood, *Kant's Rational Theology*, Ithaca: Cornell University Press, 130–45.

von Wright, Georg Henrik 1993, *The Tree of Knowledge and Other Essays*, Leiden: Brill.

Wright, Larry 1973, Functions, *Philosophical Review* 82, 139–68.

Yarus, Michael 2010, *Life from an RNA World*, Cambridge: Harvard University Press.

Zammito, John 1992, *The Genesis of Kant's* Critique of Judgment, Chicago: University of Chicago Press.

Zammito, John 2006a, Kant's Early Views on Epigenesis: The Role of Maupertuis, in: *The Problem of Animal Generation in Early Modern Philosophy*, ed. by Justin E. H. Smith, Cambridge: Cambridge University Press, 317–54.

Zammito, John 2006b, Teleology Then and Now: The Question of Kant's Relevance for Contemporary Controversies over Function in Biology, in: *Studies in History and Philosophy of Biological and Biomedical Sciences* 37, 748–70.

Zammito, John 2007, Kant's Persistent Ambivalence Toward Epigenesis, 1764–90, in: *Understanding Purpose. Kant and the Philosophy of Biology*, ed. by Philippe Huneman, Rochester: University of Rochester Press, 51–74.

Zammito, John 2012, The Lenoir Thesis Revisited: Blumenbach and Kant, *Sudies in History and Philosophy of Biological and Biomedical Sciences* 43, 120–32.

Zuckert, Rachel 2007, *Kant on Beauty and Biology. An Interpretation of the* Critique of Judgment, Cambridge: Cambridge University Press.

Zumbach, Clark Edward 1980, The Lawlessness of Living Things: Kant's Conception of Organismic Activity, Rutgers University.

Zumbach, Clark Edward 1984, *The Transcendent Science. Kant's Conception of Biological Methodology*, Den Haag: Martinus Nijhoff.

Index of Names

Index of Subjects

Contributors

Angela Breitenbach is Lecturer in Philosophy at Cambridge University. She is author of *Die Analogie von Vernunft und Natur. Eine Umweltphilosophie nach Kant* (2009). Her essays on Kant's philosophy include: Kant on Causal Knowledge: Causality, Mechanism, and Reflective Judgment (2010), Teleology in Biology: A Kantian Approach (2009), Umweltethik nach Kant: Ein analogisches Verständnis vom Wert der Natur (2009), and Two Views on Nature: A Solution to Kant's Antinomy of Mechanism and Teleology (2008).

Mark Fisher is Lecturer in Philosophy and Assistant Director of the Rock Ethics Institute at the Pennsylvania State University. He is author of *Organisms and Teleology in Kant's Natural Philosophy* (PhD 2008) and the article Kant's Explanatory Natural History: Generation and Classification of Organisms in Kant's Natural Philosophy (2007).

Hannah Ginsborg is Professor of Philosophy at the University of California Berkeley. She is author of *The Role of Taste in Kant's Theory of Cognition* (1989) and of a forthcoming collection of essays, *The Normativity of Nature*. Her articles on Kant's theory of biology include: Kant's Biological Teleology and Its Philosophical Significance (2006), Two Kinds of Mechanical Inexplicability in Kant and Aristotle (2004), and Kant on Understanding Organisms as Natural Purposes (2001).

Ina Goy is Research Fellow and Assistant Professor at Eberhard-Karls-Universität Tübingen. She is currently writing a commentary on Kant's theory of biology. Further articles include: On Judging Nature as a System of Ends: Exegetical Problems of §67 of the 'Critique of the Power of Judgment' (2013), Kant on Formative Power (2012), Kant's Theory of Biology and the Argument from Design (2014), and Die Teleologie der Natur (§§ 64–68) (2008).

Paul Guyer is Jonathan Nelson Professor of Philosophy at Brown University. His books include *Kant and the Claims of Taste* (1979), *Kant and the Claims of Knowledge* (1987), *Kant's System of Nature and Freedom* (2005) and *Kant* (2006). He is the editor of *Kant's 'Critique of the Power of Judgment': Critical Essays* (2003) as well as three Cambridge Companions on Kant. He is the editor and co-translator of Kant's *Critique of the Power of Judgment* (2000) as well as of Kant's first *Critique* (1998) and Kant's *Notes and Fragments* (2005).

Philippe Hunemann is Researcher at the Institut d'Histoire et de Philosophie des Sciences et des Techniques (CNRS/Université Paris 1 Sorbonne). He has published extensively on the philosophy of evolutionary biology and ecology. His works on Kant's theory of biology include the monograph *Métaphysique et biologie. Kant et la constitution du concept d'organisme* (2008), the edited volume *Understanding Purpose: Kant and the Philosophy of Biology* (2007), and the essays Naturalizing Purpose: From Comparative Anatomy to the "Adventures of Reason" (2006), Reflexive Judgment and Wolffian Embryology: Kant's Shift between the First and the Third Critique (2007), and Théorie kantienne des organismes et révision transcendantale du concept métaphysique de finalité (2010), and Kant vs. Leibniz in the Second Antinomy: Organisms Are Not Infinitely Subtle Machines (2013).

Luca Illetterati is Professor of Theoretical Philosophy at the Università degli Studi di Padova. He is author of *Natura e Ragione. Sullo sviluppo dell'idea di Natura in Hegel* (1995), and co-editor of the essay collection *Purposiveness. Teleology between Mind and Nature* (2008). Further essays include: Being-For, Purposes and Functions in Artifacts and Living Beings (2008), Hegels Kritik der Metaphysik der Naturwissenschaften (2008), Between Science and Wisdom. On the Kantian Notion of Philosophy (2005), and La decisione dell'idea, L'idea assoluta e il suo passaggio nella natura in Hegel (2005).

Peter McLaughlin is Professor of Philosophy at Ruprecht-Karls-Universität Heidelberg. He is author of *Kants Kritik der teleologischen Urteilskraft* (1989, engl. 1990) and *What Function Explains* (2000). His articles on the philosophy of biology and its history include: Regulation, Assimilation and Life—Kant, Canguilhem and Beyond (2008), Functions and Norms (2008), Kant on Heredity and Adaptation (2007), and Naming Biology (2002).

Ernst-Otto Onnasch is Professor of Philosophy at the Universiteit Utrecht. He is editor of the essay collection *Kants Philosophie der Natur. Ihre Entwicklung im Opus postumum und ihre Wirkung* (2009). Further essays include: Immanuel Kants Philosophie in den Niederlanden zwischen 1785 und 1804 (2010), Kants Transzendentalphilosophie des *Opus postumum* gegen den transzendentalen Idealismus Schellings und Spinozas (2009), and Die erste Rezeption der Philosophie Immanuel Kants in der Niederlande (2005).

Marcel Quarfood is Lecturer and Researcher at Stockholm Universitet. He is author of *Transcendental Idealism and the Organism: Essays on Kant* (2004). Further essays include: Discursivity and Transcendental Idealism (2011), Kant on

Biological Teleology: Towards a Two-level Interpretation (2006), and The Individuality of Species: Some Reflections on the Debate (1999).

Siegfried Roth is Professor of Developmental Biology at the Universität Köln. Beside his numerous articles on issues in developmental biology, he is author of the articles Kant und die Biologie seiner Zeit (2008) and Mathematics and Biology. A Kantian View on the History of Pattern Formation Theory (2011).

Predrag Šustar is Professor of Philosophy at Sveučilište u Rijeci. He is author of *Il problema delle leggi biologiche: Una soluzione di tipo kantiano* (2005). His articles on Kant and the philosophy of sciences include: La generazione e l'impresa critica. La costituzione della filosofia kantiana della 'biologia' (2001), The Organism Concept: Kant's Methodological Turn (2008), Nomological and Transcendental Criteria for Scientific Laws (2005), and Sistema delle scienze naturali e unità della conoscenza nell'ultimo Kant (2001).

Eric Watkins is Professor of Philosophy at the University of California San Diego. He is author of *Kant and the Metaphysics of Causality* (2005) and editor of *Kant and the Sciences* (2001). Further essays include: The Antinomy of Teleological Judgment (2009), Die Antinomie der teleologischen Urteilskraft und Kants Ablehnung alternativer Teleologien §§ 69–71 & 72–73 (2008), and Kant on Transcendental Laws (2007).

Rachel Zuckert is Associate Professor at Northwestern University. She is author of the *Kant on Beauty and Biology. An Interpretation of the 'Critique of Judgment'* (2007). Her articles on Kant's philosophy include: History, Biology, and Philosophical Anthropology in Kant and Herder (2012), Kant's Rationalist Aesthetics (2007), The Purposiveness of Form: A Reading of Kant's Aesthetic Formalism (2006), and Boring Beauty and Universal Morality: Kant on the 'Ideal of Beauty' (2005).

www.ingramcontent.com/pod-product-compliance
Lightning Source LLC
Chambersburg PA
CBHW070018100426
42740CB00013B/2542